SYMBIOSIS OF SCIENCE AND SPIRITUALITY

SYMBIOSIS OF SCIENCE AND SPIRITUALITY

Generation of Innovation in Science for Human Survival

Sampooran Singh
Dr (Mrs) Kanwaljit Kaur

Publications
KALPAZ PUBLICATIONS
DELHI-110052

Symbiosis of Science and Spirituality

Rs. 750

ISBN: 81-7835-509-4

Published in 2006 in India by
Kalpaz Publications
C-30, Satyawati Nagar, Delhi-110052
E-mail : kalpaz@hotmail.com
Website : http://www.gyanbooks.com

Laser Typesetting at : Shiva Infosystem, Delhi
Printed at : Salasar Imaging System, Delhi

Contents

PART III
SCIENCE, EDUCATION AND ETHICS

PART IV
A NEW VISION ON EARTHQUAKES AND TSUNAMIS

Dr. Karan Singh
Member of Parliament (Rajya Sabha) &
Chancellor, Jawaharlal Nehru University

3, Nyaya Marg, Chanakyapuri
New Delhi-110021

Foreword

Today, when humankind is poised at a crucial crossroad in its long and tortuous journey from the dawn of civilisation, scientists and philosophers the world over are agreed that the present paradigm of "objective science" is leading it headlong towards crass materialism apparently oblivious of its impact on the natural and spiritual worlds. The heedless pursuit of materialistic paradigm of scientific research and study will lead us ultimately to self-annihilation and the destruction of our planet. There is the need for the spiritual rebirth of humankind, a perception that the Parliament of World Religions also stressed during its centenary meeting in Chicago.

Without the resurgence of spirituality our civilization appears to be doomed as science and spirituality are unlikely to serve the cause of humankind's renaissance in a stand-alone mode. There is the need for symbiosis of science with spirituality, and that need is more urgent today than it was a couple of centuries back, when Industrial Revolution gave our lives a new direction. Dr Sampooran Singh and Dr Kanwaljit Kaur have engaged themselves, over the decades, in exploring the dimensions of science and spirituality and how to meld the synergies of both into a symbiotic whole for humankind's benefit. Their intuitive perceptions of both science and spirituality are encapsulated in this treatise. I commend both of them for their labour of love and their efforts to initiate a dialogue on this complex theme.

Karan Singh

Buddh Purnima,
23rd July 2005

NEW WORLD MOVEMENT

280, Bank Enclave, Laxmi Nagar, Delhi-110092 (India)
Phones: 011-22545424, 22545425 • E mail: n.shandilya@rediffmail.com

Swami Om Poorna Swatantra
President

Ref. No..........................

Message

With the degeneration in the life of Vedic Age – the most enlightened, dynamic and creative age of all human history – the truth of life which is naturally holistic, integral and non-divisible got fragmented into two half-truths of subjective spirituality and objective materiality. And that was the genesis of the malady with which the humanity has been afflicted ever since.

For quite some millennia the subjective half of spirituality held sway and oriented mankind towards 'other-worldliness', that is, man focused on life beyond body and earth and neglected the physical existence here. This approach led to the ethereal heights of spiritual development but degeneration of physical side which was not considered of much significance and value – all the religions of the world are various manifestations of this approach.

As a result of reaction to this subjective extremism, objective extremism represented by modern science came into being. Some 500 years ago when faith got saturated, reason asserted itself and when its entry into subjective world was not allowed by the Christian mode of spirituality, it revolted and turned outward towards Nature for its exercise – and that was the beginning of modern science. But the science also suffered from the same draw-back of one-sidedness: it became exclusively earthly and materialistic and neglected the underground and invisible aspect of life and which is the source of quality and value.

Ultimately in 20th century man realized the futility of these two mutually exclusive and contrary approaches in regard to the fulfilment of life and began to search for a third dimension and way wherein the two apparently contradictory approaches are established to be complementary aspects of a whole and healthy life where they cannot be separated and divorced from each other. This was the beginning of a New Age.

The leading edges of both the branches of science, physical as well as spiritual, are moving towards each other in a spirit of convergence and ultimate fusion. And paradoxically, the fatal crisis facing mankind today is goading it towards the acceptance of the new way.

Dr Sampooran Singh and Dr Kanwaljit Kaur are one of the pioneers of this movement for searching the Third Way. Their latest contribution

'Symbiosis of Science and Spirituality' is a landmark achievement in his ceaseless endeavour. This monograph prepares the theoretical ground for practical action: a quantum jump or mutation in human consciousness from partial or dualistic to integral and holistic one which is the bedrock foundation of a new Way and a New World.

I am sure, Dr Singh will live immortally: the posterity will build the future on his foundations.

My salutations to this modern Rishi, the moral guardian of mankind.

Swami Om Poorna Swatantra
President
New World Movement

Date: 20.06.2005

Preface

Junctures come in human history which come but rarely, when one age draws to a close and a new age begins – an age of transition, transformation, transmutation, mutation. Humanity in fact is simply, at the present moment, at the head of an evolutionary pyramid, but man is an unfinished creature, half-way between the animal and the Divine [Life Field]. But now it is clear that humanity is poised for the most crucial and the most difficult transition it has ever sought to accomplish, and that is the transition to global consciousness, a transition to a new level of awareness and being.

We are transiting into a global society, or to be more precise, a global village, or a global cottage. If one can get a television and CNN [Cable News Network], you know exactly what is happening anywhere in the world. Politics have become global. Economics has become global. Now the growth of globalization has taken place, but not the development of global consciousness. Globalization is taking place as the result of science and technology, and the thrust of evolution, but global consciousness is not keeping pace with the globalization. In five years, for all practical purposes, there is a new technological generation. So time has encapsulated, but our minds are still functioning in the old rhythms. There is an urgency to bring in mutation in our psyche.

The psychological imbalances, ecological and pollution challenges, a rising wave of violence and crime, terrorism, religious fundamentalism, ethnic conflicts, corruption, increasing unemployment, poverty, hunger, malnutrition, health crisis, are forcing man inevitably to face the realities of the human situation. All appeals to achieve full disarmament and permanent peace; all declarations by Nobel laureates; all appeals by Pugwash Conferences (since 1957); all appeals by the International Physicians for the Prevention of Nuclear War (IPPNW); by Secretaries-General of the United Nations; all proclamations of a manifesto by eminent scientists, philosophers, mystics have fallen on the deaf ears of mankind.

The events happening in the world are so fast that it is difficult to keep oneself abreast with the events and their underlying currents. This monograph investigates the cause of emergence of human challenges and answer to that paradox for human survival and excellence.

1. UNESCO International Year of Physics 2005

(a) Albert Einstein has been the greatest physicist of the age. In addition to his contribution as a scientist, he has contributed as man-cum-philosopher (Chapter 1) and his contributions are parallel to that of Lord Buddha, or the writings of Upanishads, or that of any other Eastern mystics. One is struck by the miraculous agreement between humans of different race, different religion, knowing nothing about each other's existence, separated by centuries and millennia, and by the greatest distance that are on our globe.

Albert Einstein discussed the challenges facing mankind in the years 1900-1950 in an essay entitled "Out of My Later Years". He wrote:

(i) The confidence in the stability, even the very basis of existence, of human society has largely vanished. One senses not only a threat to man's cultural heritage but also that a lower value is placed upon all that one will like to see defended at all costs .... Life in human society meant dangers of all sorts .... even standards of virtue, aspiration and practical truth were taken for granted as an inviolable heritage, common to all civilised humanity .... The sanctity of life vanished .... The lie was raised to the dignity of a political instrument. It was a manifestation of weakness of character and lack of responsibility on the part of the leaders in the affected countries.

(ii) And yet I know that, all in all, man changes but little, even though prevailing notions make him appear in a very different light at different times, and even though current trends like the present bring him unimaginable sorrow. Nothing of all that will remain but a few pitiful pages in the history books, briefly picturing to the youth of future generations the follies of its ancestors.

Einstein proclaimed that the present science is leading mankind to self-annihilation, self-destruction. Einstein's future prediction about the survival of mankind are identical to that of George Wald, Nobel Prize in Medicine, 1967, who wrote, "Our society has taken a wrong turn .... By now the problem is not transcendence but survival". Jacques Monod, Nobel Prize winning biologist, wrote in 1972, "What we see opening before us is an abyss of darkness". Our society is moving to a collective suicide. We have to take urgent steps for our survival.

2. A Renewed Scientific Attitude Regarding Human Challenges

In modern physics, a very different attitude has now developed. Physics have come to see that all their theories and natural phenomena, including the 'laws' they describe are creations of the human mind, properties of our conceptual map of reality, rather than of reality itself. The creations of the mind are parts of the map, not of the territory.

(a) The Challenges Facing Mankind: For about a million years or more, man has engaged himself in gathering fragmentary experience and partial knowledge, and storing these in the mind-brain-body system as frequency and amplitude information; and called these as memory or cultural belief structure. Man started living in a synthetic world (man-made social cosmos of time-space-causation matrices) and divorced the Reality. It constructs duality and there is conflict, which results in nervous tension and chemical pressure, and these express themselves as pain, misery and travails.

(b) The Challenges Facing Mankind: 1900-1950: Einstein perceived that the human psyche got conditioned at an accelerative phase in the first five decades of the twentieth century. Due to conditionings, the conscious mind has lost sensitivity.

(c) The Challenges Facing Mankind : 1950 – to date: The recent challenges – global terrorism, hijacking, religious fundamentalism, ethnic conflicts, organised crime, drug-trafficking, corruption, and so on – have emerged in the last five decades of the twentieth century. All these challenges are constructs of the human mind, and forms the root of intellection and is therefore next to impossible to uproot by intellection. Einstein stated, "The significant problems we face cannot be solved at the same level of thinking we used when we created them". All these challenges are constructs of the intellect, hence these are dualistic. The error of dualism forms the root of intellection and is therefore next to impossible to uproot by intellection. These challenges have now gone from the conscious level to sub-conscious level.

(d) The cause is that (i) the human psyche during the twentieth century has got heavily conditioned and has lost its sensitivity; (ii) this loss of sensitivity is leading to self-destruction and self-annihilation; (iii) intellect or logical reason cannot transform the conditioned psyche, which implies that one has to transcend the partial experience and fragmentary knowledge (information) – a symbolic-dualistic frame of reference of mind spectrum; (iv) one is driven to objective observing and understanding; (v) all discoveries are the result of common intuition and faith, and not cold logic; (vi) we have to enhance our awareness which is observation without identification, without condemnation, without denial and acceptance; (vii) we need moral education where morals emerge from the deepest dimension (subjective) of Consciousness; and (viii) this leads to a quantum jump from symbolic-dualistic frame of reference to non-dual and non-conceptual frame of reference of Mind Continuum.

(e) The accelerative conditioning of the human psyche during the twentieth century has led to a serious crisis – which is that knowledge (information) proliferates but wisdom atrophies. The crisis cannot be resolved at socio-economic-political level or by "United Nations Security Council (UNSC)". The answer is embedded in Albert Einstein's prophetic words, that mankind needs "objective observing and understanding" to attain "peace and serenity", or in other words, "Symbiosis of Science and Spirituality". In common parlance, the aspirations of mankind (and that of UNSC) is: "Creativity in Science!", or "Generation of Innovations for Freedom and Peace".

3. Einstein and Peace

(a) In 1932, after a World Disarmament Conference in Geneva, Einstein, in an unusual bout of anger, exclaimed, "they (politicians and statesmen) have cheated us, they have fooled us. Hundreds of millions of people in Europe and in America, billions of people have been cheated, traded and ticked out of their lives, health and well being". These remarks are equally valid today. In principle man is deceiving himself.

(b) Einstein wrote, "Humanity needed to reach a higher level of awareness to solve the problems of nuclear war". Martin Luther King rightly said, "Science has completely finished our spiritual power, we have guided missiles but misguided men".

4. Where is Science Going?

> Max Planck, Nobel Prize in 1931, in an essay "Where is Science Going?", wrote: It is a fact that there is a point, one single point in the immeasurable expanse of mind and matter, where science and therefore every causal method of research is inapplicable, not only on practical grounds, but also on logical grounds, and will always remain inapplicable. This is the point of [our] individual awareness.
>
> Ilya Prigogine, Nobel Prize in 1977, wrote: Science initiated a successful dialogue with nature .... The dialogue with nature isolated man from nature instead of bringing him closer to it. A triumph of human reason turned into a sad truth. It seemed that science debased everything it touched.

Alexander Koyre wrote that science created "a world in which, though there is a place for everything, there is no place for man. Thus the world of science – the real world – became estranged and utterly divorced from the world of life".

5. Answer to the Present Paradox

Robert Muller, in the Foreword to the SourceBook for the Parliament of Religions, Chicago, 1993, wrote, "Nevertheless, one dimension has been missing from the extraordinary journey, a dimension lamented by Secretaries-General Dag Hammarskjold, U. Thant and Javier Perez de Cuellar : the spiritual dimension, the highest, deepest, most common, universal and binding dimension of all". Towards the end of his mandate Dag Hammarskjold said, "I see no hope for permanent world peace. We have tried but failed miserably. Unless the world has a spiritual rebirth, civilization is doomed". Andre Malraux said, "The third millennium will be spiritual or there will be no third millennium".

Albert Einstein prophetically wrote, we need "objective observing and understanding" which leads mankind to "peace and serenity". But these words have fallen on the deaf ears of mankind.

Herman Weyl wrote, "I have a direct knowledge entirely different from the theoretical knowledge that represents the 'parallel' cerebral process in symbols. This inner awareness of myself is the basis of right understanding of my fellow men". Alan W. Watts wrote that we have to aim at the understanding of life directly, instead of in the abstract, linear terms of representational thinking.

Chinese and Japanese thinking emphasizes the "intuitive component – the importance of inspiration in scientific thinking – the important would appear to be the balance or cooperation between intuition and abstraction". The "understanding of life directly", or "direct perception", or "intuition/ insight/inspiration" is the source of all creativity in science. Nature invariably expresses itself in a non-dual frame of reference of Mind Continuum as direct perception, so nature is always creative. Mankind needs a quantum jump from intellection to intuition.

6. Our Power of Commitment Through "Objective Observing and Understanding"

(a) Life is a multi-dimensional phenomenon and human beings who are expressions of that phenomenon are themselves multi-dimensional organisms. Life is self-generated, self-sustained, self-perpetuating and self-regulated phenomenon. Life is Infinite, Eternal and Immutable. Life is an expression of Supreme Intelligence.

(b) When a system with higher quantum energy potential observes another system with a lower quantum energy potential, the lower one is knocked off or goes to abeyance. This bestows understanding and learning. The intuition (insight), spontaneity and right action is born in the Mind Continuum which blossoms to regeneration of Supreme Intelligence and bring in mutation of the conditioned human psyche.

(c) We will have to initiate educating our physical organism and psychological structure, and work for a paradigm shift in the whole system of education. We will have to work for restructuring of science, restructuring of education, awareness of human perennial values, and live that understanding every moment.

(d) The real science (restructured science-education-health) and real religion (restructured Religion-Faith-Values) are in search of truth and both approaches are non-authoritarian, non-dualistic and non-causation. We urgently need a symbiosis of real science and real religion, which are the dimensions of the wholeness of life, and as such are synonymous. Every existence/event/process/happening is permeated and surrounded by the Cosmic Consciousness, the Life Field, the Unmanifest.

(e) By offering one's life for that psychic mutation to take place, one might create a living cell of a new life style, new way of living. One might be setting into the orbit of human consciousness a spark of love and compassion. One embarks upon that enquiry, not for oneself, but on behalf of the whole human race. This is fundamental to the survival and excellence of man. This is the consummation of human life and living.

7. Relationship

(a) Life flowers and blossoms only in and through relationship. Relationship is the soil in which life can grow, the act of living can grow. The symbolic-dualistic frame of reference constructs violence, fear, confusion, disorder; and has no capacity to construct a true relationship. The non-dual frame of reference bestows true relationships. Relationship is a process of self-revelation, or self-discovery, and without understanding oneself there cannot be any relationship.

(b) Nature invariably expresses itself in a non-dual realm. The perceptions are born of non-duality, of silence; and responses are born of spontaneity. A natural response is invariably non-dual. Restructured science bestows a non-dual perception. Insight is invariably non-dual. Human being has to learn to make a quantum jump from knowledge (duality) to pure perception (non-dual). A non-dual Mind perceives the truth spontaneously. The whole Consciousness is a seamless fluid web of process structures, it is one dynamic flow, and cannot be disassembled or split into simpler and simpler parts.

(c) Modern science has brought man on the control panel of transformation of human psyche. It is hoped that world scientists will act as a beacon to the present tormented world, or perhaps a collapsing world, and guide it in ushering in a "New World Order". It advocates a radical transformation or mutation of conditioned psyche and awaken the supreme Intelligence, which has the capacity to resolve all human challenges and construct a harmonious global society, which is invariably non-violent and peaceful world.

8. Spirituality

Spirituality is the physics of Consciousness, and what we call physics is the science of matter and energy; it is science of living. It is a science that embraces the whole of life – the macrocosm (cosmos), the microcosm (atomic and sub-atomic world) and the sub-microcosm (thought with its associated volitions). Spirituality is the awareness of the wholeness around you and wholeness within you.

Spirituality proceeds from the whole and with the awareness of the whole we observe the particular. In spiritual science, the emphasis is on the act of perception, the quality of the instruments employed for perception, the quality of consciousness behind the perception. Both spirituality and science have non-authoritarian approach – an approach not conditioned by the past.

The Universal Consciousness or Spiritual Energy exists throughout the cosmos, and perhaps beyond the cosmos. Emptiness cannot be measured, silence cannot be measured, love cannot be quantified, tenderness of compassion cannot be verbalized.

We want a revolution, psychological revolution, in which the whole human race has to participate and every man is an equal partner. This revolution will regenerate or rejuvenate human Mind so that it perceives the new dimension of an integral and all-inclusive Consciousness that can guarantee human survival along with all fauna and flora.

9. The monograph is divided in four parts, but this is merely an exegetical convenience, a device to make the complex story of symbiosis of science and spirituality a little easier to tell. Part I investigates "Creativity in Science : Generation of Innovation in Science and Technology". It consists of 8 Chapters. Part II explores "Symbiosis of Science and Spirituality: Survival of Mankind". It consists of 8 Chapters. Part III deals with Science, Education and Ethics. Part IV deals with "A New Vision on Earthquake and Tsunami" and consists of 3 Chapters.

10. Aims and Objects

This monograph is "to initiate an open dialogue with an open mind" with scientists, philosophers and social activists on "Symbiosis of Science and Spirituality". It discusses the generation of innovation in science for human survival. Dialogue implies a willingness to observing (seeing) and listening (hearing) the aspirations of mankind (and that of United Nations Security Council, UNSC) which is "Creativity in Science?" or "Generation of Innovations for Freedom and Peace".

11. The subject is difficult because it is not taught in any school/college, or by parents, or our social society. The present monograph is not written for specialists in different branches in science or for spiritualists, but for layman. We have given copious references to support our findings. The occasional repetition of our insights are not oversights; they are meant to emphasise the insight where emphasis seems advisable and also with the aim that the reader may read any chapter he is interested in and each Chapter must be complete in itself.

12. Our contribution is naught as this is only a collection of 'flowers' from the great scientific truths and from the Supreme Truth handed over to humanity through insight harmoniously blended and presented as a bouquet. The conclusions are valid for all humanity and no sectarianism is inculcated. It is only an infinitesimal collection of flowers in the ambrosial hours of morning from the garden of the Lord.

13. An openness or receptivity implies that inwardly we are not pre-occupied with any activity other than the act of perception. Perhaps such a dialogue may open a new field of mutation of human psyche and may lead to order, harmony, non-violence, peace and bliss.

14. It is evident that mankind cannot afford to waste even a single second, because mankind's survival is embedded in the mutation of conditioned psyche. If man escapes his responsibility, it will be on his own peril, of India and mankind.

Chandigarh, India **Sampooran Singh**

Introduction

1. New Human Challenges in 2005

Mankind is transiting into a global society. Politics have become global. Economics has become global. Globalization is taking place as a result of science and technology and an example is the satellite technology (television with CNN), and the thrust of evolution. The globalization at the material level is happening at an accelerative pace, but global consciousness is not keeping pace with the globalization at the physical level. The events happening in the world are so fast that it is difficult to keep oneself abreast with the events and their underlying currents. We briefly discuss some of the events of 2005.

(a) The UNESCO has declared the year 2005 as the International Year of Physics. We investigate the theme of Albert Einstein's words as man-scientist-philosopher.

(b) President Dr Abdul Kalam addressed the seminar organised by the Indian Physics Association, IIT, Delhi, on the occasion of International Year of Physics, on March 31, 2005. He suggested that we should, "become innovators and creators of science and technology – intensive products". As a response to the call, we penned: "Where is Science Going? : Where is Nature Guiding Mankind!" and "Man's New Dialogue with Nature".

(c) Mr Kofi Annan, Secretary-General, United Nations gave a talk at New Delhi on May 6, 2005 and released a paper entitled, "A More Secure World : Our Shared Responsibility". This will be discussed at UN probably in September 2005.

(d) Dr Manmohan Singh, Prime Minister, on May 8, 2005 said, "A reformed and expanded UNSC has become necessary to meet the global challenges that calls for multilateral action while controlling one's own destiny".

(e) Prime Minister, Dr Manmohan Singh, has constituted a "National Knowledge Commission" vide Gazette Notification No. A.43011/17/2005-Adm.I of June 13, 2005 to "sharpen India's knowledge edge". The Commission will advise P.M. to promote excellence in the education system and creativity in science.

(f) The University of Delhi has decided to celebrate the Birth Centenary of late Prof D.S. Kothari by holding "National and International Seminars" on "Science, Ethics and Education" in 2006.

(g) Hon'ble President Abdul Kalam during his visit to St Petersburg (Russia) on May 25, 2005 called for "pooling of scientific knowledge and resources between the countries to enhance capabilities for the prediction of earthquakes, tsunamis and other natural disasters". He also stressed the need for more collective effort to address issues such as ecological imbalances.

2. Limitations of Conditionings of Human Psyche

Albert Einstein perceived an accelerative conditioning of the human psyche during the first five decades of twentieth century. The psychological imbalances, ecological crisis, a rising wave of violence and crime, terrorism, religious fundamentalism, ethnic conflicts, corruption, increasing unemployment, poverty, hunger, malnutrition, health crisis, are forcing man inevitably to face the realities of the human situation. Man has been trying to resolve these challenges for the last 10,000 years, or more, but all these efforts have proved to be sterile. All knowledge (information), thought-structures, ideals, concepts are sterile in transforming the psyche of man, or in other words, to bring radical transformation or mutation in the human psyche. Einstein prophetically wrote the following:

(a) The significant problems we face cannot be resolved at the same level of thinking we used when we created them.

(b) Humanity needed to reach a higher level of awareness to solve the problem of nuclear war.

(c) Related truths can be achieved by the process of reasoning.

This suggests that reason or intellect has no capacity to bring mutation in the human psyche, because the construct of the mind and the intellect are at the same quantum energy potential of consciousness. The construct of the mind is invariably dualistic so it always creates chaos and conflict. Einstein's "intuitive wisdom" is identical to the ancient Indian heritage or message of Upanishads, or the saints-sages of the East.

3. The Arts of Observing and Listening

(a) Everything that we experience as material reality is born in the invisible realm beyond space and time – a non-dual, a non-conceptual, and timeless Reality. The arts of Observing, Listening, Understanding, Learning, Action emerge from the non-duality. The invisible source of all that exists is not an empty void, but the womb of creation. There is a basically Unconditioned Energy and perennial knowledge. This is called Active Reality, which is powerful, highly directional, monochromatic (same frequency) and coherent (the wave is in one phase) beam of energy, or a state of highest frequency (higher frequency means higher energy), or highest quantum energy potential.

Man is an amphibian who lives simultaneously in two dimensions – the *known*, the time-space matrices; and the *Unknown*, the timeless realm. It is a quantum jump from time to timelessness, from activity to non-action, from movement to non-movement, and from speech to verbal silence. The timeless realm is hidden in man, but he has a hidden potential to energise this dimension. The known constructs conflict, duality which leads to nervous tension and chemical pressure. The known is leading man to self-destruction or self-annihilation.

Erwin Schrodinger remarked, "We cannot make any factual statement about a given natural object (or physical system) without 'getting in touch' with it – (and it involves consciousness of the object). Even if it consists only in our 'looking at the object' the latter must be hit by light rays and reflect them into the eye, or in some instrument of observation. This means that the object is affected by our observation. You cannot obtain any knowledge about an object while leaving it strictly isolated".

Sullivan put it, "We cannot observe the course of nature without disturbing it", or Andrade, "Observation means interference with what we are observing .... observation disturbs reality". It was clear to these physicists that objective measurement and verification could no longer be the mark of absolute reality because the measured object could never be completely separated from the measuring subject.

(b) Psychological Mode of Mind: There is sharing of energy between the object and the subject, so both are distorted. The subject is aware of distorted perception and the distorted memory is recorded.

The subject is invariably the past. The past is the "recalled memory", which results in coherent superpositions of different webs on the direct perception. This results in distortion of perception and modification of memory. The world is seen not as it is, but in its various relationships to the "me" of "memory". This division in the life and the flourishing of everything we call our psychological being, and from this arises all contradiction and division. The "me" is the response, in relationship, of past memory, past experience. It has no factual reality, it is a concept, an ideal. There is the superficial awareness at the psychological mind.

In the timeless realm, the "me" is non-seeing. The "me" cannot see, cannot be aware. There is no "me" in the present, so the past is the "me".

We observe whatever comes up in a non-reactional attentiveness, without accepting or rejecting it, without any measurement or evaluation, then the quantum energy potential of the subject of cognizance enhances. When such observation takes place and the past contained in us is exposed to our attentiveness, there is a living encounter with one's factual being, and the past goes to abeyance. So observation exposes the realities to ourselves. If the state of observation is sustained then a change comes about in the person spontaneously.

(c) Objective Science: The objective science is aware of the objective reality. The objective fact cannot penetrate a boundary that physicists call "the event horizon". On one side there is everything – transient, ephemeral, finite, mutable and below the speed of light – and on the other side, there is non-duality, timelessness, eternal, infinite and immutable. The speed of light is absolute, it is like a wall that no object can go through. As we approach the wall, time slows down, mass increases and space becomes curved.

There is no transference of energy between the object and the subject. It moves from *fact* to *fact*. It is a movement in the field of materialism. It is a movement from the *known* to the modified *known*, so it constructs artifacts with higher sensitivity. It constructs gadgets for human comfort, say an aeroplane, refrigerator, air-conditioners. It also leads to fabrication of war gadgets capable of destruction. It is leading mankind to self-annihilation.

(d) Subjective Science: We have to restructure subjective science so that it takes in its realm the objective and the subjective dimensions of life. The objective is a tiny wave of the subjective and there is a continuous interaction between the two which takes few microseconds. The outer (objective) and the inner (subjective) are not two separate movements, both are one whole, a unitary movement. The dimensions of the restructured science are:

[Science-Education-Ethics-Health]

It has total understanding of the wholeness of life, and no part is ignorant. The non-subjective perception bestows direct perception. We are part of the cosmic dance of energy. It is a pointer to oneness of life.

An "Open Science" bestows awareness of the Oneness and the Wholeness of life. It is Perceptive Intelligence or Discovery to enjoy order, rhythm, freedom, nonviolence, love, peace and bliss.

4. Resolution of the Present Paradox of Conditioning of Human Psyche

We have been investigating the "Symbiosis of Science and Spirituality" for the last 25 years. Nature is benevolent, and has now revealed ***Her Truths*** as "intuitive wisdom" which has been penned at the rational level of mind. There is an urgency to share this with mankind. Copious references have been added wherever necessary. We have divided the vision in three parts, but this is merely an exegetical convenience, à device to make the complex story of manifestation of Consciousness a little easier to tell. A part dealing with a vision of Earth Quakes and Tsunamis have been added.

This monograph discusses the resolution of all old and new human challenges; or in other words, it leads to a mutation of the human psyche.

5. Spirituality

Max Planck, Nobel laureate in physics and founder of quantum mechanics, in an essay entitled "*Where is Science Going*?" wrote:

> It is a fact that there is a point, one single point in the immeasurable expanse of mind and matter, where science and therefore every causal method of research is inapplicable, not only on practical grounds, but also on logical grounds, and will always remain inapplicable. This is the point of [our] individual awareness.

We intuit that the "one single point of awareness" is the non-dual and non-conceptual and timeless frame of reference of Mind Continuum. Every causal method of research or symbolic-dualistic frame of mind spectrum is inapplicable. This is the spiritual dimension in man.

Spirituality is the science of the wholeness of life. In spiritual science, the emphasis is on the act of perception, the quality of the instruments employed for perception, the quality of consciousness behind the perception. Spirituality implies exploration of the maximum quantum energy potential of consciousness. Both Open Science and Spirituality have a non-authoritarian approach.

Albert Einstein, the greatest physicist of the age, wrote about 70 years ago, that mankind needs "objective observing [and listening] and understanding" which leads to "peace and serenity". But these words have fallen on the deaf ears of mankind. This is evident from the words of Dag Hammarskjold, Secretary-General of the United Nations, which is as under:

> I see no hope for permanent world peace. We have tried but failed miserably. Unless the world has a spiritual rebirth, civilisation is doomed.

The Parliament of World's Religions, Chicago, 1993, did not touch the dimension of "spirituality" and emphasised that unless the world has a spiritual rebirth, civilisation is doomed. Many eminent scientists – say George Wald (N.L., 1967), Jacques Monod (N.L. 1972 in Biology) – stated that the present objective science is leading mankind to materialism, and this is in the direction of self-annihilation, self-destruction. The whole world is drifting towards a collective suicide.

There is an urgency to "initiate a dialogue with an open mind" and such "open sharing" may lead to a second model of mutation of human psyche. We have to restructure our educational model at University/College/School levels, ethics and health technology. The need of the hour is "exploration, observing and understanding" the wholeness of life, which regenerates the Intelligence. This has the capacity to construct a sane, ordered and harmonious society.

The greatest mystery of life is that man has probably been conditioning for about a million years [the evidence is for 10,000 years or more] and man has a hidden potential to transcend all man-made barriers in a few years; and revel in order, harmony, nonviolence, peace and bliss.

6. Human Survival: A Scientific Vision

Albert Einstein, in the prophetic vision, stated that we need "objective observing and understanding" and this bestow us "peace and serenity". He added, "What is truly valuable in our bustle of life is not the nation, I should say but the creative and impressionable individuality, the personality – he who produces the noble and sublime". He added, "A human being is a part of the whole, called by us universe .... Our task must be to free ourselves from this prison by widening our circle of compassion to embrace all living creatures and the whole nature in its beauty".

Einstein in an article entitled "*Out of My Later Years*" prophetically stated, "Nothing of all that (psychological imbalances) will remain but few pitiable pages in history books, briefly picturing to the youth of the future generations the follies of our ancestors". George Wald, Nobel prize in medicine, 1967, wrote, "By now the problem is not transcendence, but survival". Jacques Monod, Nobel prize in 1972 in biology wrote, "What we see opening before us is an abyss of darkness". P. Krishna, an eminent physicist, on April 3, 1993, said, "We are going towards self annihilation and not much time is left now". George Blaker, President of the Network, UK, put it succinctly, in 1999, "science of today cannot stop the spiritual advance, but can only delay it for a while, for such delays will automatically drag in their wake calamitous consequences, including geological disasters".

Werner Heisenberg, Nobel laureate in physics, said to one of his students, in 1975:

> You know, in the West we have built a large, beautiful ship. It has all the comforts in it, but one thing is missing: it has no compass and does not know where to go. Men like Tagore and Gandhi and their spiritual forebears found the compass. Why can this compass not be put in the human ship so that both can realize their purpose?

The aims and objects of this monograph is to "initiate a dialogue with an open mind" on the design and fabrication of the *compass* and then instal it in the human ship. This is bound to lead to ending all human challenges and usher a new era of order, harmony, non-violence and peace.

Hon'ble President Abdul Kalam wrote in the "Ignited Minds" (2003):

> Spirituality must be integrated with education. Self-realization is the focus. Each one of us must become aware of our higher Self. We are links of a great past to a grand future. We should ignite our dormant inner energy and let it guide our lives. The radiance of such minds embarked on constructive endeavour will bring peace, prosperity and bliss to this nation.

7. Human Survival : A Philosophical Vision

In 1911, Sri Aurobindo wrote in the *Song of Humanity* :

> A time will come when the Indian mind will shake off the darkness that has fallen upon it, cease to think or hold opinions at second and third hand and reassert its right to judge and enquire with perfect freedom into the meaning of its own culture and tradition.

Sri Aurobindo in the message on August 15, 1947 stated, "A catastrophe may intervene and interrupt and destroy what is being done, but even then the final result is sure .... A new spirit of oneness will take hold of the human race". The Divine Mother in 1963 wrote, "The only hope for the future is a change of man's consciousness and the change is bound to come". The message in 1967 read, "Men, countries, continents! The choice is imperative: Truth, or the abyss".

The Mother had once observed, "Three-fourth of humanity is obsolete".

J. Krishnamurti, the saint-philospher had made some prophetic remarks, which are as under:

(a) An ancient race, three to five thousand years of certain culture has been wiped overnight .... We are the history of mankind. We are the story of mankind.... I am neither individual, not that I am the collective. I am humanity, not the collective.

(b) India has lived all these centuries on ideas. She has to move from a life based on ideas, which are non-facts, to seeing of fact only.

(c) Krishnamurti spoke of the true religious spirit that probed, questioned and negated. He said, "America had not got it. Europe had not got it. In India it has been wiped out and destroyed. Yet it was there in India, waiting in the soil. Can we do something about it?

8. Concluding Comment

Fritjof Capra, an eminent physicist, wrote, "As the turning point approaches, the realization that evolutionary changes of this magnitude cannot be prevented by short-term political activities provides our greatest hope for the future".

For the past few decades, nature is working on mutation of human psyche, or in other words, a quantum jump from psychosocial evolution to spiritual evolution. We perceive that:

If we move one step towards God,
God advances a hundred steps to receive us

Our spiritual wisdom has been our strength. Our duty is "Symbiosis of Science and Spirituality". We have to enhance our awareness through "observing (seeing) and listening (hearing)" both individually and collectively. In this is embedded our survival. This only is the social, psychological, and even the political revolution. This is the only vocation of mankind.

Chandigarh, India **Sampooran Singh**

Part-I
Creativity in Science

When non-dual and non-conceptual frame of reference of Mind Continuum is bathed by the highest quantum energy potential of Consciousness (Life Field), then true science and true art emerge. Einstein wrote that we need "objective observing and understanding" to attain "peace and serenity". This observation flows to intuition or insight, this regenerates or rejuvenates Intelligence which blossoms to ending all psychological imbalances and ecological crisis and other human challenges.

When thought-structure coupled with volitions go to abeyance, all conflict, misery and pain go to abeyance and the Mind constructs a society which is free from all conflict, chaos, fragmentation, and this constructs a sane society.

Reforms need further reform, this is an endless process. Reform is in the social-political-economic field, and any action cannot bring in mutation in the psyche.

It is only ending of thought-structure coupled with volition that can bring about mutation of the mind-brain-body system. Mutation means ending of conflict and assertiveness. The mind is free from the weight of time, free from the burden of the past, free from all the actions of the will – this is the new human being. This only is the social, psychological, and even the political revolution.

1

Albert Einstein : Man, Philosopher and Scientist 14 March, 1879 – 18 April, 1955

Abstract

UNESCO has declared the year 2005 as the International Year of Physics. To celebrate Einstein's birthday, Prof S.P. Puri gave a talk entitled "*Albert Einstein: Man, Scientist and Philosopher*" mainly discussing the "Time Line of Scientific Discoveries" and also quoted some of his philosophies.

This paper discusses the distilled intuitive wisdom of Albert Einstein especially as man, philosopher and scientist. One is deeply struck by the broad agreement of Einstein's utterances (quotes) with those of the Upanishadas of India, or with his contemporary Mahatma M.K. Gandhi. This substantiates that the truth is one which is Eternal, Infinite and Immortal, and is independent of time. It discusses a string of quotations – each emerging from a non-dual and non-conceptual frame of reference of Mind Continuum, perhaps like beads on a string.

A depth exploration of Einstein's essay entitled – "Out of His Later Years" – reveals two shocking discoveries – first, the human psyche has got highly conditioned during the first few decades of the twentieth century; and second, in Einstein's words, "Nothing of all that [conditioning] will remain but a few pages in the history books, briefly picturing to the youth of the future generation the follies of its ancestors'; or in other words, the human survival is at stake.

The teachings of Einstein on education are identical to those of Gandhiji and Sri Aurobindo. Education is a deep mystery of life as it brings mutation in the human psyche. These have a special connotation because these deal with a new culture, a new dynamics and a new foundation for human relationships.

1. Introduction

"Einstein was rather a lonely man; he had few friends, few coworkers, few students. He lived in a sad time: the two World Wars. It is not surprising that for Einstein science was the road that led to victory over the turmoil of time. His world was full of observers, of scientists situated in various coordinate systems in motion with one another, situated on various stars differing by their gravitational fields. All these observers were exchanging information through signals all over the universe. What Einstein wanted to preserve above all was the objective meaning of this communication", in the words of Ilya Prigogine, Nobel laureate in physics, 1977. He was in passionate love for mankind. He was embedded in intuitive wisdom. His love for the creator (God) and his words show clearly that he was totally realised and he lived in the highest quantum energy potential of Cosmic Consciousness. We '*intuitively feel*' that Einstein was gifted by nature with Supreme Intelligence which he expressed as highly awakened intuition (insight). Here are revealing glimpses of the unique personality.

Federico Mayor Zaragoza, both as head of the UNESCO and his personal capacity as a scientist, addressed seventy-four Nobel Prize Winners in Paris in early 1988 (refer The Courier, May 1988) prophetically said, "UNESCO's task (is) paving the way for that new ethical outlook without which peace and genuine human progress are not possible, and never will be". Einstein also emphasised that man has "to attain peace and serenity" which he cannot find in the conditioned psyche. This article is also in the direction of peace – the aim set by UNESCO.

The deep understanding of life by Albert Einstein, Niels Bohr, J.A. Wheeler and many other scientists make one certain that men like Gandhiji, Buddha, Jesus, Laotse, Confucious really lived.

2. Einstein as Man

(a) Man is an amphibian, who lives simultaneously in two worlds – (a) the *known*, the time-space-causation matrices, the objective world; and (b) the *Unknown*, the timelessness, the subjective world. Science explores the objective world and religion explores the subjective world. Albert Einstein stated that "Science is lame without religion, and religion is blind without science".

(b) Brain is material and belongs to objective world. It can be investigated objectively. Mind is outside science, it is subjective. Mind is independent and perhaps controls the brain. He firmly believed that it is a Superior Mind that reveals itself in the world of experience. We see that the non-dual and non-conceptual frame of reference of Mind Continuum is the Superior Mind.

(c) We should take care not to make the intellect our God, it has, of course, powerful muscles, but no personality.

(d) Man is here for the sake of other men, for the sake of the whole mankind; and indeed for the sake of all living beings .... The chief motivating force behind all technical effort ought to be a supreme concern for the destiny of the human race .... Strange is our situation here upon earth. Each of us comes for a short visit, not knowing why, yet sometimes seeming to divine a purpose.

(e) I am absolutely convinced that no wealth in the world can help humanity forward, even in the hands of the most devoted worker in this course. The example of great and pure individuals is the only thing that can lead us to noble thoughts and deeds. Money only appeals to selfishness and irresistibly invites abuse. [1934]

(f) The most important human endeavour is the striving for morality in our action. Our inner balance and even our very existence depend on it. Only morality in our actions can give beauty and dignity to life.... To make this a living force and bring it to clear consciousness is perhaps the foremost task of education .... The foundation of morality should not be made dependent on myth nor tied to any authority lest doubt about the myth or about the legitimacy of the authority should imperil the foundation of sound judgement and action. [20 November, 1950]

(g) How strange is the lot of us mortals! Each of us is here for a brief sojourn; for what purpose he knows not .... I regard class distinctions as unjustified and in the last resort based on force .... The trite objects of human efforts – possessions, outward success, luxury – has always seemed to be contemptible. [1931]

3. Einstein and Religion

(a) The most beautiful experience we can have is the mysterious. It is the source of all true art and science. He to whom this emotion is a stranger, who can no longer pause to wonder and stand rapt in awe, is as good as dead: his eyes are closed. The insight into the mystery of life, coupled though it be with fear, has also given rise to religion. To know that what is impenetrable to us really exists, manifesting itself as the highest wisdom and the most radiant beauty which our dull faculties can comprehend only in their most primitive forms – this knowledge, this feeling, is at the centre of true religiousness. In this sense, and in this sense only, I belong in the ranks of devoutly religious men.

(b) The conversation (on Einstein's yacht) drifted from profundities about the nature of God, the universe and man, to lighter questions. Suddenly, Einstein lifted his head, looked up at the skies and said, "We know nothing about it at all. Our knowledge is but the knowledge of schoolchildren.... we shall know a little more than we do now. But the real nature of things – that we shall never know, never".

(c) "A human being", wrote Einstein "is a part of the whole, called by us 'Universe', a part limited in time and space. He experiences himself, his thoughts and feelings as something separated from the rest – a kind of optical delusion of his consciousness. This delusion is a kind of prison for us, restricting us to our personal desires and to affection for a few people nearest to us .... Our task must be to free ourselves from this prison by widening our circle of compassion to embrace all living creatures and the whole nature in its beauty. Nobody is able to achieve this completely, but the striving for such achievement is in itself a part of the liberation and a foundation for inner security".

(d) I believe in Spinoza's God, who reveals himself in the harmony of all being, not in a God who concerns himself with the fate and actions of men.

4. Einstein as Philosopher

(a) He, during a serious illness, was questioned, whether he was at all afraid of death, he replied, "I feel such a sense of solidarity with all living things that it does not matter to me where individual begins and ends". And he added, "There is nothing in the world which I could not dispense with at a moment's notice". Such utter fearlessness of death and total detachment "liberation from the self" and such complete dedication in ceaseless pursuits of truth can be achieved only on the foundation of the deepest spiritual state. This suggests that Einstein had achieved the highest quantum energy potential of Consciousness.

(b) He wrote, "The significant problems we face cannot be solved at the same level of thinking we used when we created them". He added, "Our new idea is simple: to build a physics valid for all coordinate systems" or frames of reference of mind spectrum (time-space-causation matrices), and Mind Continuum (the non-dual and non-perceptual frames of reference). He added, "I cannot conceive of a genuine scientist without that profound faith [the most incomprehensible thing about the universe is that it is comprehensble]". He also said, "Related truth can be achieved by process of reasoning".

c) It is enough for me to contemplate the mystery of conscious life perpetrating itself through all eternity, to reflect upon the marvellous structure of the universe which we can dimly perceive, and to try humbly to comprehend even an infinitesimal part of the intelligence manifested in nature.

5. Einstein and Gandhiji

(a) Einstein said (1939) on the occasion of seventieth birthday of Gandhiji that "generation to come will scarce believe that such a one as this ever in flesh and blood walked upon this earth".

(b) Revolution without the use of violence was the method by which Gandhi brought liberation of India. It is my firm belief that the problem of bringing peace to the world on a supranational basis will be solved only by employing Gandhi's method (*Satyagraha*) on a large scale .... Gandhi, the greatest political genius of our time, indicated the path to be taken. He gave proof of what sacrifice man is capable of once he has discovered the right path. His work on behalf of India's liberation is living testimony to the fact that man's will, sustained by indomitable conviction, is more powerful than material forces that seem insurmountable".

(c) In his later years at the Institute of Advanced Study in Princeton (USA), Einstein had on the almost bare walls of his room a picture of Gandhiji – whom he had never met personally. To visitors, pointing to the picture, he would say Gandhiji was the greatest man of the age. The ethical/spiritual values of Einstein and Gandhiji were very close.

6. Einstein and Peace

(a) In 1932, after a World Disarmament Conference in Geneva, Einstein, in an unusual bout of anger, exclaimed, "they (politicians and statesmen) have cheated us, they have fooled us. Hundreds of millions of people in Europe and in America, billions of people have been cheated, traded and ticked out of their lives, health and well being". These remarks are equally valid today. In principle man is deceiving himself.

(b) The famous statement, issued in 1955, jointly by Bertrand Russell, Einstein, Yukawa and others is, if any thing, even more relevant today. The declaration of the Pugwash Movement on the occasion of the 25th Pugwash Anniversary Conference in Warsaw (August 1982), signed by amongst others, nearly 100 Nobel Laureates, concludes:

> There lies before us, if we choose, continual progress in happiness, knowledge, and wisdom. Shall we, instead choose death, because we cannot forget our quarrels? We appeal, as human beings to human beings: Remember your humanity and forget the rest.

(c) Had I known that the Germans would not succeed in producing an atomic bomb, I would not have lifted a finger.

(d) I know not with what weapons World War III will be fought, but World War IV will be fought with sticks and stones.

7. Einstein as a Scientist

(a) The theory of special and general relativity is such a radical and profound departure from anything previously known to physics that it could hardly be discovered on the basis only of developments with physics and mathematics. The dethronement of the concepts of absolute space and absolute time, deeply rooted in meta-physics and Newtonian physics, came about, it seems, because of a confluence – a meeting together, as it were, in one mind – of the profound mathematical developments originating with Gauss and Riemann in Germany in the nineteenth century and a new (for the West) philosophical – epistemological outlook very similar to Buddhism and Vedanta, and possibly derived from them.

(b) Soon after recommending Einstein for the job, Max Planck, said, "His work on relativity probably exceeds in audacity everything that has been achieved so far in speculative science and even in epistemology; non-Euclidean geometry is child's play in comparison".

(c) You believe in a dice-playing God and I in perfect laws in the world of things existing as real objects, which I try to grasp in a wildly speculative way.

(d) Quantum mechanics is certainly imposing. But an inner voice tells me that it is not yet the real thing. The theory says a lot; but does not really bring us any closer to the secret of the 'old one'. I, at any rate, am convinced that *He* is not playing at dice.

(e) The Heisenberg-Bohr tranquilizing philosophy – or religion? – is so delicately contrived that, for the time being, it provides a gentle pillow for the true believer from which he cannot very easily be aroused. So let him lie there.

(f) One thing I have learned in a long life: that all our science, measured against reality, is primitive and childlike – and yet it is the most precious thing we have.

8. Einstein and Education

(a) The mind can proceed only so far upon what it knows and can prove. There comes a point where the mind takes a leap – call it intuition or what you will – and comes upon a higher plane of knowledge, but can never prove how it got there. All great discoveries have involved such a leap.

(b) The experiences of an individual appear to us arranged in a series of events, in this series the single events which we remember appear to be ordered according to the criterion of 'earlier' and 'later'. There exists, therefore, for the individual, an I-time or subjective time. This in itself is not measurable. We opine that I-consciousness has no factual reality at all. It has a conceptual reality, it has an ideational reality and it has a relevance in our lives".

(c) I believe with Schopenhauer that one of the strongest motives that lead persons to art and science is flight from the everyday life, with its painful harshness and wretched dreariness, and from the fetters of one's own shifting desires. One who is more finely tempered is driven to escape from personal existence and to the world of objective observing and understanding. .... In this image and its formation, he places the centre of gravity of his emotional life, in order to attain the *peace and serenity* that he cannot find within the narrow confines of swirling, personal experience.

(d) Appearing in time-space and causality, this Being is man, as we know him, but behind all appearance is the One Reality, the 'Mysterious', the Life Field. Einstein called the time-space-causation as an "optical delusion (a prison) of his consciousness".... we have "to free ourselves" from this prison, and to perceive the mysterious. And this freedom is possible through "objective observing and understanding". We opine that Einstein's words and concepts of mental "person", and aim of human life is identical to that of Upanishads and those of Gandhiji.

(e) The most incomprehensible thing about the universe is that it is comprehensible.

(f) Out yonder there was this huge world, which exists independently of us human beings and which stands before us like a great, eternal riddle, at least partially accessible to our inspection.

(g) The essence of a man like me lies just in what he thinks and how he thinks – not in what he does or suffers.

(h) To the sphere of religion belongs the faith that the regulations valid for the world of existence are rational, that it is comprehensible to reason. I cannot conceive of a genuine scientist without that profound faith.

(i) Physics is essentially an intuitive and concrete science. Mathematics is only a means for expressing the laws that govern phenomena. We opine that fundamental discoveries which go deep into the nature of things are the results of common intuition and faith, and not cold logic.

(j) How can it be that mathematics, being after all a product of human thought which is independent of experience, is so admirably appropriate to the objects of reality?

(k) The only rational way of education is to be an example, if one cannot help it, a warning example.

(l) Teaching should be such that what is offered is perceived as a value gift and not as a hard study. In our understanding teaching should refer to direct perception and to reject the indirect perception, and further value education is imparted in educational pattern.

(m) What is truly valuable in our battle of life is not the nation, but the creative and impressionable individuality, the personality – he who produces the noble and sublime while the common herd remains dull in thought and insensible in feeling.

(n) It is, in fact, nothing short of a miracle that the modern methods of instruction has not yet entirely strangled the holy curiosity of enquiry; for this delicate plant, aside from stimulation, stands mainly in need of freedom, without this it goes to wrack and ruin without fail.

9. Einstein to resolve Human Challenges

(a) Einstein – **Out of His Later Years** – wrote, "The confidence in the stability, even the very basis of existence, of human society has largely vanished. One senses not only a threat to man's cultural heritage but also that a lower value is placed upon all that one will like to see defended at all costs .... Life in human society meant dangers of all sorts .... even standards of virtue, aspiration and practical truth were taken for granted as an inviolable heritage, common to all civilised humanity .... The sanctity of life vanished .... The lie was raised to the dignity of a political instrument. It was a manifestation of weakness of character and lack of responsibility on the part of the leaders in the affected countries". This presents evidence that the human psyche during the first five decades of twentieth century got heavily conditioned leading to lowering of its sensitivity.

(b) Recent rise of terrorism, religious fundamentalism, drug trafficking indicates an enhancement of the conditioning of the psyche. To sum up, the twentieth century has added to conditionings of the psyche and is leading to self-annihilation, or self-destruction. And this is now threatening the extinction of all life-supporting system within a foreseeable period of time.

(c) Recently many eminent scientists – say Harlow Shaply, George Wald, Jacques Monod, P. Krishna – have discussed the conditioning of the human psyche which is leading to self-annihilation.

(d) There is an urgency to resolve the crisis of our perception or human paradox.

10. UNESCO on Education and Society

(a) The UNESCO Commission presided over by M. Edgar Faure, in 1972; the UNESCO International Commission, Chaired by Jacques Delor in 1996, have discussed educational reforms. Some other perceptive thinkers have suggested an action plan or a paradigm shift, but all these have proved to be sterile for radical transformation of the human psyche.

(b) Einstein discussed Superior Mind, which enjoys freedom; transcends wealth, desires, experience and knowledge (information); transcends intellect (reason); through objective observing and understanding; and this leads to a perception which flows to revolution without violence. It also bestows gift of intuition and moral values (spiritual). It leads to a quantum jump from a dualistic mind to non-dualistic (Superior) Mind. The Superior Mind leads to perception of the 'Mysterious', which is the source of all art and science, peace and serenity. It is unfortunate that mankind is escaping his responsibility to understand and live this intelligence/wisdom; so the present ignorance is leading man to self-annihilation or self-destruction.

(c) Albert Einstein in an article entitled "***Out of My Later Years***" pleaded that: (i) the human psyche during the first few decades of the twentieth century has got heavily conditioned and has lost its sensitivity; and (ii) this loss of sensitivity is leading to self-destruction and self-annihilation. Recently we proposed restructuring of the subject of cognizance – modified dimensions of science-education-health – which deals with the wholeness of life. When thought structure coupled with volitions (object) encounters subject-of-cognizance, the object gets exploded and is annihilated. The object goes to abeyance and dualism ends. The space-time matrices also goes to abeyance. The knowledge also stops. All this happens in less than hundred microseconds. This leads to ending of thought-structure and the subject has the awareness of Cosmic Consciousness, or Life Field, or Mysteriousness. This is briefly discussed in a recent research paper entitled, "***Peace, Security and Survival? : A Scientific Quest, July 5, 2004.***" [submitted to UNO]. A paradigm shift or action plan is also discussed in this paper.

(d) Einstein is an example that his communion with the Superior Mind or awareness of the Mysteriousness has led him to do better science. All "perennial knowledge" is structured in Cosmic Consciousness, so a creative mind is sensitive and can produce a noble and sublime creativity or originality in all art and science. We opine that discovery is the beginning of creativeness; without creativeness (intuition) there can be no peace or happiness for man.

(e) Nature bestowed the gift of intuition to Einstein. An awareness of the Mysteriousness, or in other words, intelligence highly awakened is intuition, which is the only true guide in life.

(f) The foundation of education for a global society is to perceive the '*mysterious*', the Life Field, and then live in the light of that communion with the laws of nature. It is a renewed look at the meaning and significance of the wholeness of life. It investigates the conditioned psyche (self-knowledge) and explored the Unknown (self-awareness).

(g) The need of the hour is that the UNESCO may implement this technology on the highest priority – as that of "Manhatten Project". It is suggested that the UNESCO may share this with mankind, and initiates "an open dialogue with an open mind" with every man on this globe.

(h) Man is not the product of socio-economic forces, he is fundamentally a mine rich in gems and pearls of inestimable value – called spiritual values – freedom, order, harmony, non-violence and peace. We see that the invisible force of nature – as Einstein's philosophy – is now moving to set man's vision right and establish a way of life which will again fulfil the high purpose of existence.

11. Concluding Comment

(a) Both Albert Einstein and Gandhiji had identical intuitive perception of life. Both advocated that the resolution of human challenges can only be possible through non-violent (Ahimsa) approach, called *Satyagraha* (sword of self-suffering). Both were spiritually identical and were the greatest human beings of twentieth century.

(b) Einstein advocates, "Objective observing and understanding" which leads to transcending the daily living in the common waking state of mind. This leads to non-dual and non-conceptual frame of reference of Mind Continuum, and this helps, "to attain the peace and serenity that he cannot find within the narrow confines of swirling, personal experience".

(c) "Physics is essentially an intuitive and concrete science". He added, "It is in fact, nothing short of a miracle that the modern methods of instruction (education) have not yet entirely strangled the holy curiosity of inquiry; for this delicate plant aside from stimulation, stands mainly in need of freedom; without this, it goes to wrack and ruin without fail".

(d) *Commenting on some physicist's work, he wrote, "He could not really understand how anybody could know so much and understand very little". Knowing so much is gathering of information at the intellect level, so intellectual approach cannot bring in mutation of human psyche. What we need is 'understanding' which comes with 'direct perception'.*

(e) The distilled intuitive wisdom of Einstein is an ample proof to declare, "When every desire that finds lodging in the heart of man, has been loosened from its mooring, then this mortal puts on immortality; even here he tastes God in this human body". [Sri Aurobindo's translation II-3-14]. In other words, India must embark on ending of violence (*Ahimsa*) through observation-listening *here* and *now*.

(f) We presented a report entitled, "***Peace, Security and Survival? : A Scientific Quest***" (sent to UNO). We have to restructure our science – "science-education-health". We need to restructure the society which is not based on symbolic-dualistic thought, but based on non-dual and non-conceptual thought. This also leads to arresting the fall of civilisation and flows to rise of civilisation. A pilgrimage to Supreme Intelligence is discussed. We have to understand and live according to the immutable laws of nature. It implies that there is an urgency to understand "Symbiosis of Science and Spirituality".

(g) The answer to human problems is embedded in direct perception (intuition), to communicate that intuition by verbal cognition and the words are drawn from the factual memory, and draw rational inference (called quantum logic or unfettered philosophical reason). Intuition is fundamental and rationality is a derivative. It is intuition or insight that bestows values, which leads to healthy and sane life. It is this living which leads to the perception of the evolutionary qualities of insight, creativity, spontaneity and oneness of life. This is the key to human survival and our planet Earth with its fauna and flora.

(h) The principles of education is to understand ourselves. It is not merely acquiring knowledge, gathering the correlating facts; it is to see the significance of life as a whole. Teaching is a value gift. All spiritual values emerge from the dimension of the Mysteriousness.

(i) The fundamental principles of education by Albert Einstein, Mahatma Gandhi, Sri Aurobindo, J. Krishnamurti, Vimala Thakar, Swami Om Poorna Swatantra, and many others are identical. Erwin Schrodinger also gave similar commentary on Aldous Huxley's book entitled, "The Perennial Philosophy". All these principles have fallen on the deaf ears of mankind.

(j) We have brought the human race to an explosive situation and our life is poised on razor's edge. Einstein wrote, "Man changes but little .... though current trends like the present bring him unimaginable sorrow. Nothing of all that will remain but few pitiable pages in history books, briefly picturing to the youth of the future generations the follies of its ancestors". Perhaps, the benign turn of destiny, who have faith in the mysterious will be saved.

Preview

Albert Einstein was one of the greatest sages-seers of the world and physics flowed from him as intuitive intelligence/wisdom from the "mysterious", the "Life Field".

Einstein's unique contribution to the world as man and philosopher is:

(a) the human mind gets conditioned in our infancy and we have to stretch it to fit the reality;

(b) the problem of bringing peace to the world on a supranational basis, will be solved only by employing Mahatma Gandhi's technique of "revolution without the use of violence";

(c) he advocated education to have foundation of "objective observing and understanding";
(d) we have to base ourselves on intuition (insight) and transcend experience and knowledge (information); and
(e) if man escapes his responsibility, he does so at his own peril and that of mankind in general.

The time is running out. We suggest that UNESCO may disseminate this intuitive wisdom to every person on the globe, and "initiate an open dialogue with an open mind". A second model of mutation of human psyche may emerge out of this sharing and dialogue. This is a cornerstone of human survival.

(S P Puri)
Prof. of Physics (Retd.)

Panjab University
Sector 14, Chandigarh – 160014,
India

2

Where is Science Going? : Where is Nature Guiding Mankind!

Abstract

The aim of science is to study nature and human experience objectively. Intellect gives us objective truth, as it ignores the subjective world. It leads to invention in the technological field, which is an improvement in the field of the known to the known.

We have to restructure the subject-of-cognizance, or the true-subject, which is a new dimension of "science-education-health". The dimension of science includes both the objective and the subjective world; the dimension of education deals with the sensual and the trans-sensual fields of life; and the dimension of health embraces the whole field of mind-brain-body system, or the fullness of health. The true-subject observes the objective world and the subjective world which are in the realm of the true-subject. It knocks down the thought-structure coupled with volitions. The true-subject observes the direct perception, or timeless perception. Nature invariably expresses itself as direct perception which blossoms to intuition (insight), and this energises the Intelligence/Wisdom. This expression of truth will lead the scientist to a "continuous innovator and creator of science and technology".

The Hon'ble President, Dr A P J Abdul Kalam addressed the seminar organised by the Indian Physics Association, IIT, Delhi, on the occasion of International Year of Physics, on March 31, 2005 and exhorted the nation to become a "continuous innovator and creator of science and technology". Some Indian eminent scientists – say Goverdhan Mehta on July 18, 2000 said, "We have to identify science with creativity". He also emphasised, "the connectivity of minds".[1] Again Jayant Vishnu Narlikar on April 9, 2001, said, "What is needed is a campaign for scientific temper".[2] M.G.K. Menon said, "Unless you are creative, you cannot be innovative .... Commitment and dedication, apart from intellect are required for great scientific discoveries and advancement".[3] M.M. Sharma, on July 14, 2004 said, "Science without conscience ruins the Soul .... We as a nation will have to innovate to survive".[4]

This article presents: "Where is Science Going?" and "Where is Nature Guiding Mankind!" It also explores the cause of man's scientific effort which is antagonistic to the immutable laws of Nature of guiding mankind.

Where is Science Going?

In investigating the nature of knowledge, or of truth, or of reality, logical scientific reason confines itself to the field of the 'known'; it ignores the 'knower', the subject, or the self; this explains the limitations of its knowledge, the partial character of the truths it finds, and the relative character of the reality it reveals. But the nature's gift of 'reason' is not tied to the 'known', it has the capacity of making a quantum jump from 'known' to the 'knower', if the conditioned psyche purifies itself from the attachments of the past. This has been expressed by Albert Einstein in poignant words, "*Our task must be to free ourselves from this prison by widening our circle of compassion to embrace all living creatures and the whole nature in its beauty*".[5] It is then that reason rises from the world of fact to the world of meaning and value. The logical scientific reason makes a quantum jump to quantum logic (philosophical reason), and this quantum jump does not throw overboard the earlier conclusion arrived at by scientific reason; on the other hand, these conclusions remain valid, but only for the world of the 'known', for quantum logic does not contradict scientific reason but only fulfils it.

Intellect gives us objective truth, as it ignores the subjective world. It leads to invention in the technological field which implies modifying the known technique. It is an improvement on the existing methodology, so it is an improvement in the field of the known to the known. It adds to materialism. Objective science does not deal with direct perception of the Reality, but it deals with objective values. Recently many eminent scientists -- Werner Heisenberg, Edwin Hubble, Max Planck, G.F. Chew, Bertrand Russell, Lincoln Barnett, Max Born and so on – have concluded that we cannot impart value education at reason or intellect level.[6]

Many eminent scientists – including Albert Einstein, Wernber Heisenberg, Erwin Schrodinger, Arthur Eddington, and so on – have stumbled upon another mode of perception (knowing), which is "understanding of life directly, instead of in the abstract, linear terms of representational thinking".[7] This understanding expresses itself in the non-dual and non-conceptual frame of reference of Mind Continuum. *Understanding is a movement in the non-duality, a movement of the whole being.* Understanding involves *direct* perception, while thought, memory, inheritance are *indirect.* Understanding has nothing to do with knowledge.

Albert Einstein wrote, "The most beautiful experience we can have is the mysterious. It is the fundamental emotion which stands at the cradle of true art and true science".[5] Hideki Yukawa, a Japanese Nobel laureate in Physics, wrote, "[Creativity] always presupposed the presence of the power of intuition, which both Greek and Chinese geniuses in ancient times were richly endowed with. The important point would appear to be the balance or cooperation between intuition and abstraction".[8] Creativity is a case of the suspension or transcendence of causality and not a violation of it.

Nature expressed itself as intuition (insight) in the minds of many eminent scientists. Albert Einstein, the greatest physicist of the age, presented two important explorations and wrote: (a) the human psyche during the first few decades of twentieth century got heavily conditioned; and (b) "Nothing of all that will remain but few pitiable pages in history books, briefly picturing to the youth of the future generations the follies of its ancestors".[9] These two explorations lead mankind to self-annihilation. Harlow Shapley, an eminent astronomer, wrote, "The danger is man himself. He is his own worst enemy. He is acquiring tools and studying techniques that might solve the problem we set – the complete elimination of Homo from the planet earth".[10] The final appraisal of our situation is registered by a Nobel-prize-winning biologist, Jacques Monod, in words:

> No society before ours was ever rent by contradictions so agonizing.... What we see opening before us today is an abyss of darkness.[11]

> George Wald, Nobel Prize in medicine, 1967, wrote: We who know where we are going, are now heading rapidly toward self-extinction, toward removing humankind from the cosmic scene, and for the most trivial of reasons: profit, the rivalry of the United States and the Soviet Union, the momentum of the military-industrial complex in both nations. The cosmic drama will go on elsewhere, ignorant of our demise, caring nothing; but we will have no further part in it.... Does that matter? It matters to us.[12]

Federico Mayor Zaragoza, both as head of Unesco and his personal capacity as a scientist, addressed seventy-four Nobel laureates, and stated:

> Never before, until now, has the tension between science and the human conscience, between technology and ethics, reached a point where it has become a threat to the world as a whole.... We are so dazzled that we do not perceive the threats hanging over our heads, warning us of the pressing need for a radically new and universal ethical outlook on the future of present-day science.

He quoted Bertrand Russell that "mankind is united in evil but not yet united for good; that people have learnt the technique of mutual destruction, but not the more desirable technique of worldwide cooperation.... he called for a new moral outlook".[13] P. Krishna, an eminent physicist and educationist, said in a lecture in USA, "We are going towards self-annihilation and not too much time is left now".[14]

Limitations of Rational Knowledge: Science is based on experiments and works predominantly in the rational mode of consciousness. It measures and quantifies, classifies and analyses, and in this way constructs an intellectual map of reality; a system of abstract concepts and symbols characterized by the linear and sequential structure which is typical of our thinking and speaking. The intellectual map of reality is symbolic-map knowledge, it is dualistic knowledge. The real world, however, is not linear and sequential. It is one of infinite varieties and complexities, a multidimensional world. It isclear our abstract system of conceptual thinking can never describe or understand this reality completely.

Science gives us the objective knowledge, or knowledge of the world of experience revealed through the senses. In the words of Arthur Eddington:

> ..... this knowledge is only an empty shell – a form of symbols. It is knowledge of structural form, and not knowledge of content.[15]

Great thinkers of science tell us that science gives us only a conceptual, short-hand report of what is in the universe, but we are far away from reality.

The rational knowledge cannot transform our thinking. This has been stated by Albert Einstein in phrases:

(i) The significant problems we face cannot be solved at the same level of thinking we used when we created them....
(ii) Humanity needed to reach a higher level of awareness to solve the problems of nuclear war.
(iii) Science can denature plutonium but science cannot denature evil in the heart of man.
(iv) The splitting of the atom has changed everything except our mode of thinking and thus we drift towards unparalleled catastrophe.[16]

The objective science has failed, so far, to resolve psychological imbalances, ecological crisis, fear of war and its commitment to world peace.

Man has become fanatic due to his cerebral-neurological-chemical pressures that due to squabbling over conflicting ideas he is prepared to destroy the world and also himself; this implies that man has started giving more importance to thoughts, to ideas – which is a synthetic construct of the mind – than to the fact, the challenge, the 'what is'. Carl Sagan says:

What a waste it would be after four billion tortuous years of evolution if the dominant organism contrived its own self-destruction. We are the first species to have devised the means.[17]

Werner Heisenberg, Nobel laureate in physics, wrote:

Natural science does not simply describe and explain nature, it is a part of the interplay between nature and ourselves.[18]

What we observe is not nature itself but nature exposed to our method of questioning.[19]

The crucial feature of atomic physics is that the human observer is not only necessary to observe the properties of an object, but is necessary even to define these properties. If there is no observer, there is no observed. The human observer is not aware of the true perception but aware of the mutilated perception, and this error has to be rectified on priority through "observation and understanding" which leads to 'silence' or 'quiet' Mind Continuum.

Lawrence Beynam wrote, "We are currently undergoing a paradigm shift in science – perhaps the greatest shift of its kind to date".[20] Einstein discussed the objective truth and suggested:

> Humanity has every reason to place the proclaimers of high moral standards and values above the discoverers of objective truth. What humanity owes to personalities like Buddha, Moses, and Jesus ranks for me higher than all the achievements of the enquiring and the constructive mind.[21]

But these teachings have to be seen in the light of latest discoveries of modern science.

Where is Nature Guiding Mankind!

(a) Edwin Hubble, in "The Nature of Science and Other Lectures", wrote, "The world of pure values, that world which science cannot enter, has no concern whatsoever with probable knowledge. There finality – eternal, ultimate truth – is earnestly sought".[22] Heisenberg, a Nobel Laureate in Physics, wrote, "A rationalistic play with words and concepts is of little assistance here; the most important preconditions are honesty and directness".[23] Niels Bohr, a Nobel Laureate in physics, wrote, "Clever, rational ways of thinking are not enough. The danger of mass slaughter .... can only be overcome by moral conviction, by a determination to replace national prides and prejudices with human love".[24]

(b) Max Planck, a Nobel Laureate in physics, in an essay entitled "Where is Science Going?", wrote:

> It is a fact that there is a point, one single point in the immeasurable expanse of mind and matter, where science and therefore every causal method of research is inapplicable, not only on practical grounds, but also on logical grounds, and will always remain inapplicable. This is the point of [our] individual awareness.[25]

We intuit that the "one single point of individual awareness" is the non-dual and non-conceptual frame of reference of Mind Continuum. It is evident that mankind needs a quantum jump from symbolic-dualistic frame of reference of mind spectrum to non-symbolic, non-dualistic, non-conceptual frame of reference of Mind Continuum, or in other words, we have to jettison the logical or intellectual approach and imbibe the intuitive or insight frame of reference.

(c) The fundamentals of creativity in science and technology [or "generation of innovations and support system to multiple application] are as under:

(i) When one is aware of the "objective truth", and ignorant of the "subjective world", the mind is engaged in "invention" (materialistic approach). If one is simultaneously aware of the subjective world (inner) and the objective world (outer), then the mind is engaged in discovery. Discoveries come as intuitive flashes.[26]

(ii) All knowledge is structured in Consciousness, and Consciousness bestows creativity. It is the expression of pure consciousness in a pure Mind. Creativeness happens when one lives in communion with nature. Creativeness, discovery, direct understanding of life, intuition, insight, are synonyms. Intuition over a long chronological time flowers to Intelligence or Wisdom.

The above attributes of creativity are identical to the Einstein's perception, who defined that all creativity in science flows from the mysterious, the Life Field.

(d) Scientists are now talking of "inner space labs". Houston said, "In high-level creativity, the mind can do weeks of work in a short time".[27] Fritjof Capra wrote, "The rational part of research would, in fact, be useless if it were not complemented by the intuition that gives scientists new insights and makes them creative".[28] A.N. Whitehead wrote, "It is no exaggeration to say that the future of our civilisation depends on the degree to which we can balance the forces of science and religion".[29] Bryce-Smith wrote, "We face an evolutionary challenge that requires a moral response: we shall survive only if we can show that we deserve to survive".[30]

(e) Wilder Penfield, an eminent brain surgeon concluded, "Our minds and our brains are not at all the same .... It is the mind which experiences and it is the brain which records the experience". He added, "To suppose that consciousness or the mind has a location is a failure to understand neurophysiology".[31] The mind appears to be everywhere. It is observing on the scale of atoms and molecules, neurons, cells, tissues, muscles, bones, organs.[32]

(f) A galaxy of scientists – Karl Pribram, Roger Sperry, John Eccles – has opined, "Brain science must deal with the awareness of awareness. It could no longer afford to shut out that part of the world which we call subjective".[33] We have to restructure science which takes in its ambient both the objective and the subjective dimensions of life; to restructure education which takes into consideration both the sensual and trans-sensual dimensions of life; and the dimension of health which considers a living in the present and takes the whole mind-brain-body as one system. The subject-of-cognizance is now restructured "science-education-health"; it is also called the true-subject.

(g) Ilya Prigogine, Nobel laureate, 1977, wrote, "It is a physics that presupposes an observer situated within the observed world. Our dialogue with nature will be successful only if it is carried on from within nature".[34] He added, "There is a need for new relations between man and nature and between man and man".

(h) If a thought coupled with volition comes in the field of true-subject, the I-consciousness is knocked off. If this observation is continued over a long chronological time, the 'photons' meet an impregnable wall, called "event horizon", or "frequency barrier", or "time-barrier", the thought-structure is also knocked off. We are left with the *fact*, the *true*, the *present*, the *what is*!

The true-subject sees the true-object (itself) or the true-subject perceives the truth.[35]

(i) Observation (seeing), listening (hearing), and so on are direct, whereas thinking, memory and inheritance are indirect. A direct perception bestows intuition (insight). Nature invariably guides you to direct perception, timeless perception, the awareness of the truth. A direct perception recalls past corresponding memory and there is coherent superpositions of past memory on the direct perception leading it to a distorted or indirect perception. The indirect perception leads you to pain, misery, conflict, duality and so on.

(j) In a non-dual Mind, the perceptions are born of non-duality, and responses are born of spontaneity. It awakens the awareness, the intelligence, which arises out of observation, self-knowing and compassion. In Einstein's words, it is "objective observing and understanding" that can transcend all psychological challenges and energises Intelligence.

Science and the Nature of God

(a) Einstein prophetically wrote, "We know nothing about it [the nature of God] at all. Our knowledge is but the knowledge of school children.... We shall know a little more than we do know. But the real nature of things – that we shall never know, never".

(b) One thing I have learned in a long life: that all our science, measured against reality, is primitive and childlike – and yet it is the most precious thing we have. This is the basic limitation of our science, and of man, but this ultimate truth we can know bestows us "*peace and serenity*".

(c) Nature has designed the nervous system where any action takes about five thousandths of a second. The signals from subjective to objective, and from objective to subjective also take that much time. It is hardly worth noticing anything on such a tiny time scale, yet this is the scale on which our nervous systems normally function. Fred Alan Wolf wrote, "We are beginning a new age of awareness, the age of quantum consciousness, the age of the conscious atom. By looking within ourselves, we may be able to solve the problems facing us on the final frontier – the frontier of the human spirit".[36]

The Action Plan

Without being aware of it and without being rigorously systematic about it, the logical scientific reason excludes the Subject-of-Cognizance (True-Subject) from the domain of nature that we endeavour to understand. This leads to our current collective worldview that we are taught by our parents, teachers and society – this is called past memory or conditioning of our psyche. The way of observing (seeing) things – the old paradigm – has aptly been called "the hypnosis of social conditioning". This is a world crisis and we have to rectify this on priority. The Action Plan is as under:

(a) We have to educate the educator. This is also termed as 'teachers training course'. The UGC may call for volunteers at post-graduate level from different Universities and from Teachers Training Colleges for advanced training. To educate the educator is to have him understand himself. This is one of the most difficult undertakings, because most of us are already crystalised within a system of thought or a pattern of action.

(b) To conduct courses at all levels – research scholars at doctoral level or post-graduate level especially in the science faculties of different universities and at lower levels in schools and colleges.

(c) To conduct seminars at the parents level.

(d) We have to introduce psychodynamics of mind-brain-body system in science faculty and transpersonal psychology in art faculty as a part of educational curriculum at University/College level.

(e) We have to share through open dialogues the psychodynamics of conditioned mind and the dimension of Consciousness. The aim of these dialogues is "exploration of a spiritually inspired understanding of life" and "holistic living".

(f) The principle governing restructuring of education is fundamentally based to assist nature for an evolution of life. The evolution of mankind can only happen when there is understanding and spontaneity; dynamic interplay of intuition and action; and awareness of the transcendent dimension of life. The real goal for each one of us is awaken and activise in ourselves perfect spiritual consciousness. A true seeker is the one who with his ideal in view, goes on purifying his feeling, thoughts and actions. We have to learn to empty our mind of all worldly thoughts and wait silently for the vision of the universal consciousness.

(g) The above programme is at national level and also at international level. We suggest that workshops be fundamentally based on symbiosis of modern science and spirituality. A tentative programme of a workshop is: Restructuring of Science and Education; Psychodynamics of Mind-Brain-Body System; Transpersonal Psychology of Human Psyche; Fundamental Revolution or Radical Transformation of Human Psyche; Intuition and Learning; Depth Introspection of Life, and so on.

(h) The Action Plan has to be multifaceted and multidimensional. The fundamentals are "to educate the educator or training of teacher", and writing of books both for the teachers and students; and naturally these are at different levels, and in different languages. This is a stupendous task.

(i) There will be need to create some infrastructure for conducting courses. There is need to write books for guidance of teachers and students, and for different levels. This is possible if Govt. of India approves of the basic programme. Without Government active support, any action plan will prove to be a failure. And if Government escapes its responsibility, it will jeopardize the future of India and mankind.

The Blossoming of Paradigm Shift

(a) It consists of (i) *Restructuring of Science:* which deals with objective and subjective dimensions of life; (ii) *Restructuring of Education:* which takes in its compass the sensual and the trans-sensual dimensions of life; (iii) *Restructuring of Health:* which rejects the diseased mind and imbibes healthy and attentive mind; (iv) *Awareness of Human Values:* which flows from the trans-sensual dimension of life, or it comprises of ethical/spiritual values; and (v) to *Live in the Light of one's own Understanding:* which is the living in the wholeness of life, or the oneness of Consciousness. This bestows freedom, order, harmony, non-violence and peace. This guarantees human survival and peace.

(b) The observing (seeing), listening (hearing), understanding and learning and true action are "direct perception"; while the conditionings, memory and inheritance are indirect. The paradigm shift bestows direct perception which is living in commune with nature.

(c) An insight is without remembrance, without memory, without time and that insight is itself the action. The insight self-manifests itself in a pure mind-brain-body system. Where there is insight, there is not only a total freedom but total learning. This is the consummation of human life.

(d) In the curriculum setting alone, an altered mind-set can be initiated through objective observing, understanding, learning and living, which flows to the dualistic mind and "non-dualistic and non-conceptual" frame of reference of Mind Continuum. This has to be preceded by educating the educator – that is, to have him understand himself. This, in essence, is the foundation of a new pedagogy which leads to mutation of the human psyche. We shall survive only if we can show that we deserve to survive.

(e) If we take an inner voyage of "self-observation and understanding", it will bestow intuitive mind that flows to discoveries; or in other words, the scientist becomes a "continuous innovator and creator of science and technology". This guarantees freedom, order, harmony, non-violence and peace. This is the key to the survival and excellence of mankind along with all fauna and flora.

(f) When thought structure and I-consciousness go to abeyance, the mind becomes still. A still, quiet mind is creative and provides an answer to every problem. In other words, the answer is embedded in the frame of reference of Mind Continuum, when the thought structure has totally ceased.

(g) This establishes new relations between man and nature, and between man and man. And new relations are in the realm of pure perception, according to the immutable laws of nature.[37]

Concluding Comment

Objective science confines itself to the field of the "known", it ignores the 'knower', so this process is antagonistic to the immutable laws of nature. Nature invariably expresses itself as pure perception, or immutable laws of nature. Nature bestows order, harmony, non-violence and peace. The creativity in science resolves all psychological imbalances, ecological crisis and other human challenges. It is embedded in pure perception, or immutable laws of nature. A quote from literature is, "Any nation today which is unable to appreciate the scientific mind is doomed".

Man has a hidden potential to enhance the level of awareness and understanding and live the highest quantum energy potential of Consciousness. Albert Einstein stated that it is "*objective observing and understanding*" which bestows a "*Superior Mind*" that emerges from the "***Mysterious***", or Cosmic Consciousness, or Life Field. It is the expression of the hierarchical level of Consciousness that can heal the psychic pandemic. We are at a critical stage where "If we move one step towards God [Mysteriousness], God advances a hundred steps to receive us". We shall survive only if we make a quantum jump [self-unfolding] from symbolic-dualistic frame of reference of mind spectrum to intuitive [non-dual] frame of reference of Mind Continuum, or in other words, from reason to intuition. Nature is guiding man from psychosocial evolution to spiritual evolution.

Man divorced from nature could not be happy and man's power to resist nature is hopelessly puny. Man has to understand and live according to the immutable laws of nature. It is admitted that the task of turning the age-old tide of human ambition and aggression is a formidable one. More than political-economic-social measures, we need understanding psychodynamics of the conditioned psyche, the Unconditioned Cosmic Consciousness and a way of thinking about human values. One is concerned about the superhuman task and frailties of human endeavours, but the goal seems attainable as evidenced by such eminent scientists, as Albert Einstein.

Man through "objective observing and understanding" transcends the "abstract, linear terms of representational thinking", and makes a quantum jump through discontinuities to "the understanding of life directly", or "a state of awareness". This bestows intuition or insight, which leads to a psychological revolution, in which the whole human race has to regenerate or rejuvenate human Mind so that it perceives the new dimension of an integral and all-inclusive Consciousness that can guarantee human survival along with all fauna and flora.

The need of the hour is to expand the dimensions of science [science-education-health] so that the subject views the wholeness of life, and it resolves I-consciousness, thought-structure, and other human challenges. It progresses steadily in the direction of peace. Let every scientist see himself as a vehicle of world freedom, peace and bliss.

References

1. Goverdhan Mehta, in *Global Education and Human Survival*, New Delhi: Sanbun Publishers, March 2002, p. 22.

2. Jayant Narlikar, *ibid.*, p. 22.
3. M.G.K. Menon, *The Tribune*, November 28, 2004.
4. M.M. Sharma, *XII D.S. Kothari Memorial Oration*, Defence Laboratory, Jodhpur, July 14, 2004.
5. Albert Einstein, *Reader's Digest*, July 1973; & Nature, 15 March 1979, Vol. 278, pp. 215-218.
6. in Sampooran Singh, Dr (Mrs) Kanwaljit Kaur and Paramjit Singh, *Global Education and Human Survival : A Scientific Appraisal*, Delhi: Sanbun Publishers, March 2002, pp. 46, 35-36.
7. Alan W. Watts, in *The Spectrum of Consciousness*, USA: The Theosophical Publishing House, 1982 (Question Book), p. 43.
8. Hideki Yukawa, *Creativity and Intuition*, Tokyo: Kodansha International Ltd., 1973, pp. 101-102.
9. Albert Einstein, in *Out of My Later Years*, Postscript, Greenwood Press, Connecticul, Ed. 1970, pp. 6-8.
10. Harlow Shapley, *Of Stars and Men*, London: Elek Books, 1958, p. 144.
11. Jacques Monod, in *Chance and Necessity*, New York: Vintage Books, 1972, p. 171; & in *Synthesis of Science and Religion*, Mumbai: The Bhaktivedanta Institute, June 1988, p. 127.
12. George Wald, in *The Evolution of Consciousness*, New Delhi: National Publishing House, 1983, p. 13.
13. Federico Mayor Zaragoza, *The courier*, May 1988, p.
14. P. Krishna, *Education, Science, Spirituality*, California: Institute for High Education, Research and Dialogue, August 1993.
15. Arthur Eddington, in Swami Ranganathananda, *Science and Religion*, Calcutta: Advaita Ashrama, 1979, p. 125.
16. Albert Einstein, in Sampooran Singh, *The Human Paradox (SS: Two)*, Chandigarh: Faith Publishers, 1988, p. 41.
17. Carl Sagan, in *ibid.*, p. 41; in Ken Keyes Jr., *The Hundredth Monkey*, Vision Books, Ed., February 1984, p. 69.
18. Werner Heisenberg, *Physics and Philosophy*, New York: Harper Torchbooks, 1958, p. 81.
19. *ibid.*, p. 58.
20. Lawrence Beynam, in Ken Wilber, *The Holographic Paradigm and Other Paradoxes*, Shambhala, 1982, p. 157.
21. Albert Einstein, in Renee Weber, *Dialogues with Scientists and Sages*, London: Routlege & Kegan Paul, 1986, p. 15; in Sampooran Singh, *Human Psyche and Contemporary Physics, (SS: Four)*, 1988, p. 70.
22. Edwin Hubble, in *Synthesis of Science and Religion*, *op. cit.*, p. 119.
23. Werner Heisenberg, in "*The Physicist's Concept of Nature*", in *ibid.*, p. 3.
24. Niels Bohr, in *Creativity and Intuition*, *op. cit.*, pp. 189-190.
25. Max Planck, in *Where is Science Going?*, in *Synthesis of Science and Religion*, *op. cit.*, p. 183.
26. Dr (Mrs) Kanwaljit Kaur, Paramjit Singh and Sampooran Singh, *Vedanta*, UK: Ramakrishna Vedanta Centre, November-December 2000, No 296, pp. 266, 263.
27. Houston, in *Science*, 9 March 1973, Vol. 179, No 4077, pp. 982-983.

28. Fritjof Capra, *The Tao of Physics* (Second Edition), Bantam Books, 1988, p. 18.
29. A.N. Whitehead, in *New Scientist*, 26 May 1977, Vol. 74, No. 1053, pp. 478-479.
30. Bryce-Smith, *ibid.*, pp. 478-479.
31. Wilder Penfield, in Deepak Chopra, *How to Know God*, London: Rider, 2000, pp. 215, 217.
32. *ibid.*, *The Mystery of the Mind*, The Princeton University Press, 1975, p. 109; in Fred Alan Wolf, *Taking the Quantum Leap*, San Francisco: Harper and Row, 1981, p. 244.
33. Karl Pribram, in Ken Wilber, *The Holographic Paradigm and Other Paradoxes*, Boulder: Shambhala, 1982, p. 17.
34. Ilya Prigogine and Isabelle Stengers, *Order Out of Chaos*, New York: Bantam Books, 1984, pp. 218, 312.
35. Sampooran Singh and Dr (Mrs) Kanwaljit Kaur, *Peace, Security and Survival: A Scientific Guide*, Chandigarh: July 5, 2004, pp. 1-40 (submitted to the United Nations).
36. Fred Alan Wolf, *Taking the Quantum Leap*, San Francisco: Harper and Row, 1981, pp. 236, 248.
37. Sampooran Singh, *A Scientific Vision on Nature's Evolutionary Impulse: Man Unfolding; Science of the Soul*, Dagshai: The Atam Science Trust, 2005, Vol. I, pp. 25-30.

3
Man's New Dialogue with Nature

Abstract

Man divorced from nature could not be happy and man's power to resist nature is hopelessly puny. Nature invariably expresses itself in a non-dualistic frame of reference of Mind Continuum as direct-perception, or thought-free perception; and then we proceed to thought, recognition and naming. We have to educate our mind-brain-body system through "objective observing and understanding", so that the mind makes a quantum jump from symbolic-dualistic frame to non-dualistic frame.

The non-dualistic frame bestows intuition or insight which is "direct perception" – which is free from all psychological imbalances and ecological crisis. It is free from entropy and causality. It flowers to the self-unfolding universe, which blossoms to the evolutionary impulse in universe and in man. It expresses itself as order, harmony, non-violence and peace. If direct perception is sustained for a long chronological time, the Cosmic Consciousness perceives Itself; and this is called initiating a new dialogue with nature.

Man has an evolutionary origin. The *Homo Homonid* emerged at about 4.5 million years ago and *Homo sapiens* at about 2.5 million years ago. Man is an amphibian who lives simultaneously in two worlds – (a) the *known*, the time-space-causation matrices, the logical and scientific reason; and (b) the *Unknown*, the quantum logic, the timelessness, the eternal, the infinite and the immutable. The Cosmic Consciousness, the Life Field expresses itself at the lowest quantum energy potential, the objective world, and called successive consciousness, the external world of time-space matrices. It expresses itself at the highest quantum energy potential, the inner realm, the subjective world, and called Simultaneous Consciousness.

Man started living in the external world and divorced the Reality. He started gathering fragmentary experience and partial knowledge (information). With the awakening of symbolic knowledge, there seems to arise a split in the universe between the knower and the known, the thinker and the thought, the subject and the object. Eddington put it succinctly, "Nature thus provides that knowledge of one-half of the world will ensure ignorance of the other half".[1] This subject-object split results in dualism, which is also called the conditioning of the psyche. The error of dualism forms the root of intellection and is therefore next to impossible to uproot by intellection. The dualism constructs nervous tension, chemical pressure, misery, strife, disorder and chaos.

In modern physics, a very different attitude has now developed. Capra wrote, "Physicists have come to see that all their theories of natural phenomena, including the "laws" they describe, are creations of the human mind; properties of our conceptual map of reality, rather than of reality itself".[2] This conceptual scheme is necessarily limited and approximate; it constructs duality, pain and misery, chaos and conflict.

Ilya Prigogine, Nobel Prize in 1977, wrote:

> Science initiated a successful dialogue with nature. On the other hand, the first outcome of this dialogue was the discovery of a silent world. This is the paradox of classical science. It revealed to men a dead, passive nature, a nature that behaves as an automaton which, once programmed, continues to follow the rules inscribed in the program. In this sense the dialogue with nature isolated man from nature instead of bringing him closer to it. A triumph of human reason turned into a sad truth. It seemed that science debased everything it touched.[3]

Science is leading to a loss of communication, a dialogue with nature? This article investigates the causes of rupture and opposition of science to nature and explores new relations between man and nature and between man and man. We further investigate, "where is science going?" and "where is nature guiding mankind!" This regenerates the holy curiosity of enquiry into "man's new dialogue with nature", or "science with creativity", or in common parlance, it "evolves policies, strategies and measures for regeneration of innovation of science and technologies" for human survival and excellence.

Ancient Indian Wisdom regarding the Art of Observation

(a) R. P. Kaushik, an eminent doctor-mystic, based on Patanjal yoga, wrote:

> The first stage prior to recognition and naming is a state of non-verbal or thought-free perception [*nirvikalpa jnana* or pure perception]. From this pure perception we proceed to thought, recognition and naming [*savikalpa jnana*].[4]

Kaushik added, "The observer starts either identifying itself with the observed if it is pleasant and desirable, or alienates itself from the observed if it is unpleasant and undesirable. However, in the psychological field, both these processes of identification and alienation are one and the same. Hence there is no observation but only distortion .... ***This psychological fusion of the observer and the observed is observation***". If, however, the observer goes to abeyance, the observed also vanishes. When the observer and the observed go to abeyance, the Intelligence is energised which resolves all challenges.

(b) Kaushik quoted Patanjal Yoga regarding ending of dualism or conditioning of the psyche in words:

> A mind which is dyed with the fusion of the observer and the observed is full of all the riches [*Kaivalyapada; Patanjal yoga, 23*]
>
> One who sees with this art of pure observation in him ego disappears [*Kaivalyapada; Patanjal yoga, 25*].

This implies that the I-consciousness and thought-structure must transcend before Mind regains the capacity of perceiving pure perception.

Scientific Vision of Conditioned Human Psyche

(a) When one sees an object/event, it forms an image on the retina, the retina sends an electrical pulse to the inner sensorium, the mind scans all the sensoriums and constructs a holistic perception. This is explained by Arthur Eddington in words:

> The physiologist can trace the nerve mechanism up to the brain (inner sensorium); but ultimately there is a hiatus which no one professes to fill up. Symbolically we can follow the influence of the physical world up to the door of the mind; they ring the bell and then depart.[5]

We opine that the Mind first perceives the message from the inner sensorium, that is, its perception is *direct*. It then recalls the past memory and grafts it on direct perception which reduces a perception to a concept. The Mind is not observing the object, the fact, but observes the distorted percept (concept) hence there is no understanding.

Einstein wrote that man needs "objective observation and understanding" which bestows him "peace and serenity. This regenerates man's new dialogue with nature.[6]

(b) The different steps in perception are: (i) the mind (subject) perceives the object or the external environments, the *fact*, the *true*, the *present*, the *what is*!; the perception of the fact is called thought-free perception, or pure perception, or factual perception; (ii) the mind recalls or retrieves the past memory; (iii) there is exchange of energy between the fact and recalled memory leading to distortion of both; (iv) there is coherent superpositions between the object and the subject; (v) the mutilated fact is recorded as object and the modified memory as mutilated subject; and (vi) the whole process is so fast that we are not aware of the fact but aware of the mutilated fact.

Logical Scientific Reason

(a) John Eccles perceives the sensory illusion with one startling but irrefutable assertion, "I want you to realize that there is no colour in the natural world and no sound – nothing of this kind, no textures, no patterns, no beauty, no scent ....".[7] In short, none of the objective facts upon which we usually base our reality is fundamentally valid.

(b) Erwin Schrodinger wrote, "Without being aware of it and without being rigorously systematic about it, we exclude the Subject of Cognizance from the domain of nature that we endeavour to understand …. Mind has erected the objective outside world of the natural philosopher out of its own stuff. Mind could not cope with this gigantic task otherwise than by the simplifying device of excluding itself – withdrawing from its conceptual creation. Hence the latter does not contain its creator".[8] Charles Sherrington wrote, "The material world has only been constructed at the price of taking the self, that is, mind, out of it, removing it; mind is not part of it; obviously, therefore, it can neither act on it nor be acted on by any of its parts".[9] Schrodinger quoted C.J. Jung and B. Spinoza, who hold the identical views. This implies that in the scientific mode of mind, the subject of cognizance is at the immanent quantum energy potential while the object is at the successive consciousness level and there is no energy transfer between the subjective world and the objective world. The spirit of science has nothing to do with individual conditions, with nationalism, with race, with prejudice. It is free from any psychological imbalances and ecological crisis. It deals with the "known" and rejects the "knower", the subjective dimension of life. The objective world has only an objective value and cannot bring in mutation in the human psyche.

(c) Many eminent scientists – including Albert Einstein, Werner Heisenberg, Erwin Schrodinger, Arthur Eddington, and so on – have stumbled upon another mode of knowing (perception), which is "understanding of life directly, instead of in the abstract, linear terms of representational thinking". The understanding comes only when the mind is quiet; or in other words, the mind is free from conditioning. Only the truth can liberate the mind from its own ideation; to see the truth the mind must realize the fact that as long as it is agitated it can have no understanding. Understanding flows when I-consciousness and thought-structure go to abeyance. It is only a quiet mind that can be creative.

(d) It is physics that presupposes an observed world within the observer (true-subject). Our dialogue with nature will be successful only if it is carried on from within nature. Galileo wrote, "He asked questions addressed to nature and succeeded in receiving answers directly from nature".[10] It implies that it is the true-subject which asks the questions, and the true-observed world – which is within the true-subject – will provide the answer. This in common parlance is called intuition or insight.

(e) It is intuition alone that can regenerate science with creativity. This is possible if we take an inner voyage of self-discovery from "objective observing and understanding" to non-dual and non-conceptual frame of reference of Mind Continuum. We want a revolution, a psychological revolution, in which the whole human race has to participate and every man is an equal partner. This revolution will regenerate or rejuvenate human Mind so that it perceives the new dimension of an integral and all-inclusive

Consciousness, that can guarantee human survival along with all fauna and flora.

(f) Quantum Jump from "disorder and chaos" to "Order and Harmony": In the psychological mode of mind or common waking state, the distorted subject observes the distorted object, so there is "disorder and chaos". A rigorous, consistent and persistent "objective observation" is capable of pursuing dualism to its limits; and in that state the dualism ends or there is quantum jump from "disorder and chaos" to "order and harmony", and this quantum jump is in the field of discontinuity. In other words, the "order and harmony" is at a higher quantum energy potential of consciousness than "disorder and chaos". It is the potential of consciousness that governs transformation of "disorder and chaos" to "order and harmony", and this process is called the self-integrative potential of the system. We have to heal the "epidemic" of disorders of the mind through "objective observation and understanding".

The symbolic-dualistic frame of reference is "far-from-equilibrium" condition, so very small perturbations or fluctuations can become amplified into gigantic structure-breaking waves. The non-dual and non-conceptual frame of reference of Mind Continuum are "near equilibrium", so it is a movement in the direction of non-violence, peace and security.

(g) Entropy: In a closed system, the Second Law of Thermodynamics states that disorder can never spontaneously give rise to order. More precisely, this law generates nature's trading account in the conditioned psyche, is in a quantity called entropy, which is, roughly, the degree of disorder in a physical system. When ordered energy becomes disordered in this fashion, the entropy rises. The Second Law forbids the entropy of a complete system from falling.

The symbolic-dualistic frame of reference of mind spectrum is a closed system, so every activity in this field adds to entropy. The scientific mode of mind is an open system embedded in the subject (society) and linked to it by very dense feed back loops. It is powerfully influenced by external environment, and in a general way, its development is shaped by cultural receptivity to its dominant idea.

The scientific frame of reference moves from fact to fact, from one web of objective values to another web of objective values, so it is free from entropy and causality. If the scientific mind starts condemning or identification, the scientific mind lapses to psychological mind and entropy and causality enters the field.

The intuitive mode of mind is also free from entropy and causality. At the substratum of scientific mind and intuitive mind is a unified order, but it is not causality and entropy; it is purpose, and not the purpose of the universe and man, but the purpose *in* the universe and *in* man. In other words, we seem to inhabit patterns of self-integrative potentials which govern the evolutionary purpose of nature.

The Wholeness of Life

(a) The wholeness of life is defined by David Bohm, an eminent physicist, as "unbroken wholeness in flowing movement".[11] Its dimensions are homogeneous, undifferentiated and dimensionless.

(b) We have to restructure the subject of cognizance – which implies: (i) ***Restructuring of Science:*** which deals with objective and the subjective dimensions of life; (ii) ***Restructuring of Education:*** which deals in its compass the sensual and trans-sensual dimensions of life; and (iii) ***Restructuring of Health:*** which rejects the diseased mind and imbibes healthy and attentive mind. The subject of cognizance now is "Restructured Science-Education-Ethics-Health", it is also called the true-subject. The true-subject is at the highest quantum energy potential of consciousness, or the highest hierarchical expression of consciousness.

(c) Ending of I-consciousness: The attention is at a higher quantum energy potential than I-consciousness. When I-consciousness encounters the attention, it goes to abeyance. This suggests that the movement of thought structure is independent of your volition. This discovery has a dual effect – the pride and vanity that 'I think' goes away and the mental movement has a momentum independent of myself. Now I look upon my thoughts without my volition, or I observe my thoughts in utter freedom.

(d) Ending of Thought: Thought-structure knows no freedom, it is bound to the past or future projections. The movement of thought-structure is within me, it is independent of my volition, it is independent of the movement of the ego (I-consciousness). My thoughts, my memory, my feelings have a movement independent of me. There is a realisation of the activity of volition and the immensity of the thought-structure.[12,13]

If thought-time is in the inner field and it encounters true-subject, the photons meet an 'impregnable wall', or 'event horizon', or 'frequency barrier', the thought is dissipated, because the true-subject is at the highest quantum energy potential and thought is at the lowest quantum energy potential of consciousness. In other words, thought is within the speed of light and true-subject is beyond the speed of light. The speed of light is absolute; it is like a wall that no object can crash through. As thought approaches the wall, time slows down, mass increases, and space becomes curved. As the time approaches zero, thought gets dissipated and ends.[14]

There is nothing like an individual mind, there is nothing like an individual consciousness. Thought is the product of collective human effort through centuries. When it is seen that there is a racial consciousness and a racial momentum of knowledge, experience and thought and there is nothing like an individual ego, the self, the me, there is a sense of great freedom within.

(e) Ending of the Fact: If we continue observing or listening to the *fact*, without justification, or condemnation, or motivation, we observe that there is only the *fact* and nothing else. The fact itself is the operator and not you operating on the fact. If you leave the fact alone, it has no time. The fact is free from fear. The fact grows, which implies that there is release of tremendous energy and the fact moves to higher quantum energy potential. This is expressed as Pure Consciousness → Fact.

(f) The search for "Who am I?" is the foundation of all morality and religion. "Who am I?" belongs to the inner world. While "What is an electron?" is a question which belongs to the external world. There is a reciprocal question of the electron: "Who are you to know me?"[15] It is only non-dual and non-conceptual frame of reference of Mind Continuum that can understand, "What is an electron?".

States of Mind

(a) Psychological State of Mind: The mind cuts itself up into at least two states – the subject and the object. In this severed and mutilated condition, whatever it sees is only partially itself. The subject and the object are at the same quantum energy potential of successive consciousness, so there is transfer of energy. It is a closed systems and constructs entropy and causality. It constructs nervous tension, chemical pressure, misery, strife, disorder and chaos, so it is a pointer to self-destruction or self-annihilation.

(b) Scientific Mode: The subject is at the immanent quantum energy potential, it observes the *fact*, the objective values. Logical and scientific reason confines itself to the field of the 'known', it ignores the 'knower', the subject. There is no energy interaction between the subject and the object. The spirit of science has nothing to do with individual psychological imbalances. It moves from fact to fact, from one set of human values to another set of values; this is due to transfer of energy from the Unmanifest; so it is an open system.

Scientific mode of mind gathers fragmentary experience and partial knowledge, and stores it in the memory. It enables man to enjoy material pleasures and comforts. It contributes to materialism, which is used to construct better artifacts with higher sensitivity and efficiency. It adds to human comfort and leisure. The scientific mode cannot transform or bring mutation in the human psyche.

(c) The Intuitive Mode of Mind: A rigorous, consistent and persistent objective observation and understanding leads to "the understanding of life directly, instead of in the abstract, linear terms of representational thinking". "The ultimate perception (understanding of life directly) does not originate in the brain or any material structure, although a material structure is necessary to manifest it. The subtle mechanism of knowing the truth does not originate in the brain".[16] The nature or life expresses itself as intuition or insight, when nature responds to your question. Intuition or insight is fundamental and reason is a derivative.

(d) Holistic Perception: Nature invariably expresses itself as direct. All human challenges – such as psychological imbalances, ecological crisis, poverty, hunger – vanish, and there is order, harmony, non-violence and peace. It is the direct perception or holistic perception of Cosmic Consciousness to Itself (pure mind).

Holistic perception implies that the mind is totally free, to be totally unattached to everything that man has put together. It implies to be aware of the wholeness of life, of the Fundamental Vibration. It dissolves the fact. Perception and understanding are not of time.

(e) Perception in the State of Meditation: It is also called the ultimate perception. In the ocean of Silence, the vibration of the cosmic energy also ends, and this bestows the perception of one Self of all, the *Brahman*, the Life Field. Man becomes a co-worker with the Divine. This establishes man's new dialogue with nature.

Aims and Objects of Human Life in the Words of Albert Einstein

(a) Albert Einstein, the greatest physicist of the age, discussed the aims and objects of human life in words:

A human being is a part of the whole, called by us 'Universe', a part limited in time and space. He experiences himself, his thoughts and feelings as something separated from the rest – a kind of optical delusion of his consciousness. This delusion is a kind of prison for us, restricting us to our personal desires .... Our task must be to free ourselves from this prison by widening our circle of compassion to embrace all living creatures and the whole nature in its beauty.[17]

What is truly valuable in our bustle of life is not the nation, I should say, but the creative and impressionable individuality, the personality – he who produces the noble and sublime while the common herd remains dull in thought and insensible in feeling .... The most beautiful thing we can experience is the mysterious. It is the source of all true art and science .... To know that what is impenetrable to us really exists, manifesting itself as the highest wisdom and the most radiant beauty which our dull faculties can comprehend only in their most primitive forms – this knowledge, this feeling, is at the centre of true religiousness.[18]

I believe with Schopenhauer that one of the strongest motives that lead persons to art and science is flight from the everyday life, with its painful harshness and wretched dreariness, and from the fetters of one's own shifting desires. One who is more finely tempered is driven to escape from personal existence and to the world of objective observing and understanding .... (Man) has to attain peace and serenity.

One thing I have learned in a long life: that all our science, measured against reality, is primitive and childlike – and yet it is the most precious thing we have.

One is deeply struck by the broad agreement of Einstein's utterances (quotes) with those of Upanishadas or Vedic Wisdom of India. But some aspects of the traditional wisdom have to be reviewed in the light of modern scientific discoveries.

(b) There is a growing realisation that the existing models of education have failed in India and in the rest of the world. Einstein wrote:

> It is in fact, nothing short of a miracle that the modern methods of instruction have not yet entirely strangled the holy curiosity of inquiry; for this delicate plant, aside from stimulation, stands mainly in need of freedom; without this it goes to wrack and ruin without fail .... The rational way of education is to be an example – if one can't help it, a warning example.
>
> Who would have thought around 1900 that in fifty years time we would know so much more and understand so much less.

The present education leads us to gather more and more information, but understanding decreases. In other words, knowledge grows but wisdom atrophies. This suggests a serious crisis of human mind-brain-body system, and has to be rectified *here* and *now*.

Man's dialogue with nature implies the awareness of the "mysterious", and this blossoms to "Creativity in Science", or "Generation of Innovation of Science and Technology".

Concluding Comment

In life, as well as in relationships, thought creates havoc. Thought in terms of belief systems is destroying man. Essentially we have to understand, learn with a Mind that is free to observe, to listen. We have to educate ourselves with the psychodynamics of the conditioned psyche (called self-knowledge) and to explore the Unconditioned Energy (called self-awareness). It is an exploration from self-knowledge through self-observation to self-awareness. The restructuring of education also includes to educate the educator. The wisdom of listening and observing is born with the epistemological evolution.

It is only "restructured science-education-ethics-health" that helps man to grow from knowledge to understanding, from confrontation to cooperation, from rationality to intuition, from intellect to intelligence. Intelligence self-expresses itself in a non-dual frame of Mind Continuum and blossoms to peace and security. The non-dual and non-conceptual frame of reference of Mind Continuum bring in mutation in the brain. Mutation is possible in the brain through insight, not through knowledge, not through accumulation. That insight is the ability to see a truth directly, a direct perception of the truth, and education must foster that.

Observation (seeing), listening (hearing), understanding, learning, and nature's expressions are direct; while thinking, memory and inheritance are indirect. Nature invariably expresses itself as direct, pure perception, thought-free perception. If pure perception is sustained for a long chronological time, then the Cosmic Consciousness perceives Itself and this is called that man initiates a new dialogue with nature.

However there is a large gap between the superhuman task of new dialogue with nature and the frailties of human endeavours; it implies that the human mind starts evaluation, identification of the pure perception due to recall of the past memories, and the mind relapses to a lower state of attentiveness. This flows to coherent superpositions of old memory on pure perception and transforms it to conceptual model leading to breakdown of all new dialogue with nature.

The conceptual model leads to nervous tension, chemical pressure, strife, disorder and chaos. The new dialogue bestows order, harmony, non-violence and peace.

It is the higher hierarchical expression of Cosmic Consciousness that flows to opening a new dialogue with nature.

References

1. Arthur Eddington, in Ken Wilber, *The Spectrum of Consciousness*, USA: The Theosophical Publishing House, 1982 (Quest Book), p. 29.
2. Fritjof Capra, *The Tao of Physics*, Bantam Books (Revised Edition), 1988, pp. 277, 147.
3. Ilya Prigogine and Isabelle Stengers, *Order Out of Chaos*, Toronto: Bantam Books, 1984, p. 6.
4. R.P. Kaushik, *Towards a New Consciousness*, Darshan Yoga Sansthan, January 1974, pp. 63-71.
5. Arthur Eddington, *The Nature of the Physical World*, Cambridge, 1928; in Sampooran Singh, *The Holistic Education*, New Delhi: Siddharth Publications, 1987, p. 41.
6. Albert Einstein, in *Nature*, 15 March 1979, Vol. 278, pp. 215-218.
7. John Eccles, in Deepak Chopra, *Ageless Body, Timeless Mind*, New York: Three Rivers Press, 1998, p. 12.
8. Erwin Schrodinger, *What is Life? & Mind and Matter*, Cambridge University Press, 1967, pp. 127, 131.
9. Charles Sherrington, in Erwin Schrodinger, *ibid.*, p. 128.
10. Galileo, in Hideki Yukawa, *Creativity and Intuition*, Tokyo: Kodansha International Ltd., 1973, p. 111.
11. David Bohm, *Wholeness and the Implicate Order*, London: ARK Paperbacks, ARK Edition 1983, p. 172.
12. Vimala Thakar, *Through Silence to Meditation*, Ahmedabad: Vimal Prakashan Trust, 1998, pp. 22-27.
13. ibid., Silence, Holland: Bookfund Vimala Thakar, 1989, p. 6.
14. Deepak Chopra, *How to Know God*, London: Rider, 2000, pp. 270-271.
15. D.S. Kothari, *Atom and Self*, in *The Evolution of Consciousness*, New Delhi: National Publishing House, 1983, pp. 208, 213.

16. David Bohm, *Lecture given April 6, 1977*, University of California at Berkeley; in Gary Zukav, *The Dancing Wu Li Masters*, Fontana/Collins, 1982, p. 327.
17. Albert Einstein, *The Merry Mind of Albert Einstein*, Readers Digest, July 1973, p. 134.
18. *ibid., I believe: Nineteen Personal Philosophies*, London: Unwin Books, 1962, p. 28.

4

Consciousness and Spirituality : Widening the Scientific Perspective

Abstract

The objective science mainly deals with material progress of life, it investigates modified use of the existing materials, which is an invention and not a discovery. It is limited because it investigates the objective (outer) dimension and ignores the subjective (inner) dimension. Reason or intellect is inadequate because reason emerges from the same quantum energy potential of consciousness as memory. It rejects intuition because intuition is beyond our observations, perceptions and logic.

A galaxy of scientists have opined that, "Brain science could no longer afford to shut out that part of the world which we call subjective". We have to "restructure science-education-health" which takes in its ambient the wholeness of life – called true-subject.

When the true-subject observes the conditioned psyche and the Unconditioned Energy it leads to mutation in consciousness. This leads to ending of I-consciousness and thought-structure. This rejuvenates Intelligence which resolves all psychological imbalances, ecological crisis and other challenges. It results in manifestation of order, rhythm, non-violence and peace. This is the consummation of human life as it guarantees human survival and excellence.

Man has an evolutionary origin. The *Homo sapiens* emerged at about 2.5 million years ago. He started living in the external world, the time-space matrices. About 10,000 years ago, or more, some saints and sages in India started exploring the unfoldment of man and found that man is an amphibian who lives simultaneously in two dimensions – the given (inner, subjective) and the home-made (outer, objective). The objective is in the field of time, it is at the lowest quantum energy potential of successive consciousness and is limited, transient and finite. The subjective is eternal, infinite and immutable and it is at maximum quantum energy field of Consciousness. Modern science has now recognised that the whole cosmos (and beyond cosmos) is full of high frequency energy (higher frequency means higher energy) which is the womb of existence.

This essay explores the limitations of objective science and technology, inadequacy of logical reason, consciousness, spirituality and widening of the scientific perspective to move from intellect to intelligence, from knowledge to understanding, from "successive consciousness" to "Simultaneous Consciousness". The salient explorations have already been discussed in two monographs.[1]

The Limitations of Science and Technology

Science is the search for truth in the objective world, the 'without'. A selfless, unfaltering devotion to truth and the use of knowledge for man's upliftment are the ideals of a scientist. Truth is what all earnest scientists seek. Free enquiry, objectivity, precision, autonomy, freedom, experience are the pillars of the inductive (experimental data) and deductive (theoretical) logic. Discovery is its mission, its perception. It is precise, clear, unprejudiced, which does not condemn, which observes, which sees. It moves from fact to fact. Science is a dialogue between Nature and the universe (man-made cosmos) where questions are asked and Nature responds with the answer, it implies that the insight is born.

A scientific mind takes up "the role of a non-concerned observer", in the words of Schrodinger;[2] it is "unrelated to our beliefs, our ambitions, our hopes", in the words of Prigogine;[3] it must not construct a pseudo-observer, it must see in the present and reject the past remembrances or future projections. The values in science are objective truth, integrity, and honesty.

Science aims at making true and adequate statements about its object. Science and technology mainly focusses attention on the material progress of life and on technical solution to human challenges; it explores a modified use of the known – and this is called invention (not discovery). Although science is a human activity, yet all values are missing and quite particularly meaning and end are missing. Science is an intellectual endeavour, and it rejects all intuition (insight). It is a non-authoritarian approach.

Science makes progress largely by a combination of two aspects: open-mindedness and the following of abstract concepts. Abstract concepts produce narrow-mindedness, so the whole question of moral, spiritual and other dimensions is pushed out of science.[4]

Schrodinger wrote, "Our science – Greek science – is based on objectivation whereby it has cut itself off from an adequate understanding of the Subject of Cognizance, of the mind". Our sensations, perceptions and observation have a strong, personal, subjective tinge and do not convey the nature of the "thing-in-itself", to use Kant's terms. Kant landed us with a complete resignation: "never to know anything at all about his "thing-in-itself",[5] the Supreme Reality, the transcendent quantum energy potential of Consciousness.

Inadequacy of Logical Reason

(a) All knowledge is structured in consciousness. The creative process involves a quantum jump from logical and scientific reason to philosophic reason or non-dual and non-conceptual frame of reference of Mind Continuum; in other words, from a relative frame of reference to an absolute rest frame of Mind. Knowledge emanates from it as quantas depending on the frame of reference of mind. D.S. Kothari, an eminent physicist, in prophetic words, wrote:

> The new theorems, unless trivial, are of course first "discovered" and then proved. Discovery is an inspired guess, a flash of insight. It is not a step by step process, but a sort of quantum jump.... A rational account of a discovery, the proof, follows and not precedes the discovery.[6]
>
> Fundamental discoveries which go deep into the nature of things are the result of uncommon intuition and faith, not cold logic.[7]

The discovery presupposes a systematic interaction between theoretical concepts (deductive, intellectual approach) and observation (inductive, experimental results). Intuition, or insight, or discovery is a timeless perception. Intuition results in the discovery of truth in nature hitherto unknown in man. Intuition is fundamental and rationality is a derivative.

(b) All great discoveries have arisen from the realm of the timelessness, so scientific enquiry is not a cerebral activity and gathering information, but a quest of the Timelessness, the Divinity, the Life Field. A selfless, unfaltering devotion to truth and use of knowledge for man's uplift – higher hierarchical expression of Consciousness – are the only vocation of a scientist.

(c) Ilya Prigogine wrote, "It is physics that presupposes an observer situated within the observed world. Our dialogue with nature will be successful only if carried on from within nature".[8] We have to initiate a dialogue with nature from within nature, and nature responds to this experimental interrogation. If a scientist,. who is aware of "restructured science-education-health" and lives in that awareness, asks questions addressed to nature, he invariably succeeds in receiving answers directly from nature. Nature is an "open-book" to him. Galileo Galilei taught us that "he asked questions addressed to nature and succeeded in receiving answers directly from nature".[9]

(d) Recently, many eminent scientists – Werner Heisenberg, Edwin Hubble, Max Planck, G.F. Chew, Bertrand Russell, Lincoln Barnett, Max Born, and so on – have concluded that we cannot impart value education at reason or intellect level.[10]

(e) The symbolic-dualistic frame of reference of mind spectrum is mechanical and repetitive, it is the past; hence it has no creativity. The non-dual and non-conceptual frame of reference of Mind Continuum is creative. The logical reason is inadequate for human creativity and progress. It has to be further developed to quantum logic (intuition). Intuition is fundamental and reason is a derivative.

Consciousness

There is One Reality, the Unmanifest-Manifested. The nature of the Unmanifest is 'Deep-Silence', 'Inactivity (Non-Vibration)', 'Passivity', 'Immutability', 'Eternal'; and also Activity (Vibration), dynamism and illimitable potency to manifest as many. Activity (Active Reality) expresses which is self-hidden in the Silence (Passive Reality); both are different states of the Supreme Reality, the obverse and converse affirmations; both are Pure Existence and conscious Force, or Self-Existent Bliss and Supreme Intelligence, and are One. It is the Active Reality which manifests as the perfect freedom and omnipotence of an eternal divine activity in innumerable galaxies as inanimate world and plant and animal kingdoms.

In a scientific conference at Moscow in 1996, the Russian scientists "have come up with the idea of a Life Field beyond our physical concepts of time, space and the known electro-magnetic and gravitational fields". In a subsequent Conference at Cambridge University, entitled "Beyond the Brain", it was concluded, "A new scientific approach would include subjective experience as valid information and address all forms of human experience – including intuition, spirituality and mystical encounters". Wills Harman argued that "we need to develop a different form of science to truly understand consciousness".[11]

Spirituality

(a) Spirituality is the physics of Consciousness and what we call physics is the science of matter and energy. Spirituality is the awareness of the wholeness around you and wholeness within you.

(b) Spirituality proceeds from the whole and with the awareness of the whole we observe the particular. In spiritual science, the emphasis is on the act of perception, the quality of the instruments employed for perception, the quality of consciousness behind the perception. Both spirituality and science have non-authoritarian approach – an approach not conditioned by the past. Spirituality explores the trans-psychological dimension of consciousness.

(c) In spirituality, the act of investigation, exploration is conducted on behalf of the whole human race. *Spirituality explores what the truth is, what is fact is, and what the Reality is behind the truth. Spirituality teaches us to live with the fact and transcend the non-fact.*

Widening the Scientific Perspective

The recent discoveries of modern science are compelling man to widen the scientific perspective, and the salient features are as under:

(a) Restructuring of Science: Many eminent scientists – including Albert Einstein, Werner Heisenberg, Erwin Schrodinger, Arthur Eddington, and so on – have stumbled upon another mode of perception (knowing), which is "understanding of life directly, instead of in the abstract, linear terms of representational thinking". This implies understanding the wholeness of life.

Another galaxy of scientists – say, Karl Pribram, Roger Sperry, John Eccles, Wilder Penfield, and so on – have opined, "Brain science could no longer afford to shut out that part of the world which we call subjective". We have to "restructure science-education", which takes in its ambient both the objective and the subjective dimension of life and is called the "true-subject". The time interval between the subjective and the objective is less than a hundred microsecond. The objective science must restructure itself and consider the subjective and the objective as one harmonious movement. The restructured science deals with the wholeness of life.

(b) Restructuring of Education: The dimension of value education is embedded in the trans-sensual dimension of human personality. We have to restructure education so that it takes in its ambient the sensual and the trans-sensual dimensions of life.

(c) Restructuring of Medical Education: The theme of the Indian Science Congress – 2005 is Health Technology. A living in the *present*, the *now* heals, and living in the *past* and *duality* suppresses man's sensitivity and enhances disease. There is an urgency to restructure medical education which takes in its ambient the psychodynamics of mind-brain-body system and the Unconditioned Energy; or in other words, the wholeness of life.

(d) The Oneness of the Cosmos: Ervin Laszlo stated, "The parts of this totality are not radically separate: they are integral elements of the whole. Every 'thing' in the cosmos is a coherent 'event' and every event is the summed result of all the other events".[12]

Mutation or Radical Transformation of the Psyche

We need restructuring of science-education-health to *observe* truthfully and sincerely the conditioned and the Unconditioned Energy, and this will flow to mutation of the psyche. The different steps in observation, which are expressed to understand and learn, are as under:

(a) Ending of I-consciousness: If a thought coupled with volitions encounters the true-subject, the I-consciousness is knocked off because it is at the lowest quantum energy potential. This suggests that: (i) the movement of thought-time is independent of I-consciousness; (ii) it is the movement of collective human effort; (iii) both the I-consciousness and thought-structure have a conceptual reality.

(b) Ending of Thought-Time: When thought-time comes in the realm of true-subject, the photons meet an 'impregnable wall', or "event-horizon", or "frequency-barrier", and are dissipated, which implies that thought-time goes to abeyance and what is left behind is the *fact*, the *true*, the *present*, the *what is*![13] When thought-time subsides, it is a discontinuity in the mental stream, and in philosophical language, it is called "no-mind" state, which comes gracefully and voluntarily. Suddenly there is an inner stillness. This state is highly alert and aware, but no thinking.

(c) Ending of the Fact: If we continue observing or listening to the *fact*, without justification, or condemnation, or motivation, we observe that there is only the *fact* and nothing else. The fact itself is the operator and not you operating on the fact. If you leave the fact alone, it has no time. The fact is free from fear. The fact grows, which implies that there is release of tremendous energy and the fact moves to higher quantum energy potential. This is expressed as Pure Consciousness ↔ Fact.

(d) Intuition: The "understanding of life directly, instead of in the abstract, linear terms of representational thinking" bestows intuition, spontaneity, sensitivity and creativity. These are metaphysical (quantum logic) insight, falling outside the realm of duality, or the intellectual construction. *Understanding, learning and action are a cyclic process, which results in a spiral going upwards in a vertical dimension; it is the growth in vertical dimension. The higher the understanding, the higher the expression of consciousness in the human mind-brain-body system.* The higher sensitivity or creativity leads to a perception of pure consciousness.

(e) Intelligence: Intelligence perceives life in its wholeness, and responds to it out of its own wholeness. The perception through intelligence, through sensitivity is a perception in non-duality. In love, beauty there is instantaneous perception. The perception and the response have no time lag between them.

The communion with the Divine, the Eternal, the Infinite is a happening; it is not the result of my doing. The uncovering of the secret is not a human achievement. It is something that happens, when the human being puts all his conditionings into relaxation, and remains before the mystery of Life in utter humility, in utter nakedness of his psyche. The consummation of human growth is the manifestation of the Divine, the Self-Unfoldment of man. Life is consummation of human growth. Then the mystery unfurls itself, uncovers itself. It uncovers itself into us.

(f) Is Man Individual or Collective or Humanity?: J. Krishnamurti wrote, "Man is the conclusion of all the experiences of mankind. He is the history of mankind. He is neither individual, nor collective, but he is humanity".[14] He does not have anything of his own both in the objective and the subjective dimensions of life. Voltaire expressed this in prophetic words, "*.... everything is governed by immutable laws .... everything is prearranged .... everything is a necessary effect .... we are toys in the hands of destiny*".[15] If this *absolute truth* is perceived in its *wholeness*, all psychological imbalances go to abeyance and man lives in communion with nature.

Concluding Comment

The humanity is poised at a crucial cross-roads in its long and tortuous history on Planet Earth. We stand at a bifurcation where one road is leading to Buddha (Intelligence, Wisdom) and the other road to bomb (self-annihilation, self-destruction). The aim of human life is to perceive and live in that wholeness, which flows into harmony, order, non-violence and peace. Let us work jointly for that consummation of life.

We want a revolution, psychological revolution, in which the whole human race has to participate and every man is an equal partner. This revolution will regenerate or rejuvenate human mind so that it perceives the new dimension of an integral and all-inclusive consciousness, that can guarantee human survival along with all fauna and flora.

The mutation in Consciousness happens when the true-subject (restructured science-education-health) observes the conditioned psyche and the Unconditioned Energy, or in other words, the wholeness of life. This resolves the age-old tide of I-consciousness, thought-structures, psychological imbalances, ecological crisis, and other human challenges. This guarantees human survival and excellence.

An exploration of the mutation in consciousness is the only consummation of life. This is the only vocation of mankind. If we escape our responsibility, it will jeopardize our future, the future of India and of mankind.

References

1. *Science and Technology – New Challenges*; and *Science, Education, Health and Consciousness: Man Unfolding*; unpublished.
2. Erwin Schrodinger, *What is Life? & Mind and Matter*, London: Cambridge University Press, Ed. 1967, p. 128.
3. Ilya Prigogine and Isabella Stengers, *Order Out of Chaos*, Bantam Books, 1984, p. 5.
4. Maurice H. Wilkins, in *Synthesis of Science and Religion*, Mumbai: The Bhaktivedanta Institute, June 1988, p. 30.
5. Erwin Schrodinger, *op. cit.*, pp. 140, 136.
6. D.S. Kothari, in *The Odyssey of Science, Culture and Consciousness*, New Delhi: Abhinav Publications, 1990, pp. 48-49.
7. *ibid.*, in *The Evolution of Consciousness*, New Delhi: National Publishing House, 1983, p. 227.
8. Ilya Prigogine, *op. cit.*, p. 218.
9. Galileo Galilei, in Hideki Yukawa, *Creativity and Intuition*, Tokyo: Kodansha International Ltd., 1973, p. 111.
10. Sampooran Singh, Dr (Mrs) Kanwaljit Kaur & Paramjit Singh, *Global Education and Human Survival: A Scientific Appraisal*, New Delhi: Sanbun Publishers, March 2002, pp. 46, 35-36.
11. Joan Walton, *New Science*, Winter 1996.
12. Ervin Laszlo, *Emerging Concepts of the Cosmos*, The Scientific and Medical Network, Newsletter No. 49, August 1992, p. 28.
13. Deepak Chopra, *How to Know God*, Rider, 2000, pp. 270-71.
14. J. Krishnamurti, KFI, 1982, No 1, p. 40, & 1983, No 3, pp. 27-28.
15. Voltaire, in Ilya Prigogine, *op. cit.*, pp. 257-58.

5

Symbiosis of Science and Spirituality: The Eternal Renewal by Nature

Abstract

The psychosocial evolution of man, during the last million years, or more, has resulted in his living in the external world, the time-space matrices. He trained his ears (listening) and eyes (observing, seeing) in an antagonistic process to the laws of nature, which implied that he was engaged in coherent superpositions of different webs of conceptual world over the perceptual world and this led to his awareness of the distorted perceptual world. This has resulted in a crisis in human consciousness. There is decrease of the sensitivity or perceptivity of mind. The true arts of understanding, learning and living were forgotten. He is now living in a chaotic world of pain, misery, conflict and violence.

This has resulted in fundamental questions as "Has humanity taken a wrong turn which has created this chaos in the world?" and "What needs to be done within the consciousness of man to preserve our very existence and bring about a new culture?"

This essay explores the conditioned mind-brain-body system through observation by true-subject (restructured science-education-ethics-health). This leads to rejuvenation of the mind to its pristine sensitivity, receptivity and clarity. It leads to a perception of the wholeness. It heals the diseased psychic mind. It blossoms to listening, observing, understanding, learning and living direct. There is an urgency that every man takes up this exploration so that he lives in communion with nature's laws.

Man is an amphibian who lives simultaneously in two dimensions – (a) the given, the absolute field of Consciousness; and (b) man-made social cosmos, the time-space-causation matrices. The absolute field of Consciousness is at the highest quantum energy potential, or the highest frequency, of Consciousness; and man-made social cosmos is at the lowest quantum energy potential, or lowest frequency, of consciousness, and this is called the *successive consciousness*. Man is aware of his mutilated mind-brain-body construct; but he is not aware of his absolute dimension of Life Field. Nature, out of its benediction, has endowed every man with the hidden dimension of undertaking a journey from psychosocial evolution to spiritual dimension, or the eternal renewal by nature, or the self-unfoldment of man. The man has an in-built capacity to rejuvenate his inner potential of Cosmic Consciousness, or of Supreme Intelligence, which can end all human challenges.

Many eminent scientists, philosophers-mystics, politicians, social activists, lovers of mankind have opined that "unless our consciousness undergoes a radical transformation psychologically, there is no hope for man". We have to take a journey together into the whole problem of our daily existence, understanding our lives, our daily miserable, conflicting unhappy lives.

Nature bestowed man five senses – seeing, listening, taste, touch and smell. This essay explores the arts of observation, listening, understanding, learning, action and living; which are all *direct*. We explore the very structure of our thinking, of our acting, of our behaviour and our outlook. This also explores the conditioned psyche and the Unconditioned Energy which is revealed by a rigorous, consistent and persistent scientific mode of mind (restructured science-education-health). This leads to annihilating edge of subject and object dualism. Man has to learn to live in synchronisation with the immutable laws of nature, and not antagonistic to the laws of nature. A living in communion with nature widens the scientific perspective of living.

Recently we discussed the quantum jump of psychosocial evolution to spiritual evolution which points out to widening the scientific perspective, peace and survival.[1,2] This essay explores the dimensions of consciousness of man to preserve the very existence and bring about a new culture.

March of Psychosocial Evolution

Man's psychosocial evolution, over uncountable aeons, has led to man's living in the external world, the time-space-causation matrices; and divorcing the Reality. The mind-brain-body system constructed a vast web of psychological imbalances and stored the subtle matter content as frequency and amplitude information; and this is called the memory, or cultural belief structure, or conditioning of the human psyche, or psychic pandemic. The memory on disintegration gave conceptual structures -I-consciousness and thought-structure. The conceptual world of different webs was superpositioned over the perceptual world which resulted in distorted perception. This has led to a crisis of perception. Every activity of the conditioned human psyche is dualistic, so it adds to entropy of the system which is a pointer towards self-destruction or self-annihilation.

Modern physics has confirmed that all the concepts we use to describe nature are limited and creations of the mind, the conditioned psyche, parts of the map, not of the territory. It is evident that the solution is not separate from the problem; the answer is embedded in the problem.

The psychological imbalances, ecological and pollution challenges, a rising wave of violence and crime, terrorism, religious fundamentalism, ethnic conflicts, corruption, increasing unemployment, poverty, hunger, malnutrition, health crisis, are forcing man inevitably to face the realities of the human situation. All appeals to achieve full disarmament and permanent peace; all declarations by Nobel laureates; all appeals by Pugwash Conferences (since 1957); all appeals by the International Physicians for the Prevention of Nuclear War (IPPNW); by Secretaries-General [Dag Hammarskjold, U. Thant, Javier Perez de Cuellar] of the United Nations; all proclamations of a manifesto by eminent scientists, philosophers, mystics have fallen on the deaf ears of mankind.

Resolution of Human Paradox : Transformation in Consciousness

Sri Aurobindo wrote, "All will change if man once consents to be spiritualized". The Divine Mother (Madame Mirra Alfasa) declared, "No human organisation can change radically unless human consciousness itself changes".[3] J. Krishnamurti stated, "We are concerned with the transformation, psychologically, inwardly of human beings. Unless our consciousness undergoes a radical transformation psychologically, really, there is no hope for man".[4] Robert Muller wrote, "Nevertheless, one dimension has been missing from the extraordinary journey, a dimension lamented by Secretaries-General Dag Hammarskjold, U. Thant and Javier Perez de Cuellar, the spiritual dimension, the highest, deepest, most common, universal and binding dimension of all".[5] Deepak Chopra wrote, "We are the only creatures on earth who can change our biology by what we think and feel".[6] Fritjof Capra wrote, "The patterns scientists observe in nature are intimately connected with the patterns of their minds – with their concepts of thoughts and values .... (These are) leading to the emergence of a new vision of reality that will require a fundamental change in our thoughts, perceptions and values".[7]

The Art of Observation

Erwin Schrodinger wrote, "We cannot make any factual statement about a given natural object (or physical system) without 'getting in touch' with it. This 'touch' is a real physical interaction. Even if it consists only in our 'looking at the object' the latter must be hit by light-rays and reflect them into the eye, or into some instrument of observation. This means that the object is affected by our observation. You cannot obtain any knowledge about an object while leaving it strictly isolated".[8] Sullivan put it, "We cannot observe the course of nature without disturbing it",[9] or Andrade, "Observation means interference with what we are observing .... observation disturbs reality".[10] George Wald wrote "The very crux of twentieth-century physics is the recognition that the observer cannot be excluded from his observation; he is an intrinsic participant in them".[11]

The object is first recorded as pure perception in the brain. Then there is recall of past memory and an interaction happens between the pure perception and recalled memory which leads to distorted perception. We introduce psychological time between pure perception and distorted perception. However we are not aware of pure perception but aware of distorted perception. This process ends all observation.

The process of distortion of observation was well known in Patanjal Yoga. R.P. Kaushik wrote, "The stage prior to recognition and naming is a state of non-verbal or thought-free perception (*nirvikalpa jnana* or pure perception). From this pure perception we proceed to thought, recognition and naming (*savikalpa jnana*)".[12] The concepts are grafted on the perception, which reduces a perception to a concept. The psychological fusion of the observer and the observed is called observation. Osho stated, "Seeing (observation) is *direct*, touching is *direct*, thinking is indirect".[13] Man has to learn to observe, to see direct without interference of the thinking process. Every activity in time is *indirect*, and every happening in the timelessness is *direct*. Observation, listening imply a communion with the immutable laws of nature, but interference of thinking in the observation is antagonistic to nature.

A deformed perception leads to damage to the brain, the neurological and chemical pressures on the brain. The brain loses the sensitivity. Ambition, violence, fear, anger, propaganda, traditions are all tremendous pressures on the brain. Reason, logic have a certain value, but they are limited. Reason cannot end the pressure in the brain. Therefore a brain that lives in pressure is distorted, deformed and damaged. A *pure perception* leaves the brain pristine, pliable, young, fresh, innocent.

The act of observation implies: (a) to see non-subjectively, to see without any evaluatory comparisons or judgements creeping in; (b) observation without an observer is attentivity, the whole being gets tuned into a state of attentivity, where there is neither subject, nor object, it is a subject-free and object-free attentivity; and (c) it blossoms to ending of I-consciousness and thought structure.

The Art of Listening

The moment we hear a word, we interpret it and translate it to a synonym to understand it. When we come to words, which have a psychological import, such as nation, prestige, God, love, and we respond differently to such words. The human mind at once loses its equanimity and responds by resistance or identification, and psychological pleasure and pain come into operation. This disturbance distorts listening. This resistance creates a chain reaction and ultimately destroys our peace and freedom.

Something which does not conform to the patterns of our conditioning becomes unpleasant and very disturbing and we react very strongly. However, in the psychological field, both these processes of identification and alienation are one and the same.

When we listen to a word it produces an automatic reaction. We should watch and feel this reaction with our whole being, patiently and effortlessly, and not interpret this automatic reaction into another set of words or symbols. To listen in such a way implies that we listen directly, without having images about ourselves. It implies complete attention; when the past conditionings encounter attentivity, all conditioning goes to abeyance and mind is rejuvenated, it becomes innocent and is fresh to understand and live an eternal life.

The Art of Understanding

Understanding comes only when the mind is very quiet. So long as the mind is agitated, it can have no understanding. Quietness of the mind, tranquillity of mind, is not a thing to be produced by will-power, by any action of desire. Stillness of mind comes only when there is no process of isolation through accumulation, but complete understanding of the whole process of relationship.

The symbolic-dualistic frame is invariably agitated and cannot give any understanding. It is only the non-dual and non-conceptual frame of Mind Continuum that is quiet and hence bestows understanding. Understanding flowers to intuition.

The Art of Learning

Understanding blossoms to learning. The learning is not recorded, not accumulated; so it changes to energy of being, and the quantum energy potential of the system rises.

Thought process follows the rules of classical logic. In the world of symbols, or experience in the common waking state, everything is either this or that. In the world of pure perception by non-dual frame of reference, there is only the *fact*, which is awareness of the wholeness. Perception follows the rules of quantum logic. The perception is an integral part of quantum logic. Further up the ladder of evolutionary growth, there is coherent superpositions of different webs, or hierarchical expression of Universal Consciousness.

Human being seems to be born with the faculty of self-awareness and understanding, and this faculty gets energised through the arts of observation and listening. Awareness is the perfume of the learning.

The Art of Action

There is no psychological time gap between learning and action, so learning and action are spontaneous. This action arises from the immanent quantum energy potential of consciousness. The action is not based on an idea. An action without idea is without conflict, without duality.

Thought is action at the lowest quantum energy potential of consciousness; it is in the outer field or time-space-causation matrices. We reject the action based on thought.

The Art of Learning-Acting-Meditating

To learn there must be complete freedom – both the conscious and the subconscious mind. And in freedom, the movement of learning, acting and meditating is one harmonious whole. Meditation is not a state; it is a movement, an action is a movement. There is no psychological time gap between learning, acting and meditating; so learning, acting and meditating are one spontaneous movement; there is no wastage of energy, but this energy becomes a state of being, or in other words, the system takes up the role of higher quantum energy potential of consciousness.

Meditation implies to be aware without any choice of the world outside you, and to the immense conflict of the world inside you. In meditation the act of will has to come totally to an end. Will is the essence of desire, heightened form of desire. J. Krishnamurti wrote, "It is *only* the action of will and assertiveness that needs to be changed at all, because the only mischief in relationship is conflict, between individuals or within individuals, and conflict is will and assertiveness".[14]

The Silence and the Space

When I-consciousness and thought-structure go to abeyance, there is silence, quietness – which is called *no-mind* state. The silent mind bestows understanding and learning which is aware of the *fact*. When the mind is totally free from all matter content, then the mind is said to have space, because pure Consciousness exists in a vibrational state. It is the manifestation of pure Consciousness in a pure mind that dissolves all human challenges.

The Art of Living

System implies the comprehension of the whole through the part. That means, give your heart, your mind and all your senses completely to something; it implies that the subject of cognizance is "restructured science-education-health", and the true-subject sees all the parts, without I-consciousness and thought-structure, and each part bestows a perception of the wholeness. The true-subject is at the transcendent quantum energy potential of Consciousness and it observes Itself in each part. There is no centre (cultural belief structure, I-consciousness), there is no limitation to space. The whole process is a function of the quantum energy potential of the subject and one with a lower potential drops away of its own accord. The art of living is embedded in silence, space and energy; and to be aware of this is meditation. Such a life would result in a holistic growth of an individual and show him an alternative way of living.

Out of this comes compassion: that is passion for all. This energises love, which is not desire, which is not pleasure, which is not the action of thought. When there is compassion, there is clarity – clarity of your mind to think clearly objectively, non-personally.

Meditation implies a state where there is absolute discontinuity of the conditioned mind. No expectation, no tension of waiting, no tension of moving in some direction – it is total relaxation. It is awakening of the intelligence that is born out of compassion and clarity.

Transformation is not in the future. It can only be now, from moment to moment. It is seeing the false as the false and the true as the true. One has to learn to live in the light that you understand. *Nature's evolutionary impulse is Eternal Renewal of man, the Self-Unfoldment of man.* This is the sacredness which is not touched by thought. This blossoms to love, compassion and clarity.

Concluding Comment

(a) P. Krishna, an eminent physicist, wrote, "We are going towards self-annihilation and not much time is left now".[15] David Bohm, a renowned physicist, and J. Krishnamurti, an eminent philosopher-mystic, discussed "*The Future of Humanity*" and raised fundamental questions as, "Has humanity taken a wrong turn which has created this chaos in the world?" and "What needs to be done within the consciousness of man to preserve our very existence and bring about a new culture?".

(b) There are arts of observation, listening, understanding, learning, acting, and all these are *direct*; but thinking is indirect. Direct implies to live in communion with the immutable laws of nature, and *indirect* means to live antagonistic to the laws of nature. Every *direct* act of living is an act of worship, because it enhances the quantum energy potential of consciousness of the system.

(c) Man does not have anything which is his personal – including brain and memory, which is of humanity and the realm of the timelessness. Everything is a grace of nature. We are only dancing dolls, where someone else (Cosmic Consciousness, Life Field) pulls the strings; we are toys in the hands of destiny. A living in communion with nature is free from all imbalances, challenges, strife, conflict, and so on.

(d) Intellect, being limited, cannot perceive what is the limitless, the timelessness – eternal. The highest realisation the intellect can have is to realise that it is limited and conditioned, therefore it is incapable of seeing truth. This is the supreme realisation for the intellect and realising this it becomes silent and still.

(e) The chaos is in our mind-brain-body system – man-made social cosmos. It is a crisis in our consciousness. The transformation/mutation in our consciousness can happen through observation-listening-learning. These acts of observation and listening bestow reaction-free perception, unpolluted perception. This bestows a higher level of awareness that has the capacity to resolve our all challenges and bestow non-violence and peace.

(f) The mutation in consciousness happens when the true-subject (restructured science-education-ethics-health) observes/listens the conditioned psyche and the Unconditioned Energy, or in other words, the wholeness of life. It blossoms to compassion; that is, passion for all. Passion bestows love. Love is not different from truth. Love is that state in which the thought process, as time, has completely ceased. Love can be found only in action, which is relationship. Where love is there is transformation. Transformation and love lead to human survival. It is the awareness of the Wholeness, the Oneness of Life that heals the psychic pandemic. It is the Supreme Consciousness/Intelligence that resolves all the human challenges, and bestows order, harmony, non-violence and peace.

References

1. Sampooran Singh and Dr (Mrs) Kanwaljit Kaur, Peace, *Security and Survival: A Scientific Quest*, July 5, 2004.
2. Sampooran Singh, *Consciousness and Spirituality: Widening the Scientific Perspective*, unpublished.
3. Sri Aurobindo and Divine Mother, in Sampooran Singh, Dr (Mrs) Kanwaljit Kaur and Paramjit Singh, *The Masterplan Paradigm for Human Survival and Excellence : The Science of Life*, Chandigarh: Faith Publishers, February 1998, p. 246.
4. J. Krishnamurti, *KFI*, 1982, No. 3, pp. 8-14.
5. Robert Muller, *Foreword to the SourceBook for the Parliament of Religions*, Chicago, 1993.
6. Deepak Chopra, *Ageless Body and Timeless Mind*, New York: Three Rivers Press, 1994, p. 1.
7. Fritjof Capra, *The Tao of Physics*, Bantam Books (Second Edition), 1988, Preface, p. xvii.
8. Erwin Schrodinger, *What is Life*? and *Mind and Matter*, Cambridge University Press, 1967, p. 135.
9. J.W.N. Sullivan, *The Limitations of Science*, New York: Mentor Books, 1949, p. 140.
10. E.N. da C. Andrade, *An Approach to Modern Physics*, New York: Doubleday Anchor Books, 1957, p. 255.
11. George Wald, in *Synthesis of Science and Religion*, Mumbai: The Bhaktivedanta Institute, June 1988, p. 19.
12. R.P. Kaushik, *Towards a New Consciousness*, Delhi: Darshan Yoga Sansthan, 1974, pp. 63-71.
13. Osho, *The Book of Nothing*, Pune: Rebel Publishing House Pvt Ltd., 1975, p. 53.
14. J. Krishnamurti, *The Urgency of Change*, London: Victor Gollancz Ltd., 1977, p. 191.
15. P. Krishna, in *Education, Science, Spirituality*, USA: Institute for Higher Education, Research and Dialogue, August 1993, pp. 12, 13, Cover Page.

6
Science with Creativity

Abstract

This article investigates, "Has humanity taken a wrong turn that has created this chaos in the world?", and "What needs to be done within the consciousness of man to preserve our very existence and bring out a new culture".

The answer to this paradox has been provided by Albert Einstein in prophetic words: "objective observing and understanding" leads to "peace and serenity", but these pearls of wisdom have fallen on the deaf ears of mankind. The United Nations on 28-31 August 2000 proclaimed that "Humanity stands at a critical juncture in history, one that calls for strong moral and spiritual leadership to help set a new direction for humanity".

The present-day science is leading mankind to self-destruction and self-annihilation. Mankind has to make a quantum jump from intellect (reason) to intuition (insight); that insight will regenerate Intelligence which has the capacity to resolve all challenges; and usher a realm of order, harmony, non-violence and peace.

The twentieth century was a century dominated by science and technology. Since the destiny of mankind is largely determined by modern science and its offspring, technology, so seventy-four Nobel Prize winners met in Paris from 18 and 21 January, 1988, and the Conference was titled, "*Facing the 21st Century: Threats and Promises*". The Director-General of Unesco, Mr Federico Mayor Zaragoza, who writes both as head of Unesco and his personal capacity as a scientist, wrote:

> Never before, until now, has the tension between science and the human conscience, between technology and ethics, reached a point where it has become a threat to the world as a whole .... We are so dazzled that we do not perceive the threats hanging over our heads, warning us of the pressing need for a radically new and universal ethical outlook on the future of present-day science.[1]

The Union of Concerned Scientists (UCS) issued a "*World Scientists' Warning to Humanity*" in April 1993, signed by over 1670 scientists including

104 Nobel laureates, concluding that "Human beings and the natural world are on a collision course .... A new ethic is required – a new attitude towards discharging our responsibility for caring for ourselves and for the earth".[2] This ethic "must motivate a great movement, convincing reluctant leaders and reluctant Governments and peoples themselves to effect the needed change". As every event/existence is interrelated and interpenetrated, so the responsibility is not towards survival of one's self but for the human race.

Ilya Prigogine wrote, "In the twentieth century, an attempt has been made to seek behind all conflicts a more universal conception of human dignity that is less subject to the constraints of history. In this respect too, the twentieth century has marked an irreversible turning-point in the history of mankind – a turning-point that heralds a new age, a bifurcation. Will this be a new point of departure towards Karl Popper's dream of an open society".[3] Karl Popper's dream is expressed in words, ".... the aim is a picture of the world in which there is room for biological phenomena, for human freedom and human reason".[4] Jean Dausset wrote, "The possession of this knowledge and this power imposes new duties on man .... he is now capable of guiding his destiny towards a well thought-out future – in other words, of assuming control of his destiny".[5] John C. Polanyi wrote, "*Conventional* thinking on nuclear weapons [including the human challenges of conditioned psyche] is a recipe for mutual suicide".[6] The French sociologist Raymond Aron put it, the world is increasingly one of "virile weapons and impotent men".[7]

Bertrand Russell wrote that thanks to science and technology, mankind is united in evil but not yet united for good; that people have learnt the technique of mutual destruction, but not the more desirable technique of worldwide cooperation. Russell believed that wisdom becomes increasingly necessary with every advance of knowledge and technology, and that although our age has surpassed all others in knowledge, it has not enjoyed a correlative increase in wisdom and he called for a "new moral outlook".[8]

A panel of scientists published "Science with Creativity".[9] Goverdhan Mehta on July 18, 2000 said, "We have to identify science with creativity". He also emphasised, "the connectivity of minds".[10] Again Jayant Vishnu Narlikar on April 9, 2001, said, "What is needed is a campaign for scientific temper".[10] M.G.K. Menon said, "Unless you are creative, you cannot be innovative .... Commitment and dedication, apart from intellect are required for great scientific discoveries and advancement".[11] M. M. Sharma wrote, "Science without conscience ruins .... We as a nation have to innovate to survive".[12]

This article investigates, "Has humanity taken a wrong turn which has created this chaos in the world?" and "What needs to be done within the consciousness of man to preserve our very existence and bring out a new culture?" It discusses a scientific mode of mind, the art of scientific observation, understanding of life directly, intuition and truth. It explores science with creativity, and has its foundations in consciousness and spirituality.

A Scientific Mode of Mind

A scientific mind takes up "the role of a non-concerned observer", in the words of Schrodinger;[13] it is "unrelated to our beliefs, our ambitions, our hopes", in the words of Prigogine;[14] it must not construct a pseudo-observer, it must see in the present and reject the past remembrances or future projections. The values in science are objective truth, integrity, and honesty. Eckhart Tolle wrote, "The moment you start watching the thinker, a higher level of consciousness becomes activated".[15]

Schrodinger wrote, "Mind has erected the objective outside world of the natural philosopher out of its own stuff. Mind could not cope with this gigantic task otherwise than by the simplifying device of excluding itself – withdrawing from its conceptual creation. Hence the latter does not contain its creator".[16] Charles Sherrington wrote, "The material world has only been constructed at the price of taking the self, that is, mind, out of it, removing it; mind is not part of it; obviously, therefore, it can neither act on it nor be acted on by any of its parts".[17] The objective world is at the *successive consciousness*, while the subjective world is at the *immanent* quantum energy potential of consciousness, hence the objective world can neither act on the subjective nor be acted on by any of its parts.

Science is a dialogue with the universe in which questions are asked and insights born. Prigogine wrote, "It is physics that presupposes an observer situated within the observed world. Our dialogue with nature will be successful only if it is carried on from within nature".[18]

David Bohm wrote that "to observe nature with a certain kind of objectivity that could be called the germ of a 'scientific' attitude". He added that "the implication of science is …. that it teaches us to look at facts in an unbiased way, *whether we like it or not*, and that is meaningless to do otherwise …. such a scientific spirit is not only necessary, not only in what is commonly called 'scientific research', but also in art and every phase of life".[19] Maurice H. Wilkins, Nobel Prize in Medicine, 1962, wrote, "The main point is that the open-minded enquiry of the scientist is not something peculiar to science itself but is a characteristic of the good way of living for human beings in general. It is equivalent to the religious concept of love".[20]

Scientific Observation: A rigorous, consistent, and persistent scientific observation implies to see an event/process/happening/object without interfering in its flow, or to see without any transfer of energy to it, or to observe non-subjectively which means any evaluatory comparison or judgement creeping in. It is to perceive without naming, identifying, without emotionally reacting.

Observing (watching) without a single thought, observing without any reaction, observing without any identity, only endlessly observing. Observing the thoughts that arise and then fade away, thought after thought, thought itself becoming aware of itself. Thought is always discontinuous. So you observe the discontinuity. In this observation there is learning, not accumulation of knowledge.

Understanding the Life Directly: Many eminent scientists – including Albert Einstein, Werner Heisenberg, Erwin Schrodinger, Arthur Eddington, and so on – have stumbled upon another mode of perception (knowing), which is "understanding of life directly, instead of in the abstract, linear terms of representational thinking".[21] Understanding expresses itself in the non-dual and non-conceptual frame of reference of Mind Continuum. *Understanding is a movement in the non-duality, a movement of the whole being.* Understanding comes only when the mind is very quiet, tranquil, silenced. Understanding leads to a perception of the whole. Perception has nothing to do with knowledge (information). *The perceptions are born of non-duality and responses are born of spontaneity.*

Ultimate perception happens in a trans-sensual dimension, which is timeless (beyond brain). As all knowledge is structured in Consciousness, so ultimate perception is a happening when Universal Consciousness bathes the brain cells. Perception bestows understanding the wholeness of life, which flowers to intuition, spontaneity and right action; it is a happening in the non-dual and non-conceptual frame of reference of Mind Continuum.

Intuition and Creativity: All knowledge is structured in consciousness. This recognizes that consciousness alone can be totally self-referral, so the pure mind or non-dual mind can perceive the consciousness. When non-dual state of mind is presented as an object to pure consciousness, then it reveals itself. In the non-dual mind, the intelligence is born of the inner silence, of the inner space. The intelligence is uncluttered by the movement of thought. The dimension of intelligence, of unconditioned energy, of Fundamental Vibration, has no past, so it is free of all conditioning and divisive loyalties, etc. It bestows insight. Insight implies something which you perceive without any logic, without any motive. It is called immediate perception. *The perceptions are born of non-duality, of silence, and responses are born of spontaneity.* Perception has nothing to do with knowledge. Intelligence begins to use the human brain and help the human being to look at the world in quite a different way.

Nature, out of Its benediction, expresses Its truth in a non-dual and non-conceptual frame of reference of Mind Continuum. Hideki Yukawa wrote, "Creativity implies discovering something that nobody has hitherto known about or inventing something new .... (It is) the discovery of a truth in nature hitherto unknown to man".[22] The discovery, or creativity, or intuition, or insight is the self-expression of pure consciousness in a pure mind. Our psychosocial evolution over uncountable aeons has resulted in a loss of instincts and intuition.

Truth: Only the truth can liberate the mind from its own ideation; to see the truth, the mind must realize the fact that as long as it is agitated it can have no understanding. Stillness of the mind comes only when there is no process of isolation through accumulation but complete understanding of the whole process of relationship.

Seeing the false as the false and the true as the true is transformation, because when you see something very clearly as the truth, that truth liberates.[23]

Consciousness and Spirituality

Jawaharlal Nehru wrote, "It is for the human mind and human spirit to control it (the Buddha or the bomb) and direct it in the right direction. The great path of science is to be tempered by something and that something is spirituality which gives us to some extent the right measure, the right perspective and right direction to look at". The great scientist, Albert Einstein, said, "Caring for man and his destiny must always constitute the principal interest of all technical efforts".

Some important characteristics of spirituality are as under:

(a) Spirituality is the physics of Consciousness. It is the awareness of the wholeness of life around you and wholeness within you; and its dimensions are "restructured science-education-health". It proceeds from the whole and with the awareness of the whole, we observe the parts.

(b) In spiritual science, the emphasis is on the act of perception, the quality of the instruments employed for perception, the quality of consciousness behind the perception. The dimension of spirituality is at the transcendent potential. When I-consciousness or thought-structure encounters the dimensions of spirituality, these go to abeyance and pure dimension of Consciousness expresses Itself as order, harmony, non-violence and peace.

(c) In spirituality, the act of investigation, exploration is conducted on behalf of the whole human race. Spirituality explores what the truth is, what the *fact* is, and what the Reality is behind the truth. Spirituality teaches us to live with the *fact* and transcend the non-fact.[24,25]

(d) The symbolic-dualistic frame of reference of the mind spectrum (dualistic thought) makes a quantum jump across a boundary, that physicists call the "event horizon",[26] or "frequency barrier" to "absolute rest frame of reference" (non-dual and non-conceptual thought) of the Mind Continuum. At the moment of making a quantum jump across the "frequency barrier", all solidity of dualistic thought (time) dissolves into a bundle of high frequency vibrating energy (timelessness realm). The frequency barrier, in a sense, represents a limiting condition which is nearing the velocity of light or quantum energy potential equivalent to the velocity of light.

(e) When there is scattering of the frequencies – which are between the frequency barrier and the lowest frequency (immanent quantum energy potential of the mind-brain-body system) at which human beings exist – then it is called the psychosocial evolution. It constructs the time-space matrices in quick succession. *Spiritual evolution happens when all frequencies are higher than the threshold frequency at the frequency barrier, that is, between the frequency barrier and the ultimate frequency in the Life Field.* These express themselves as creativity, intuition and spontaneity. These bestow perfect freedom, non-violence, order, harmony and peace.

When one crosses the frequency barrier, or in other words, when one makes a quantum jump across the frequency barrier, we enter a realm of timelessness, where there is perception of the wholeness of life; and this bestows creativity, intuition and spontaneity, fellow feelings of love and compassion – this is the truth of life.

Discovery: A discovery or an intuition is an expression of truth in nature hitherto unknown to man. It happens if there is timeless perception. It happens when there is quantum jump from a symbolic-dualistic frame to a non-dual and non-conceptual frame of reference of Mind Continuum. D.S. Kothari, an eminent scientist, wrote:

> The new theorems, unless trivial, are of course first 'discovered' and then proved. Discovery is an inspired guess, a flash of insight. It is not a step by step process, but a sort of quantum jump.... A rational account of discovery, the proof, follows and not precedes the discovery.[27] .... Fundamental discoveries which go deep into the nature of things are the result of common intuition and faith, not cold logic.[28]

All great discoveries have arisen from the realm of timelessness, so scientific enquiry is not a cerebral activity and gathering information but a quest of the Timelessness, the Divinity, the Life Field.

When one is aware of the "objective truth", and ignorant of the "subjective world", the mind is engaged in "invention". If one is simultaneously aware of the subjective world (inner) and the objective world (outer), then the mind is engaged in discovery. Invention implies a modified use of the known material.

Science with Creativity

Karl Pribram wrote, "Brain science must deal with the awareness of awareness. It could no longer afford to shut out that part of the world which we call subjective".[29] Roger Sperry spoke of the importance of a neglected area of science, namely the study of the human subjective experience".[30] John Eccles also emphasizes the need to deal scientifically with the neglected area of inner conscious awareness.[31] The above referred galaxy of scientists have opined that the outer (objective) and the inner (subjective) are not two separate movements, but one dynamic whole; or in other words, the outer and the inner are one, a unitary movement, not separate, but whole.

Restructuring of Science: We have to restructure science, which takes in its ambient both the objective and the subjective dimensions of life. It is a harmonious movement. The time interval between the subjective and the objective is less than a hundred microsecond. The restructured science deals with the wholeness of life.

Restructuring of Education: The dimension of value education is embedded in the trans-sensual dimension of human personality. We have to restructure education so that it takes in its ambient the sensual and the trans-sensual dimension of life.

Restructuring of Health Technology: The theme of the Indian Science Congress – 2005 is Health Technology. A living in the *present*, the *now* heals. And living in the *past*, and the *duality* suppresses man's sensitivity and enhances disease. There is an urgency to restructure medical education which takes in its ambient the psychodynamics of mind-brain-body system and the Unconditioned Energy, or in other words, the wholeness of life.

To sum up, we have to restructure science-education-health, and the subject of cognizance is called the true-subject.

The Ending of Human Challenges or Healing the Psychic Pandemic

We need restructuring of science-education-health to *observe* truthfully and sincerely the conditioned and the Unconditioned Energy, and this will flow to mutation of the psyche. The different steps in observation, which are expressed to understand and learn, are as under:

(a) Ending of I-consciousness: If a thought coupled with volitions encounters the true-subject, the I-consciousness is knocked off because it is at the lowest quantum energy potential. This suggests that: (i) the movement of thought-time is independent of I-consciousness; (ii) it is the movement of collective human effort; (iii) both the I-consciousness and thought-structure have a conceptual reality.[32]

(b) Ending of Thought-Time: When thought-time comes in the realm of true-subject, the photons meet an 'impregnable wall', or "event-horizon", or "frequency-barrier", and are dissipated, which implies that thought-time goes to abeyance and what is left behind is the *fact*, the *true*, the *present*, the *what is*!

Man has to learn through the art of observation to make a quantum jump across the frequency barrier – from lower frequencies (symbolic-dualistic frame of reference) to higher frequencies (non-dual and non-conceptual frame of reference of Mind Continuum). Below the frequency barrier, there is disorder, chaos and disharmony, and this is called the sensual world; and above the frequency barrier there is perfect order, harmony; and this is called the natural world, the Quantum World. Below the frequency barrier is the time field and above the frequency barrier is the timeless realm, the Unknown, the Reality. When Reality bathes the brain cells, there is no pain and misery, and one revels in freedom, harmony, non-violence, peace and bliss. To live in communion with the natural world, Reality, is the consummation of human life.

(c) Ending of the Fact: If we continue observing or listening to the *fact*, without justification, or condemnation, or motivation, we observe that there is only the *fact* and nothing else. The fact itself is the operator and not you operating on the fact. If you leave the fact alone, it has no time. The fact is free from fear. The fact grows, which implies that there is release of tremendous energy and the fact moves to higher quantum energy potential. This is expressed as Pure Consciousness → Fact.

(d) The Art of Living: The arts of observing, listening, understanding, learning, action are direct; and thinking is indirect. The indirect processes are antagonistic to the immutable laws of nature, so the conditioned psyche constructs conflict, pain, and sorrow. The direct happening implies an expression of Cosmic Consciousness in a non-dual and non-conceptual Mind Continuum, so it is beyond the brain and intellect. "Science with creativity" is direct expression of Cosmic Consciousness in the pure mind.[33,34] David Bohm, an eminent physicist, wrote:

> The ultimate perception does not originate in the brain or any material structure, although a material structure is necessary to manifest it. The subtle mechanism of knowing the truth does not originate in the brain.[35]

The art of living emerges from direct perception or ultimate perception; and this is free from any psychological imbalances, conflict, chaos. The art of "science (restructured science-education-health) with creativity" leads to an expression of the wholeness of life.

Fred Alan Wolf wrote, "We are beginning a new age of awareness, the age of quantum consciousness, the age of the conscious atoms. By looking within ourselves, we may be able to solve the problems facing us on the final frontier – the frontier of the human spirit".[36] It appears that the order of the universe is the order of our minds and reveals a relationship.

Concluding Comment

(a) Under the auspices of the United Nations, an International Conference discussed, "Humanity stands at a critical juncture in history, one that calls for strong moral and spiritual leadership to help set a new direction for society".[37] The challenges facing mankind are not in the external field, but are embedded in the human psyche, and have assumed a global dimension.

(b) Science has now recognized that the whole cosmos (and beyond cosmos) is full of high frequency energy (higher frequency means higher energy) which is the womb of existence. The aim of human life is to perceive and live in that wholeness, which flows into harmony, order, non-violence and peace. Let us work jointly for that consummation of life.

(c) There are arts of observation, listening, understanding, learning, acting, and all these are *direct*; but thinking is indirect. Direct implies to live in communion with the immutable laws of nature, and *indirect* means to live antagonistic to the laws of nature. Every *direct* act of living is an act of worship, because it enhances the quantum energy potential of consciousness of the system.

(d) Man does not have anything which is his personal – including brain and memory, which is of humanity and the realm of the timelessness.

Everything is a grace of nature. We are only dancing dolls, where someone else (Cosmic Consciousness, Life Field) pulls the strings; we are toys in the hands of destiny. A living in communion with nature is free from all psychological imbalances, challenges, strife, conflict, and so on.

(e) The chaos is in our mind-brain-body system – man-made social cosmos. It is a crisis in our consciousness. The transformation/mutation in our consciousness can happen through observation-listening-learning. These acts of observation and listening bestow reaction-free perception, unpolluted perception. This bestows a higher level of awareness that has the capacity to resolve our all challenges and bestow non-violence and peace.

(f) The mutation in consciousness happens when the true-subject (restructured science-education-health) observes/listens the conditioned psyche and the Unconditioned Energy, or in other words, the wholeness of life. This resolves the age-old tide of I-consciousness, thought-structures, psychological imbalances, ecological crisis, and other human challenges. This guarantees human survival and excellence. It blossoms to compassion; that is, passion for all. Passion bestows love. Love is not different from truth. Love is that state in which the thought process, as time, has completely ceased. Love can be found only in action, which is relationship. Where love is there is transformation. Transformation and love lead to human survival. It is the awareness of the Wholeness, the Oneness of Life that heals the psychic pandemic. It is the Supreme Consciousness/Intelligence that resolves all the human challenges, and bestows order, harmony, non-violence and peace.

(g) We want a revolution, psychological revolution, in which the whole human race has to participate and every man is an equal partner. This revolution will regenerate or rejuvenate human mind so that it perceives the new dimension of an integral and all-inclusive consciousness, that can guarantee human survival along with all fauna and flora.

(h) The "science with creativity" is an experiment of creating a "New Man" – a human being who lives in harmony with the inner and the outer, with himself and his environment, and who is free from all ideologies and conditionings that now divide humanity. The human survival is embedded in this gigantic experiment. And if man escapes his responsibility, he does so at his own peril, of India and of mankind.

References

1. Federico Mayor Zaragoza, in *The Courier*, May 1988, p. 4.
2. Union of Concerned Scientists, *World Scientists' Warning to Humanity*, USA: 26 Church Street, Cambridge MA 022 38, April 1993.
3. Ilya Prigogine, in *The Courier*, May 1988, p. 13.
4. Karl Popper, in *ibid.*, p. 10.
5. Jean Dausset, in *ibid.*, p. 16.
6. John C. Polanyi, in *ibid.*, p. 31.

7. Raymond Aron, in *ibid.*, p. 33.
8. Bertrand Russell, in *ibid.*, p. 4.
9. Dr (Mrs) Kanwaljit Kaur, Paramjit Singh and Sampooran Singh, *Science with Creativity*, Vedanta, UK: Ramakrishna Vedanta Centre, No. 296, November-December 2000, pp. 256-267.
10. Goverdhan Mehta, in *Global Education & Human Survival*, New Delhi: Sanbun Publishers, March 2002, p. 22; and Jayant Vishnu Narlikar, *ibid.*, p. 22.
11. M G K Menon, *The Tribune*, November 28, 2004.
12. M.M. Sharma, XII D.S. Kothari Memorial Oration, Defence Laboratory Jodhpur, July 14, 2004.
13. Erwin Schrodinger, *What is Life*? and *Mind and Matter*, London: Cambridge University Press, Ed 1967, p. 128.
14. Ilya Prigogine and Isabelle Stengers, *Order Out of Chaos*, Bantam Books, 1984, p. 5.
15. Eckhart Tolle, *The Power of the NOW*, Hodder, 2004, p. 17.
16. Erwin Schrodinger. *op. cit.*, p. 131.
17. Charles Sherrington, in Erwin Schrodinger, *ibid.*, p. 128.
18. Ilya Prigogine, *op. cit.*, p. 218.
19. David Bohm, *New Scientist*, 11 November 1982, p. 364.
20. Maurice H. Wilkins, *Synthesis of Science and Religion*, Mumbai: The Bhaktivedanta Institute, June 1988, p. 29.
21. Alan W. Watts, *The Ways of Zen*, New York: Vintage Books, 1957, p. 10.
22. Hideki Yukawa, *Creativity and Intuition*, Tokyo: Kodansha International Ltd., 1973, pp. 123, 113.
23. J. Krishnamurti, *The First and Last Freedom*, London: Victor Gollancz Ltd., 1977, pp. 278, 279, 286.
24. Dr (Mrs) Kanwaljit Kaur, Paramjit Singh, and Sampooran Singh, *Symbiosis of Science and Spirituality*, Vedanta: UK, op. cit., No. 301, September-October 2001, pp. 224-231; and No 302, pp. 261-270.
25. Sampooran Singh, *A Review on Symbiosis of Science and Spirituality*, Dagshai: The Atam Science, October 7, 2002, pp. 16, 17.
26. Deepak Chopra, *How to Know God*, London: Rider, 2000, pp. 270-271.
27. D. S. Kothari, *The Perception of Truth in Science and Philosophy*, in *The Odyssey of Science, Culture and Consciousness*, New Delhi: Abhinav Publications, 1990, pp. 48-49.
28. *ibid.*, *Atom and Self*, in *The Evolution of Consciousness*, New Delhi: National Publishing House, 1983, p. 227.
29. Karl Pribram, in Ken Wilber, *The Holographic Paradigm and Other Paradoxes*, Boulder: Shambhala, 1982, p. 17.
30. Roger Sperry, *Annual Review of Neuroscience*, 1981.
31. John Eccles, in *Synthesis of Science and Religion*, The Bhaktivedanta Institute, June 1988, p. 199.
32. Sampooran Singh and Dr (Mrs) Kanwaljit Kaur, *Peace, Security and Survival*: *A Scientific Quest,* July 5, 2004.
33. Sampooran Singh, *Consciousness and Spirituality*: *Widening the Scientific Perspective*, unpublished.
34. *ibid.*, *Symbiosis of Science and Spirituality*, *The Eternal Renewal by Nature*, unpublished.

35. David Bohm, in Gary Zukav, *The Dancing Wu Li Masters*, Fontana/Collins, 1982, p. 327.
36. Fred Alan Wolf, *Taking a Quantum Leap*, San Francisco: Harper and Row, 1981, pp. 236, 248.
37. *The Millennium World Peace Summit of Religious and Spiritual Leaders* (*United Nations*), New York: 28-31 August, 2000.

7
The Foundations of a New Human Society

Abstract

We explore the foundations of a new society with a view to end all human challenges and conflict. It presents the recent findings of Indian sages and Western thinkers. It discusses knowledge and intuition (insight), invention and discoveries, life and death, and mysterious Consciousness.

We have to restructure science which implies that science takes in its fold the objective (outer) and the subjective (inner) dimensions of life, which embraces the wholeness of life. The I-consciousness (ego) and thought structure go to abeyance when these come into field of restructured science-education.

Every existence/event/process/happening is permeated and surrounded by Universal Consciousness, the Life Field, the Unmanifest.

Man has an evolutionary origin. The *Homo sapiens* appeared at about 2.5 million years ago. Perhaps the concepts of "in her" and "out there" arose in a rudimentary form with the beginning of stone age – "I (subject) go for search of food (object)". The evolution of the nervous system hastened this growth of awareness. It appears that man alone has awareness of the self, as the *subject* of experience, along with awareness of the not-self, as the *object* of experience, of both the *within* and the *without* of nature. The *within* is called the subjective and the *without* is called the objective dimension of man. Man is unique in developing and nourishing his subjective dimension of life.

Modern physics has confirmed that all the concepts we use to describe nature are limited and transient, that they are not features of reality, but creations of the human mind; properties of our conceptual map of reality, rather than of reality itself; parts of the map, not of the territory. The world extended in space and time is but our representation.

The human brain has evolved through thousand and thousand years. It has been gathering partial experience and fragmentary knowledge (information) for uncounted aeons and storing these as memory or cultural belief structure – which is frequency and amplitude information – in the brain. The consciousness is shared by all humanity, you are the entire humanity. The memory or thought-structure is shared by mankind. We are the rest of mankind. We live in the past, which has conditioned the psyche. The conditionings have constructed an aggressive and violent society. There is global disorder, and we are responsible for that.

Spencer Brown, in a most spectacular passage, wrote, "In order to do so (see), evidently it must cut itself up into at least one state that sees, and at least one other state which is seen. In this severed and mutilated condition, whatever it sees is *only partially itself*".[1] Eddington put it succinctly, "Nature thus provides that knowledge of one-half of the world will ensure ignorance of the other". There is energy exchange (photon transfer) between the seer and the seen, and this leads to mutilation of both. We are aware of the mutilated seen and the mutilated seer gets recorded. This leads to a crisis of perception. This results in seer-seen or subject-object dualism. We live in pain, misery and conflict. Alexandre Koyre stated, "Two worlds: this means two truths. Or no truth at all. This is the tragedy of modern mind which solved the riddle of the universe, but only to replace it by another riddle: the riddle of itself".[2]

We have built a society – a synthetic man-made construct which is divorced from reality – which is violent. The violence is disorder, it has destroyed every form of relationship, and has brought deep agony in oneself. The foundations of our society are built on a sand dune, or on a ripple of a lake, or on the waves of an ocean, or on patterns of clouds. It is man's signature as "*distorted perception*" and recorded in the brain as memory or "cultural belief structure" over uncountable aeons of time. This is called the conditioning of the human psyche, which is shared by the whole mankind. The memory is finite, transient and ephemeral. The memory is meta-stable, it spontaneously disintegrates to thought-time. Thought is invariably dualistic; and it gives rise to conflict, confrontation, violence, pain, misery and travails.

It has become evident that a non-violent and peaceful society cannot come to existence by merely wishing it, by past behaviour patterns, by thought-time structures, by beliefs, ideals, concepts and so on. If human mind constructs conflicts, then to end conflict we have to transform human psyche. But all traditional approaches over millennia of years have proved to be sterile to transform the mind. As the cause of our conflict is thought-time or psychological imbalances, so we have to end (or resolve) the cause *here* and *now*; and if we ignore this, we will remain violent and miserable as we are today even after one thousand years. The answer to this paradox is: (a) investigating the conditioned psyche – which is seeking out inwardly and finding out oneself a tremendous realm beyond thought, and (b) explore the Unconditioned Energy which is impersonal, unconditioned and beyond the known sensual realm. This exploration will lead to an inner journey from rationality to intuition (insight) and energise the Intelligence which will take hold of man and guide him to non-violent and peaceful world.

Man wants to live a really peaceful life in which there is order, rhythm, non-violence and inner equipoise, in which there is deep abundance of love. This article explores the foundations of a new human society with a view to end this tremendous violence and conflict. It discusses the Universal Consciousness, the Life Field, the *Brahman*, the One Unmanifest-manifested. It explores the psychodynamics of the conditioned psyche; to educate our physical body and psychological structure so that these perceive the wholeness of life and this implies higher awareness; we will have to "understand the life directly"; perceive true perception of the *fact*, and live in that wholeness of life. This is to live in communion with the laws of Nature. This lays the foundation of a non-violent and peaceful New Human Society.

Intuitive Findings of Indian Sages

India is the cathedral of human civilisation. It has a unique cultural heritage for the last 10,000 to 30,000 years. Swami Vivekananda declared, "Shall India die? Then from the world all spirituality will be extinct, all moral perfection will be extinct .... Such a thing can never be".[3] Sri Aurobindo, in a prophetic vision, wrote:

> The East alone has some knowledge of the truth, the East alone can teach the West, the East alone can save mankind.... This is the reason of India's resurgence, this is why God has breathed life into her once more.... India cannot perish, our race cannot become extinct, because among all the divisions of mankind it is to India that is reserved the highest and the most splendid destiny, the most essential to the future of the human race.[4a]
>
> The evolution of mind working upon life has developed an organization and use of matter which can no longer be supported by human capacity without an inner change. The problem is fundamental and in putting its evolutionary Nature is confronting man with a critical choice which must one day be solved in the true sense if the race is to arrive or even to survive.[4b]
>
> If this evolution (spiritual) is to take place, since it must come through the growth of the spirit and the inner consciousness, the initiative can come from India and although the scope must be universal, the central movement may be hers.... A catastrophe may intervene and interrupt and destroy what is being done, but even then the final result is sure.... A new spirit of oneness will take hold of the human race.[4c]

Sri Aurobindo added, "All will change if man once consent to be spiritualized". The Divine Mother (Madame Mirra Alfasa) declared, "No human organisation can change radically unless human consciousness itself changes".

Man is caught in two crises – the civilisational crisis and crisis due to the thrust of evolutionary impulse. The Divine Mother in 1963 wrote, "The only hope for the future is a change of man's consciousness and the change is bound to come". The message in 1967 read, "Men, countries, continents!

The choice is imperative: Truth, or the abyss". The Mother had once observed, "Three-fourth of humanity is obsolete".[5]

Vimala Thakar, the greatest philosopher-mystics, prophetically stated in November 2000, "The time has become ripe for the next phase of mutation, in which the quality and content of consciousness will become unconditionally free – from psychological structure". She again stated on March 17, 2003, "Before the old order (conditioned psyche) gets completely dismantled, by the onslaught of new globalized trends or withers away by the weight of out of date philosophies, mankind will have to focus on the Vision of the New". We will have to purify our physical body and psychological structure to perceive the *Nature's Grace*.

Recent Findings of Western Thinkers

Vaclav Havel, President of Czech Republic, stated that the only reliable path to peaceful co-existence is, "It must be rooted in self-transcendence .... Transcendence as the only alternative to extinction".[6] He added that the genuine universal experience that could, "unite all of humanity is the experience of transcendence in the broadest sense of the word .... modern man has lost his transcendental anchor".[7] George Wald, Nobel Prize in Medicine, 1967, wrote, "By now the problem is not transcendence, but survival".[8] The former Governor of Colorado, Richard Lamm, quoted Stuart Udall, and concluded, "The whole human enterprise is a machine without brakes, for there are no indications that the world's political leaders will deal with the realities until catastrophes occur .... We have already passed the point of no return, and tragic human convulsions are at hand".[9] Schweitzer warned us: "We have lost the ability to foresee and forestall. We shall end by destroying the Earth".[10]

Michael Burghley concludes, "There is an intelligent order and beauty to life already, and it does not need our interference .... We must resign from trying to manage the universe. Until we do, as individuals and as a race, we are doomed to disaster. Even as it stands, we will not be spared the repercussions of the way we have conducted our lives .... Sages and teachers down through the centuries have told us that our value and fulfilment spring from within".

Many eminent thinkers have opined that we are passing through a transitional period – it seems that the psychic age is on the way out and spiritual age is painfully being born.

Mysterious Consciousness?

Man simultaneously lives in two dimensions: the time-space matrices, the *known* and the timeless realm, the Divine Potential, the Life Field, the *Unknown*. The *known* is an expression of successive consciousness at the lowest frequencies. The *Unknown* is an expression of the Simultaneous Consciousness at the maximum frequency, or at the highest quantum energy potential of Universal Consciousness. As a symbol, we may compare the *known* to a thimble, and the *Unknown* to a fathomless and boundless ocean. The *Unknown* is immutable, infinite and eternal.

In a recent (1996) scientific conference at Moscow, the Russian scientists proposed, "A Life Field, beyond our physical concepts of time, space and the known electromagnetic and gravitational fields". In another conference at Cambridge, it was proposed, "A new scientific approach would include subjective experience as valid information and address all forms of human experience – including intuition, spirituality and mystical encounters".[11]

George Wald wrote:

> One can put together everything that we have learned, and add to it everything that workers in this area hope to learn; and none of it comes anywhere near, or even aims in the direction of *what it means to see* …. Seeing – the event in consciousness – seems to lie in another universe, unapproachable by science…. Science offers no way of testing for the mere presence or absence of consciousness, let alone its content…. Science has no way of dealing with consciousness, not even to identify its presence or absence …. Since consciousness presents us with no identifying signals, nothing measurable, no apparent energy expenditure, it would be a comfort for a scientist to deny that it exists, or to declare it irrelevant …. Science tells us nothing about consciousness, nor does it promise ever to do so …. Science does not only discover, it creates, makes real what it asserts to be real …. science becomes a Tao, a Way: the way to the boundaries.[12]

There is no problem more mysterious than the mind-brain-body interaction. Modern science has not made it less intractable. On the other hand, it has added new urgency and also a new poignancy.

Hideki Yukawa concluded, "Unless something rises into the consciousness it cannot become an object of rational thought, of rational consideration".[13] Eugene Wigner stated, "The measurement is not completed until its result enters our consciousness".[14] Roger Sperry wrote, "Current concepts of the mind-brain relation …. gives full recognition to the primacy of inner conscious awareness as a causal reality".[15]

Mystery of Man: Arthur Eddington hinted at the emergence of the mystery of man from the study of the mystery of physical nature:

> Knowledge is only an empty shell – a form of symbols. *It is knowledge of structural form, and not knowledge of content.* All through the physical world runs that unknown content, *which must surely be the stuff of our consciousness*.[16]

Teilhard de Chardin concludes:

> Finally, it breaks down completely with man, in whom the existence of a within can no longer be evaded, because it is the object of a direct intuition and the substance of all knowledge.[17]

> *The true physics is that which will, one day, achieve the inclusion of man in his wholeness in a coherent picture of the world.*

Lincoln Barnett wrote:

> Niels Bohr puts it, "We are both spectators and actors in the great drama of existence". *Man is thus his own greatest mystery .... Least of all does he understand his noblest and most mysterious faculty: the ability to transcend himself and perceive himself in the act of perception.*[18]

The great neurologist, Charles Sherrington, wrote:

> Today, Nature looms larger than ever and includes more fully than ever ourselves. It is, if you will, a machine, but it is a partly mentalised machine and in virtue of including ourselves, it is a machine with human qualities of mind. It is a running stream of energy – mental and physical – and unlike man-made machines, it is actuated by emotions, fears, and hopes, dislikes and love.[19]

Richard Feynman wrote:

> Living things, masses of atoms, DNA, protein .... Dancing a pattern ever more intricate .... Out of the cradle onto the dry land .... Here it is standing .... Atoms with consciousness .... Matter with curiosity .... Stands at the sea .... Wonders at wondering .... I .... A universe of atoms .... An atom in the universe.[20]
>
> That is what it means when one discovers how long it takes for the atoms of the brain to be replaced by other atoms, to note that the thing which I call my individuality is only a pattern or dance. The atoms come into my brain, dance a dance, then go out; always new atoms but always doing the same dance, remembering what the dance was yesterday.[20]

Man is a miniaturised cosmos. Since atoms with consciousness are looking at atoms with consciousness, we have the experience of wonder and mystery whenever we look at the universe. We are looking at ourselves; mind looking at mind. It is this process which we call the universe (time-space matrices).

Consciousness performs a dual role in the universe. In the world of the quantum, it is both the awareness and the creation of experience. It is being and the knowing of experience. Awareness implies that the Consciousness is expressing itself at the highest quantum energy potential, or at the highest frequency (higher frequency means higher energy); it is also called the Unmanifest. Experience implies that the 'successive consciousness' expresses itself at the lowest quantum energy potential as time-space- matrices, or matter – it is called the manifest. Consciousness may be written as Unmanifestmanifest. With the stroke of the twentieth-century quantum eraser, the dividing line between ontology (theory of being) and epistemology (theory of knowing) is rubbed out. The fusion of ontology, epistemology, psychology, philosophy, and physics is beginning to emerge.

The One Unmanifest-Manifested: There is One Reality, the Unmanifestmanifested. The nature of the Unmanifest is 'Deep-Silence',

'Inactivity (Non-Vibration)', 'Passivity', 'Immutability', 'Eternal'; and also Activity (Vibration), dynamism and illimitable potency to manifest as many. Activity (Active Reality) expresses which is self-hidden in the Silence (Passive Reality); both are different states of the Supreme Reality, the obverse and reverse affirmations; both are Pure Existence and conscious Force, or Self-Existent Bliss and Supreme Intelligence, and are One. It is the Active Reality which manifests as the perfect freedom and omnipotence of an eternal divine activity in innumerable galaxies as inanimate world and plant and animal kingdoms.[21]

The Active Reality is a powerful, highly directional, monochromatic (same frequency) and coherent (the wave is in one phase) beam of energy; or a state of highest frequency (higher frequency means higher energy), or high quantum energy potential. At the level of man, the intense energy of insight expresses itself as direct perception of the fact, the Reality. The insight acts upon the brain matter and transmutes it, so the brain matter transforms and attains a higher order through insight, or higher quantum energy potential of Consciousness. The insight or the understanding permits the powerful subtlest energies to penetrate the known and transform the known. The subtlest energy (higher potentials) knocks the thought out of commission, not by thinking, not by reasoning, but rather a direct transformation due to its contact with the subtlest energy. The insight is the supreme Intelligence.

The Universal Consciousness (highest quantum energy potential) heals, because it has the dimensions of creativity and spontaneity, nonviolence, beauty and harmony, love and compassion. Man has to choose between healing and annihilation, between the *Buddha* (Wisdom) and the *bomb* (annihilation). Obviously the decision is to work honestly, sincerely and truthfully for mutation of one'' own psyche.[22]

New Meaning of the Word 'Order': Physicists have showed a new meaning for the word *order*. This new order, the basis of the new physics was not found in the particles of matter. Rather it was found in the minds of the physicists. The order of the universe may be the order of our own minds.

Deepak Chopra wrote, "To know God personally, you must penetrate a boundary that physicists call '*the event horizon*', a line that divides reality sharply in half [*or to be more precise, in two parts – the subject and the object*]. On this side lies anything that remains within the speed of light; on the other side is anything faster than the speed of light".[23] We call the boundary, or the wall, between symbolic-dualistic frame and non-dual and non-conceptual frame of reference of mind spectrum as 'the frequency barrier' or 'the event horizon'. We assume the mass of thought as electronic rest mass, m_0 [$m_0 = 9.1091 \times 10^{-31}$] and c (velocity of light) is 3×10^8 meters/sec. Based on this assumption, the energy of the frequency barrier comes out to be 8×10^{-14} joules, or 0.5×10^6 electron volt. It means that the frequency barrier represents half-a-million electron volts (0.5 MeV). On this basis, the energy comes out to 8×10^{-14} joules, or 0.5×10^{6} electron volts. The frequency of vibration comes out to be 10^{20} vibrations/sec. This has not been measured, but we take it as a first approximation. It gives an order of magnitude. Perhaps, it is the energy level of the "frequency barrier", which makes it difficult to make a quantum jump from time to timelessness. It is limited, an intellectual approach, and requires further exploration.[24]

Whenever any species is in danger of extinction, some evolutionary impulse/urge within the race itself pushes to a new mode of consciousness which would enable it to overcome the danger. This is known as self-unfoldment of consciousness. This survival mechanism has in fact been largely responsible for evolution so far.

> Ervin Laszlo discussed the emerging concepts of the cosmos in words: The parts of this totality are not radically separate: they are integral elements of the whole. Every "thing" in the cosmos is a coherent "event" and every event is the summed result of all the other events.... The cosmos is a system of interacting waves: it is a sea where the effect of the smallest perturbation propagates throughout the entire field.... the new physics and the new cosmology do not recognize discrete things and independent events in the universe, only ripples upon ripples, waves upon waves, interpenetrating and propagating in a seamless sea.[25]

Every thing in the cosmos is a coherent event and every event is the summed result of all other events. The observable world appears to have emerged from the "quantum vacuum (potentialities)" that constitutes the "ground energy state" of the universe. The ground energy state of the cosmos is the Eternity.[26]

Mankind needs a new "instrument of thought", in the words of David Bohm[27]; "the poetic intuition.... which the intellect can never fully grasp", in the words of Max Planck[28]; "a completely new form of human intellectual endeavour", in the words of G.F. Chew.[29] These physicists point out that the symbolic-dualistic frame of reference or relative frame of reference of the mind spectrum cannot understand the 'intuition' or 'new form of human intellectual endeavour', but it is the non-dual frame of reference of mind spectrum that has the capacity to perceive the truth.

David Bohm said, "Man is potentially infinite".[30] The potential of the human race is infinite as life is infinite. It appears that the potential contained in the human beings to peel off layers after layers of untruth and reach the core of truth is infinite. *Spirituality is defined to maximise the potential of the human being and to live in that highest potential.*[31]

Knowledge and Intuition (Insight)

Knowledge: The conditioned human psyche is invariably engaged in gathering experience and knowledge and storing these in the brain as frequency and amplitude information – called memory or cultural belief structure. The memory is meta-stable and spontaneously disintegrates into a fragment –

called thought-time. Memory gives adequate consideration to skills, beliefs, presuppositions, psychological imbalances and ecological challenges.

Objective science is a movement of knowledge gathering more, and more. The more is the measurement and thought can be measured because thought is a material process. Thought can never be creative because creativity is in the realm of the Unknown and thought can never touch the Unknown.

Knowledge and experience are always fragmentary, divisive, dualistic, and constructs conflict, pain, misery and travails. So knowledge is always limited whether now or in the future, and where there is limitation there is a conflict. Thought has brought out tremendous conflict between human beings – not only between the individuals, but also collectively. Knowledge can never be creative, though knowledge can bring about a better physical world.

We share the consciousness of all humanity, for our suffering, our pleasures, beliefs, conclusions, opinions, and all the religious dogmas and faiths are shared by all human beings on this earth.

Insight: Many eminent scientists – Heisenberg, Schrodinger, Einstein, Eddington – have stumbled upon another mode of knowing. They called it intimate, or non-dual, or direct, or intuitive mode of knowing. Eddington wrote:

> We have two kinds of knowledge which I call symbolic knowledge and intimate knowledge .... [The] more customary forms of reasoning have been developed for symbolic knowledge only. The intimate knowledge will not submit to codification and analysis; or, rather, when we attempt to analyze it the *intimacy is lost and it is replaced by symbolism*.[32]

Allen W. Watts wrote, "It is understanding of life directly instead of in the abstract, linear terms of representational thinking".[33] David Bohm discussed:

> The ultimate perception does not originate in the brain or any material structure, although a material structure is necessary to manifest it. The subtle mechanism of knowing the truth does not originate in the brain.[34]

Intuition appears to be a direct input of the truth into the brain cells without mediation of the sensual apparatus. Intuition happens when the subject of cognizance is *non-variable attentivity*, *reaction-free attentivity*, or *objectless attention* – this is called a Mind with a scientific temper. It is tuned to perceive the *fact*, the *true*, the *what is*!, the *non-dual*. It bestows a perception of intuition (insight) and spontaneity. The pure perception bestows non-violence, order, harmony, truth, love and compassion, so it is in communion with the Universal Consciousness, the Life Field.

A mind that has set aside seeking in the future, and has denied its past is ethical – this is also called a mind with scientific temper. A non-dual and non-conceptual frame of reference is ethical. The ethical nature is beyond conflict and psychological imbalances. An ethical mind is synergistic to psychological mode of mind.

Seeing is direct, touching is direct, action is direct, thinking is indirect.

Thinking is a vicious circle, it is repetitive and is the past, it never leads anywhere. Only renounce thinking, but you have to become more conscious, aware and alert.

The epistemological revolution is the theory of knowing. It is a quantum jump from symbolic-dualistic frame of reference to non-dual frame of reference. The symbolic-dualistic frame of reference is in the spatio-temporal dimension, while the non-dual frame is a happening in the timelessness. It is a quantum jump from intellect to intuition (insight). The intuitive mode is synergistic to the intellectual mode. The non-dual frame of reference is the basic substratum for the right kind of education, or value education.

Human being seems to be born with the faculty of self-knowedge (*apara vidya*, lower knowledge, or logical reason), and understanding and self-awareness (*para vidya*, higher science, or the faculty of perception or quantum logic). We define '*apara vidya*' as empirical, rational and analytic mode of knowing by symbolic-dualistic frame of reference; and '*para vidya*' as mystical and intuitive modes of perceiving (not knowing) by non-dual and non-conceptual frame of reference of mind spectrum. The *para vidya* is also termed as 'intuitive vision of non-duality'. It is a metaphysical insight, falling outside the realm of duality. The noblest and mysterious faculty of man is to transcend himself and perceive himself (*para vidya*) in the act of perception. It implies that when consciousness at the higher quantum energy potential bathes the brain cells, it knocks down the experience and knowledge at the lower quantum energy potnetial of consciousness and bestows the ability to perceive himself in the act of perception. The present educational structure deals with the sensual world, the *without*, and does not touch the *inner* world, the subjective dimension of man.

It is understanding that flows to learning. Understanding and learning blossoms to spontaneity and right action. It flowers to ending the movement of pain, strife, conflict, and so on. The intuition or insight can decondition the mind-brain-body system. Insight is perception of the wholeness and from that completeness there can be logical, sane, rational action. Only then can you have a complete relationship with the mind. The consummation of human life is a pilgrimage to the sacred temple of understanding, learning and right action.[35]

Intuition or insight is the timeless perception. It is insight alone that can decondition the brain. Insight is complete and from that completeness there can be logical, sane and rational action. Only then there can be flowering of the brain. Pure clear insight is perception without any doubt. *Understanding, learning and action are a cyclic process, which results in a spiral going upwards in the vertical dimension; it is a growth in the vertical dimension.* The higher the understanding, the higher the expression of the Consciousness in the human brain-body system.

Invention and Discovery

Invention: Knowledge has its own limited insight. It is responsible for invention – not discovery. Invention is to use the known in a new mode for better performance.

Discovery: Kothari wrote, "The new theorems, unless trivial, are of course first 'discovered' and then proved. Discovery is an inspired guess, a flash of insight. It is not a step by step process, but a sort of quantum jump. A rational account of a discovery, the proof, follows and not precedes the discovery".[36] He again wrote, "Fundamental discoveries which go deep into the nature of things are the result of uncommon intuition and faith, and not cold logic".[37] Maurice H. Wilkins wrote, "The logical and rational is a very important element in science. But emphasis on that tends to make people not notice the essential role of intuition. In fact, in any creative activity you can never analyze the whole process completely because psychological and other dimensions are involved. One cannot hope to understand completely the nature of the creative process".[38]

All perennial knowledge is structured in Universal Consciousness or Life Field. David Bohm described the perennial knowledge as "an undivided whole in flowing movement, an ongoing process, an inseparable part of our overall reality". He defined insight as "an act of perception, permeated with intense energy and passion, that brings out great clarity".[39]

Our activity introduces discontinuities in whatever we are observing. These discontinuities are fundamental to the new physics of twentieth-century. To Bohr, discontinuity was a fundamental truth.[40] In an atom, there is a quantum jump from one orbit to another, and the quantum jump is a discontinuity, a timeless realm. So in an atom there is an evidence of time → timeless → time, or continuity → discontinuity → continuity. The dynamics of the mind is governed by the immutable laws of nature – manifest → Unmanifest → manifest. In every atom, the immutable laws of nature are embedded. Voltaire wrote:

> .... everything is governed by immutable laws .... everything is prearranged .... everything is a necessary effect .... (We are) both toys in the hands of destiny.[41]

We are like puppets where someone else (Life Field) pulls the strings, and we dance to Its tune. This dance is ever repeatable and it is called perfect Silence.

As we think – in any sequence of thought – there is always a sudden, discontinuous jump from one thought to the next. We are not aware at all of any intermediate stages. If we were so aware, no thinking would be possible. There would be an almost infinite sequence of sub-thoughts between two thoughts, like an infinity of points between the two ends of a line. *Quantum jumps for thought appear to be as much a necessity for the conscious self as these are for the atoms.*[42]

A Critique on Life and Death

Life and Death: Life begins with breathing-in (a set of atoms) and death begins with breathing-out (a set of modified atoms). Life is inhalation, death is exhalation. Breathing-in is at higher quantum energy potential than breathing-out, so there is discontinuity. Discontinuity happens when there is the unmanifest. Breathing-in and breathing-out are not two opposite things, they are parts of one whole. Similarly love is breathing-out and hate is breathing-in. So the unmanifest transforms death to life, hate to love.

Between the waking state and deep sleep, there is the Unmanifest. Sri Ramana Maharshi also says: "The non-manifest equally exists in your waking state".[43] But our *particular* attention or successive consciousness splits the unmanifest and the manifest and we are ignorant of the unmanifest. One can move to a higher quantum energy potential, not by human effort or will, but by the grace of the Unmanifest, and we have to purify our instruments of perception to accept the grace. HIS grace is flowing everywhere in the cosmos and beyond cosmos.

The perception that *Atman* is *Brahman* is not an experience but a metaphysical insight, falling outside the realm of duality. Faith is a perception through "understanding of life directly"; it gives us the perception of the wholeness of life. Belief is an intellectual construction. The metaphysical insight is not a term of mental cognition, or intellectual construct, but it emerges from a perception of the whole. The notion of unreality, or of illusions, or of ignorance is felt only in relation to a foregoing experience. With the realisation of metaphysical insight all rapport collapses into the non-dual one. *Duality is but an artifice of the memory.*[44] The non-dual and non-conceptual awareness reveals the Reality. There is only *Atman-Brahman*, the eternal subject, the Self of all our experiences. The Brahman is unmanifest and the universe is the manifest; so "*all that exists is Brahman*".

Unmanifest → Manifest: The Unmanifest is fundamental and manifest is a derivative; this is written as Unmanifest → manifest; Timelessness → time; Highest Frequency → lowest conscious frequencies; Simultaneous Consciousness → successive consciousness; Eternity (Reality) → transient (false). It is summed up as under:

> In the art of seeing there is the Unmanifest.
> In the art of hearing there is the Unmanifest.
> In the art of breathing there is the Unmanifest.
> In the art of sensual perceptions there is the Unmanifest.
>
> In the waking and deep sleep states, there is the Unmanifest.
> In the deep sleep state, there is the Unmanifest.
>
> *Atman-Brahman* is the Self of all our experiences.
> The experiencing is the awareness of the Unmanifest.

Duality is but an artifice of the memory.
All that exists is the Unmanifest.

The occurrence of seeing or hearing, smelling or breathing is a movement in the sphere of non-duality. In the dimension of Awareness – which is an energy born of the emptiness of Silence – an energy free of volitions is born. It operates without our choice. Silence or emptiness is the total unconditioned energy. Nature (Life Field) invariably expresses itself in the non-dual realm and it obeys the laws of non-causality. Whatever happens (intuition, insight) it happens for the best, and it is always the right action. The Unmanifest transforms waking state to deep sleep state, death to life, violence to non-violence, hate to love. Each moment when you breathe in you are reborn; each moment when you breathe out you are dead, because breath is life.

The crux of the twenty-first century physics is as under:

In an atom there is the Unmanifest.
Between two consecutive thoughts there is the Unmanifest.
Between conditioned mind and fact there is the Unmanifest.
When fact dissolves there is the Unmanifest.

If you want to be with life, with the whole, you have to be choiceless. We need a laboratory, an experiment in creating a new man, a new society – a human being who lives in harmony with the inner and the outer, with himself and his environment, and who is free from all ideologies and conditionings that now divides humanity.

Science and Consciousness

We have to restructure science which implies that science takes in its fold both the objective (outer) and the subjective (inner) dimensions; to restructure education, which deals both with the sensual and supersensual world, and these lead to a pure perception of pure Consciousness. The subject of cognizance (true-subject) of "restructured science" is the "non-variable attentivity", or "reaction-free attentivity", or "objectless attention", which is synergistic to thought-time. The true-subject bombards thought-time and annihilates it, which implies that I-consciousness goes to abeyance, or the subject-object dualism ends. George Wald stated, "The very crux of twentieth-century physics is the recognition that the observer (subject) cannot be excluded from his observations; he is an intrinsic participant in them".[45] Ilya Prigogine opined, "It is physics that presupposes an observer situated within the observed world. Our dialogue with nature will be successful only if it is carried on from within nature".[46] The observer and the observed are not two separate worlds, but one dynamic whole. This was highlighted by Erwin Schrodinger in words:

> The world is given to me only once, not one existing and one perceived. Subject and object are only one. The barrier between them cannot be said to have broken down as a result of recent experience in the physical

> sciences, for this barrier does not exist.[47]
> The world is given but once. Nothing is reflected. The original and the mirror-image are identical. The world extended in space and time is but our representation.[48]

Observation/Inquiry: A scientific mind takes up "the role of a non-concerned observer", in the words of Schrodinger.[49] It is unrelated to our beliefs, our ambitions, our hopes", in the words of Prigogine.[50] It must not construct a pseudo-observer; it must reject the past remembrances or future projections. It bestows understanding of life directly, which is perception of the fact; it has the awareness of the inner perceptual image, the values, the significance. When you observe, there is no observer; when you dance, there is no dancer; when you love, there is no lover; when you walk, there is no walker inside, it happens. The perceptions are born of non-duality, of silence, and responses are born of spontaneity. Intelligence begins to use the human brain and help the human being to watch the world as one wholeness. The future of mankind is embedded in understanding, learning and living in the perception of the pure Consciouseness. You have to leave it to Him, the Unmanifest.

Spirituality: It is the science of the wholeness of life. Spirituality implies exploring the non-cerebral through the non-action of the conditioned mind, it is the growth to a new dimension of Consciousness. It is the awareness of the wholeness around you and the wholeness within you.

Radical Transformation or Mutation: Modern science has brought man on the control panel of mutation or transformation of human psyche. Scientists have discovered Life Field or Universal Consciousness, which is the substratum of all existences, and which is at a very high frequency (higher frequency means higher energy) throughout the cosmos. We are on the threshold of extending this energy into human dimension.

It is the integral dimension of Consciousness that can resolve all challenges facing mankind. It is to heal the diseased mind-brain-body system. It is an open call to express higher hierarchical level of expression of consciousness eloquently through restructured science and restructured education for a global society, rising civilisation and our daily living in freedom, peace and bliss. This is the only guarantee to human survival and excellence.

We urgently need radical transformation or mutation of the mind-brain-body system which implies abeyance of the ego, the subject-object interaction; and simultaneously it activates the Universal Consciousness, the perceptive Intelligence centered in life. It is the transformation of the foundation of perception – from conditioned mind to unconditioned Mind, from intellect to intuition, from part to wholeness, from lowest frequencies of consciousness to the highest frequency of Consciousness. This is the basis of holistic education, or to inculcate ethical-spiritual values in daily living.

The whole cosmos is pulsating with an eternal, infinite, immutable, beginningless and endless Universal Consciousness, the Unmanifest, the highest frequency Conscious Energy. Every existence (physical, biological and psychological) – atom, event, happening – is a cosmic dance of the play of the manifest and Unmanifest, limited and Limitless, measurable and Immeasurable. Only the Homo sapiens – man – is endowed with the capacity to perceive the Unmanifest, the highest expression of Consciousness; and to live in the light of the Divine Energy. This is the only vocation of man. This is the consummation of human life.[51]

A Vision on Evolutionary Impulse of Nature: Teilhard de Chardin, the eminent paleontologist, divided the evolution into four stages – geosphere, biosphere, noosphere and christosphere. Broadly geosphere deals with inorganic (physical) evolution, biosphere with biological (organic) evolution, noosphere with psychosocial evolution and christosphere with spiritual evolution.

Cosmology talks of Red Giant – a dying star. They are constantly distilling their stuff into space. Every now and then the entire star threatens to blow up – a nova, or a supernova. George Wald wrote, "In its dying (supernova) it has made the elements of which life is composed. It is a moving realisation that stars must die before organism can live".[52] In other words, the physical evolution of stars must end before biological evolution of organisms starts.

Homo sapiens appeared at about 2.5 million years ago. According to generally accepted anthropological findings, the anatomical evolution of human nature was virtually completed some fifty thousand years ago.[53] The growth of subjective dimension of man led to transcendence of the biological evolution and imbibed the psychosocial evolution.

Many eminent scientists and philosophers-mystics have opined that man will transcend the psychosocial evolution and imbibe spiritual evolution. Vimala Thakar, on February 27, 2003, prophetically predicted that "he twenty first century is destined to see the disappearance of conflicting Civilisations and the emergence of a new harmonious, global civilisation".

The immutable law of nature is that the old order gets completely dismantled or is transcended before the new order emerges. It is a happening – the higher hierarchical expression of Awareness (Consciousness) burns the old dross and new order emerges – the physical evolution is transcended before the biological evolution emerges; the biological evolution is transcended before the psychosocial evolution emerges; and the psychosocial evolution is transcended before the spiritual evolution emerges. The evolutionary impulse of Nature supports the radical transformation or mutation of the evolution from one stage to the next and all processes are going on simultaneously in Nature. The human race rejoice that the divinity of the manifest (successive consciousness) is being transcended slowly and it is being replaced by the Divinity of the Unmanifest, the Life Field. The seeds of a new culture (New Human Society) have to be sown in the atmosphere of freedom and energy of love and compassion.

The Urgency of Mutation of Human Psyche

(a) Awareness is the movement of Intelligence. It is the response of the whole being to the communion with the fact.

(b) To live is to be related, and for getting related the awareness of '*what one is*' is absolutely necessary.

(c) An inquirer has the humility to face the global problems in his own life, in his own being; in his relationship with others to face them there.

(d) It is the understanding of life directly that resolves the inheritance.

(e) The awareness of the infinite and eternity makes you extremely sensitive in your relation to the moment *here* and *now*. The awareness of the whole puts you in correct perspective of the particular. We have to deal with the *present*, the *now*, the *here*.

Concluding Comment

All that exists is the Reality, the *Brahman*, the Universal Consciousness, the Life Field. The Reality is the inner Consciousness in every existence/ event/process. It is at the maximum frequency, the highest quantum energy potential of Consciousness.

The successive consciousness constructs the time-space matrices in quick succession. It constructs conflict, pain, misery and travails. The past memory – thought-time structure or cultural belief structure is responsible for all human psychological imbalances and ecological challenges. The Unmanifest, when it bathes the brain cells, brings radical transformation from violence to non-violence.

The Committee on Science and Technology, USA, submitted in June 1981 a lengthy report to the House of Representatives, USA, which encouraged serious assessment of further research on "*the physics of consciousness*", and concluded that recent experiments:

> .... that there exists an 'interconnectiveness' of the human mind with other minds and with matter; .... that the human mind may be able to obtain the information independent of geography and time.[54]

The Union of Concerned Scientists (UCS) issued a "*World Scientists' Warning to Humanity*" in April 1993, signed by over 1670 scientists including 104 Nobel laureates, concluding that "Human beings and the natural world are on a collision course .... A new ethic is required – a new attitude towards discharging our responsibility for caring for ourselves and for the earth".[55] As every event/existence is interrelated and interpenetrated, so the responsibility is not towards survival of one's self but for the human race.

Albert Einstein, the greatest physicist stated, "The significant problems we face cannot be solved at the same level of thinking we used when we created them". He himself provided an answer to this human paradox in words, "Our new idea is simple: to build a physics valid for all coordinate systems",[56] or frames of reference of the Mind Continuum. Einstein wrote, "The chief motivating force behind all technical effort ought to be a supreme concern for the destiny of the human race", or in other words, we have to work sincerely and honestly for our own survival and progress. We must lay the foundation of a new human society based on non-violence and peace.

The restructured science and the restructured education are fundamental for constructing a sane global society and the perception of the highest hierarchical level of expression of Consciousness (Life Field, *Brahman*, the Absolute Subjectivity). This bestows non-dual Subjectivity, or non-dual and non-conceptual awareness. It resolves all human psychological imbalances and ecological challenges facing mankind, and usher in a new era of non-violence, peace and bliss. It is the greatest vocation in the world. This is the consummation of human life. This only is the social, psychological, and even the political revolution.

We want a revolution, psychological revolution, in which the whole human race has to participate and every man is an equal partner. This revolution will regenerate or rejuvenate human mind so that it perceives the new dimension of an integral and all-inclusive Consciousness, that can guarantee human survival along with all fauna and flora.

The cultural transformation or the evolutionary mutation of this magnitude and depth cannot be prevented by short-term socio-economic-political activities. This provides mankind the strongest faith in the Order, the Wisdom, the Intelligence, and future of mankind along with all fauna and flora.

Man is his own greatest enemy. Knowing fully well, the path of *Light*, yet he wants to live in ignorance and darkness. If man escapes his responsibility at this critical juncture, he does so at his own peril.

References

1. Spencer Brown, *Laws of Form*, New York, Julian Press, 1972, pp. 104-05.
2. Alexandre Koyre, in Ilya Prigogine and Isabelle Stengers, *Order Out of Chaos*, New York: Bantam Books, 1984, p. 36.
3. Swami Vivekananda, in *All India Magazine*, Pondicherry, Sept-Oct 1993, p. 48.
4. Sri Aurobindo, (a) in *All India Magazine*, ibid., pp. 15, 16, 13; (b) in *The Odyssey of Science, Culture and Consciousness*, New Delhi: Abhinav Publications, 1990, p. 41; and (c) *A Message on August 15, 1947*, in The Tribune, August 15, 1997.
5. The Divine Mother (Madame Mirra Alfasa), in *CIRHU* (*Centre of International Research in Human Unity*, Auroville), Papers 4, February 1997, p. 50.
6. Vaclav Havel, *Speech at Philadelphia, July 4, 1994.*
7. *Ibid., Speech at Stanford University, September 29, 1994.*
8. George Wald, in *The Evolution of Consciousness*, New Delhi: National Publishing House, 1983, p. 14.

9. Richard Lamm, in Michael Burghley, *My World, My Responsibility*, USA: Loveland, 1987, p. 10.
10. Schweitzer, *ibid.*, p. 10.
11. Joan Waldon, *Scientists Tune into Consciousness*, New Science, Winter 1996.
12. George Wald, *Synthesis of Science and Religion*, Bombay: The Bhaktivedanta Institute, June 1988, pp. 20, 18.
13. Hideki Yukawa, *Creativity and Intuition*, Tokyo: Kodansha International Ltd., 1973, p. 203.
14. Eugene Wigner, *Two Kinds of Reality*, Monist, 1964, pp. 248-64.
15. Roger Sperry, in *Synthesis of Science and Religion*, *op. cit.*, p. 199.
16. Arthur Eddington, in Swami Ranganathananda, *Science and Religion*, Calcutta: Advaita Ashrama, 1979, p. 125.
17. Pere Teilhard de Chardin, *The Phenomenon of Man*, London: Collins, 1959, p. 55.
18. Lincoln Barnett, in Ranganathananda, *op. cit.*, p. 27.
19. Charles Sherrington, *Man on His Nature*, Pelican Edition, p. 38; in Swami Ranganathananda, *op. cit.*, p. 32.
20. Richard Feynman, in Fred Alan Wolf, *Taking a Quantum Leap*, Harper and Row, 1981, pp. 227, 228.
21. Sampooran Singh, *Dynamic Interplay Between Science and Religion*, Jodhpur: A Jainsons Publication, 1978, pp. 357-386.
22. Sampooran Singh, AVM Manjit Singh (Retd), Dr (Mrs) Kanwaljit Kaur and Paramjit Singh, *God in Science*, Chandigarh: Vishvas Publication Pvt Ltd, July 1999, p. 2.
23. Deepak Chopra, *How to Know God*, London: Rider 2000, pp. 270-271
24. Sampooran Singh, *Global Education and Human Survival*, New Delhi: Sanbun Publishers, March 2002, p. 77.
25. Ervin Laszlo, *Emerging Concepts of the Cosmos*, *The Scientific and Medical Network*, Newsletter No. 49, August 1992, p. 28; in Sampooran Singh, *The Pilgrimage to the Third Millennium*, Chandigarh: Faith Publishers, 1995, p. 31.
26. Sampooran Singh, *God in Science*, *op. cit.*, p. 124.
27. David Bohm, in Gary Zukav, *op. cit.*, p. 326.
28. Max Planck, *The Philosophy of Physics*, New York: Norton, 1936, p. 83.
29. G.F. Chew, in Gary Zukav, *op. cit.*, p. 331.
30. David Bohm, in *PsychoScience*, Spring 1994, Vol. 1/No. 1, p. 27.
31. Sampooran Singh, Dr (Mrs) Kanwaljit Kaur and Paramjit Singh, *Global Values Education*, Chandigarh: Faith Publishers, January 2001, pp. 63-64.
32. Arthur Eddington, in A. Commins and R.N. Linscott, ed., *Man and the Universe*, *The Philosophers of Science*, New York: Washington Square Press, 1969, p. 453.
33. Alan W. Watts, *The Ways of Zen*, New York: Vintage Books, 1957, p. 10.
34. David Bohm, in Gary Zukav, *op. cit.*, p. 327.
35. Sampooran Singh, *A Review on Symbiosis of Science and Spirituality*, Dagshai Cantt: The Atam Science Trust, October 7, 2002, pp. 14-15.
36. D.S. Kothari, *The Perception of Truth in Science and Philosophy*, in Kishore Gandhi, *The Odyssey of Science and Culture and Consciousness*, New Delhi: Abhinav Publications, 1990, pp. 48-49.

37. *Ibid.*, *Atom and Self*, New Delhi: National Publication House, 1983, p. 227.
38. Maurice H. Wilkins, in *Synthesis of Science and Religion*, *op. cit.*, p. 31.
39. David Bohm, *Insight, Knowledge, Science and Human Values*, *Within the Mind: On J. Krishnamurti*, KFI, 1982, pp. 40, 47.
40. Niels Bohr, in Fred Alan Wolf, *op. cit.*, p. 85.
41. Voltaire, in Ilya Prigogine and Isabelle Stengers, *Order Out of Chaos*, New York: Bantam Books, 1984, p. 257.
42. D.S. Kothari, *Atom and Self*, *op. cit.*, p. 8.
43. Sri Ramana Maharishi, in *Vedanta*, UK: Ramakrishna Vedanta Centre, Issue No 305, p. 115.
44. Swami Siddheswarananda, *Seeing Brahman with Open Eyes*, Vedanta, *ibid.*, p. 122.
45. George Wald, in *Synthesis of Science and Religion*, *op. cit.*, p. 19.
46. Ilya Prigogine and Isabelle Stengers, *Order Out of Chaos*, *op. cit.*, p. 218.
47. Erwin Schrodinger, *What is Life*? and *Mind and Matter*, Cambridge University Press, Ed 1967, p. 137.
48. *Ibid.*, p. 146.
49. Erwin Schrodinger, *op. cit.*, p. 128.
50. Ilya Prigogine, *op. cit.*, p. 5.
51. Sampooran Singh, *Waking, Dreaming and Deep Sleep States*, unpublished.
52. George Wald, *op. cit.*, p. 14.
53. Fritjof Capra, *The Turning Point*, London: Wildwood House, 1982, p. 325.
54. *Report to the House of Representatives*, USA: Committee on Science and Technology, June 1981, in R.G. Jahn, February 1982, Proceedings of the IEEE, Vol. 70, No. 2, p. 165.
55. *Union of Concerned Scientists, World's Scientists' Warning to Humanity*, USA: 26 Church Street, Cambridge, MA 02238, April 1993.
56. Albert Einstein, in Sampooran Singh, *Foundations of Human Survival*, Chandigarh: Faith Publishers, 1996, p. 5.

8
The Emergence of the New Human Being

Abstract

The psychosocial evolution of man – for a million years or more – has resulted in mutilated perception of the whole human race. We have inherited the conditioned psyche (thought patterns coupled with volitions) which is leading us to self-destruction or self-annihilation.

We explore the challenge of thought, observation of the psychosocial field and trans-psychological realm, discontinuous world, freedom and mutation, understanding and learning, and harmonious interrelatedness in nature.

We will have to initiate educating our physical organism and psychological structure, and work for a paradigm shift in the whole system of education. We have to work for restructuring of science, restructuring of education, awareness of human perennial values, and live that understanding every moment. This will lead to the emergence of the new human being.

Humanity is standing on the threshold of a new era. We are living in extremely explosive times. Mind has been invariably engaged in gathering partial experience and fragmentary knowledge (information) for the last one million years or more. He has stored this as frequency and amplitude information in the brain, and this store house is called 'memory' or 'cultural belief structure', so it is the repository of total experience and information of humanity. Memory is meta-stable, it spontaneously disintegrates as thought-time coupled with variety of urges and drives – say, psychological imbalances and an urge for security. The desire for acquisition and accumulation is born of an urge for security. It appears that the desire to identify oneself with a nation, with a race, with a religion, with an ideology, may also be a result of that desire for security. The whole human psyche is conditioned. The conditionings express themselves as psychological imbalances, pollution and other environmental disasters, a rising wave of violence and crime, high inflation and unemployment, energy crisis, a crisis in health care, poverty, malnutrition, and so on. All these crises are different facets of one and the same crisis and this crisis is essentially a crisis of perception.

Relationship of the mind to acquisition, accumulation and preservation of wealth; relationships to ownership, property and prestige; relationship to experience, knowledge and authority; have not been changed. Knowledge has failed to end his selfishness, his aggressive, competitive recklessness, to end the past thought structures. He is inwardly very primitive, almost barbarous. He is responsible for this violence. This is basically due to that the mind is mechanistic, repetitive, divisive and deals with a part, a fragment. The human reason or intellect is sterile in transforming the human psyche. Many eminent scientists – Werner Heisenberg, Edwin Hubble, Max Planck, G.F. Chew, Bertrand Russell, Lincoln Barnett, Max Born – have concluded that we cannot impart value education at reason or intellect level.[1] All legislation, all reforms of the socio-political-economic system, any tinkering of the time-space-causation matrices, all traditional approaches, cannot transform the psyche. All education of the intellect or reason has no significance regarding understanding the psychodynamics and limitations, and ending the causes of conflict of human psyche or mind-brain-body system.

This article explores the meaning of life: (a) the facts of psychological life, and (b) the transpsychological truth behind these facts and the absolute ground of our being, the ground of existence. It is a quantum jump from the psychological life to the absolute ground of existence. It deals with the wholeness of life.

A Critique on Mind-Brain-Body System

Wilder Penfield concluded, "Our minds and our brains are not the same.... It is the mind which experiences and it is the brain that records the experience".[2] Erwin Schrodinger concluded, "Mind is by its very nature a *singulare tantum*. I should say: the over-all number of minds is just one. I venture to call it indestructible since it has a peculiar time-table, namely mind is always *now*. There is really no before and after for mind. There is only a now that includes memories and expectations".[3] Gregory Baterson concluded, "Mind is the essence of being alive".[4] He added:

> The individual mind is immanent (successive consciousness) but not only in the body. It is immanent also in pathways and messages outside the body; and there is a larger Mind of which the individual mind is only a subsystem. This larger Mind is comparable to God and is perhaps what some people mean by 'God'.[5]

The immanent mind "is neither individual, nor collective, but humanity".[6] The Mind Continuum is just one; so the non-dual frame of Mind Continuum is shared by the whole mankind.

What we call mind (with a lower case 'm') is a conditioned energy, moving through the body, monitored by brain, operating through the senses. The mind is in the relative frame of reference, or self-interacting dynamic state at immanent quantum energy potential of consciousness. The mind is at the lowest frequency of consciousness, or the minimum quantum energy potential of consciousness, or the lowest hierarchical expression of consciousness. When we use the word mind, the brain is also included. The mind expresses itself as the ego, the self, the "I", the "me"; the memory or cultural belief structure; and the conscious, sub-conscious and the unconscious faculties. We may call conscious mind as short-range memory and the sub-conscious as the long-range memory.

What we call Mind (with upper case 'M') is the non-dual frame of reference of Mind Continuum. The Mind is in the self-referral state, or the highest frequency of Consciousness, or the highest quantum energy potential. Higher frequency means higher energy. The psychic energy makes a quantum jump beyond 'frequency barrier' or 'event horizon', where the energy does not transform to matter, so in that state there is no recording of perception as conative (memory). In this state, the original (Simultaneous Consciousness) and the mirror-image (pure perception) are identical.

The Challenge of Thought

There is nothing like an individual mind, nothing like an individual consciousness. The consciousness is shared by the human race. It is the human violence, human anger, conditioned in various ways. The human being can measure things, with the help of rationality it can build up concepts and ideas about life, about reality. And we have been busy on the racial scale conceptualizing, ideating, for thousands upon thousands of years.

The world is a construct of the human mind. We graft the concepts on the reality and are aware of the distorted perception. We live as biological vegetation and psychological mentation. We are missing the communion with the real because we are every busy with the unreal, the unfactual. It is this isolation from reality which is the source of all suffering. It is the absence of awareness about the ever vibrating, all permeating life which causes misery, suffering and constructs strife and conflict. The whole process of conceptualization and ideation and identification, which has been necessary in building up the culture and civilization, is the greatest obstacle in communion with the reality, communion with life.

The challenges of thought are as under:

(a) No thought is my original thought. No emotion is my personal emotion. It is the production of collective human activity, or human race. Mental action is a mechanistic action, it is not going to help us in creating a new society.

(b) The inner and the outer are complementary to each other – a unitary dynamic flow.

(c) If we allow the thought to come in the field of attentivity, the momentum of thought loses its grip and goes to abeyance.

(d) The attentivity bestows intuition (insight), spontaneity and right action.

(e) The challenge is to be free from the clutches of irrationality and rationality and grow into a non-subjective dimension of consciousness, and make a quantum jump from rationality to intuition (insight).

Enquiry or Observation

The neuro-chemical system is loaded with tensions so there cannot be relaxation, which is the content of silence. So enquiry or observation is a non-authoritarian approach which implies that you observe the movement of every thought without supplying any energy to it for its sustenance; you do not look to the past where nothing is interpreted or identified or recognised which implies freedom to observe at the first step; and such observation bestows humility that the ego or thought is not your and this results in their blossoming and withering away.

The state of observation which is a reaction-free attentiveness, which is a non-evaluatory perception is absolutely necessary. Observation is watching the movement of mind without accepting or rejecting, without identifying oneself with it or condemning it. We have to learn to take everything – the wholeness of life – in one sweep of attention.

Everything that the mind brings up gets exposed to that silence; to the humility; to the all inclusive attention; and the quality of your response changes. This is a happening and does not require your conscious effort. So one learns to put oneself in the state of watching, looking innocently, defencelessly.

Knowledge or organised information have no dynamism. They are sterile and do not result in understanding. And it is only encounter with facts that results in understanding. Freedom implies acquaintance with the conditioning. If one is attentive and alert, the cause and the source of conditioning is understood and these go to abeyance.

Life is maximum quantum energy potential of creative energy, it is inexhaustible creativity and infinity, immeasurableness, and learning has no end. An inquirer is a learner, and life is a movement of learning. Learning is to perceive the truth, and the false gets dropped by itself with the realisation that it is not the truth. Sensitivity is the perfume of truth. Truth is its own defence, and truth is its own sensitivity.

Freedom and Mutation

We urgently need radical transformation or mutation of the conditioned mind-brain-body system. Mutation is radical qualitative change in the quality of consciousness. Mutation is a decrease of matter content of the consciousness, or total change of the mutant in the consciousness, in the whole being. Mutation is a happening in the realm of total freedom, which happens when there is ending of the urge for security in the psychological realm. The urge for security is embedded in the symbolic-dualistic field of mind spectrum, and the freedom is characteristic of the non-dual Mind

Continuum. Mutation implies direct contact with reality, communion with reality, without the intervention of the brain. The urge to be free of the fear of freedom is in contradiction to the psychological wants. This endows a realisation that all the acquisition, accumulation, and preservation, in which mind has been collecting through untold centuries, is a hindrance to the psychological transformation, to freedom.

Freedom has no motive, no direction. It is not an intellectual action, nor an emotional expression. In freedom, there cannot be any experience. Freedom transcends psychological barrier. Freedom happens when we make a quantum jump across 'frequency barrier' or 'event horizon'. Freedom is existing as a *fact* (the *true*, the *what is!*, the *present*, the *non-dual*) within me, it is related to perception and understanding.

Thoughts, emotions, feelings and memories are mechanical actions. A mind that has its roots in the past, is the known, and it cannot come directly into contact with reality. A conditioned mind does not know, how to perceive without a motive, without comparing, without trying to recognize or identification. The very act of perception is contaminated by the process of naming, recognizing, comparing and judging. Mind becomes silent only, when it is confronted with something which has staggeringly huge dimensions, or higher quantum energy potential of consciousness, or deep sleep state of consciousness, which it cannot interpret.

Understanding

The objective science sees only a part of life, so it adds to conditioning of the psyche. Nature has provided an inner dimension to man, called subjective. The outer (objective) and the inner (subjective) are a single dynamic flow, a unitary movement. This is called ***restructured*** science, which sees the wholeness of life, or is aware of the objective and the subjective dimensions of life simultaneously. The restructured science is synergistic to the objective science, so the subject of cognizance is non-variable attentivity, or reaction-free attentivity, or objectless-attention. When thought comes in the realm of restructured science, it bombards thought and thought gets annihilated. Understanding and learning results in instantaneous dropping the false. This bestows understanding of life directly, spontaneity and total action.

Understanding of mind awakens a new quality of attention, all inclusive attention, observes the objective challenge, the nature of subjective reactions and the causes of the subjective reactions simultaneously. The all inclusive attention has the awareness of the objective and the subjective dimension of life – the wholeness of life – and this flows to freedom. Then the momentum of reaction will not carry me over with it, but I will be ahead of the reaction. Total silence of mind in action is freedom. Action out of silence is freedom in movement.

Discovery of truth is an inward journey. It is a journey, a voyage which one has to launch alone. Taking the inward journey and arriving at silence, is a voyage in solitude.

Love

Love is not different from truth. Love is that state in which the thought process, as time, has completely ceased. Love is a state of being in which the movement of life is unsoiled by the touch of mind. Life is moving and the movement of life untouched by the mind, is the state of love. Truth, love, reality, beauty, harmony, death are phenomena which are outside the realm of duality. There is no subject-object relationship. There is a movement without any cause and motive. Love is a state of being in which one has arrived at the humility to act without the ego.

Emergence of the New Human Being

All social-economic-political reforms are in the field of duality, these are divisive, and cannot bring in radical transformation or mutation of the psyche. In the religious formula of action based on belief, idealism, dogmatism, conformity are embedded authority and acceptance, obedience and the utter denial of freedom, so these add to conflict.

J. Krishnamurti wrote, "It is only the action of will and assertiveness that needs to be changed at all, because the only mischief in relationship is conflict between individuals or within individuals, and conflict is will and assertiveness".[7] If one is attentive and alert throughout the day, the mischief of weaving psychological patterns go to abeyance. The new human being can emerge if man understands and lives in the higher hierarchical expression of Consciousness. J. Krishnamurti adds, "A man who is passionate about the world and the necessity for change, must be free from political activity, religious conformity and tradition – which means, free from the weight of time, free from the burden of the past, free from all the action of will: this is the new human being. This only is the social, psychological, and even the political revolution. By allowing that mutation to occur in our life, we contribute to the total human race, we enrich the orbit of global human consciousness.

Integral (Holistic) Education

R.D. Laing, an eminent psychiatrist, emphatically reminds us:

> Out go sight, sound, taste, touch and smell and along with them has since gone aesthetics and ethical sensibility, values, quality, form; all feelings, motives, intentions, soul, consciousness, spirit. Experience as such is cast out of the realm of scientific discourse.[8]

According to Laing, hardly any thing has changed our world more during the past four hundred years than the obsession of scientists with measurement and quantification. Laing gave a critique on educational system in UK in words:

A child born today in the United Kingdom stands a ten times greater

> chance of being admitted to a mental hospital than to a university.... This can be taken as an indication that we are driving our children mad more effectively than we are genuinely educating them. Perhaps it is our way of educating them that is driving them mad.[9]

Vimala Thakar stated, "In ancient India, the system of Education included the science of mind as one of the subjects".[10] She added, "When a person born in East becomes alert and sensitive to this inquiry of Truth, if he is earnest, if he is honest, if he is not trying to borrow conclusions and theories, but trying to probe within the depth of his being, then the residue of the experience of his race might help him to understand the mind quickly". J.Krishnamurti prophetically stated:

> India has lived all these centuries on ideas. She has to move from a life based on ideas, which are non-facts, to seeing the fact only.... In India it has been wiped out and destroyed. Yet it was there in India, waiting in the soil .... Something had been destroyed in India, but a presence waited in the land. The benediction of the timeless is waiting, the ground is ready. Can we create something that is of this benediction".[11]

Educational system of the last hundred and fifty years in India has uprooted the minds of people. We have to regenerate this educational model but have to reject the traditional approach and replace that by modern scientific thought.

The recent discoveries of modern science are: (a) all the concepts we use to describe nature are limited, that they are not features of reality; parts of the map, not of the territory; (b) the outer and the inner are not two separate movements, but are one, a unitary movement, not separate, but whole; (c) many eminent scientists – including Albert Einstein, Werner Heisenberg, Erwin Schrodinger, Arthur Eddington – have stumbled upon another mode of knowing, called intimate, intuition (insight) and understanding of life directly; and (d) human being seems to be born with the faculty of self-awareness and understanding of life directly.

Harmonious Interrelatedness in Nature: Fritjof Capra wrote, "I believe that the world-view implied by modern physics is inconsistent with our present society, which does not reflect the harmonious interrelatedness we observe in nature".[12] The mystery of interrelatedness is called Divine; it is not the creation of human thought or knowledge or language; and it is self-created and self-sustained. To achieve such a state of dynamic balance, we need a quantum jump from symbolic-dualistic frame to a non-dual and non-conceptual frame of reference of Mind Continuum. Krishnamurti adds, "But to be psychologically free and original can only come about when you are aware of your own inward activities, watch what you are thinking and never let one thought escape without observing the nature of it, the source of it".[13]

Discontinuous World: The art of observation by an attentive and alert mind introduces discontinuities in whatever we are observing. These discontinuities are fundamental to the new physics of the twentieth century. To Niels Bohr, "discontinuity was a fundamental truth .... The world was fundamentally a discontinuous and quantum jump world".[14]

An attentive mind during observation of the movement of thought introduces a discontinuity from conditioned energy to fact (unconditioned energy), which is the substance of silence. So silence is an expression of the highest quantum energy potential of consciousness (Unmanifest). Silence is absolute discontinuity of the conditioned mind – no expectation, no tension of waiting, no tension of moving in some direction, total relaxation, so life immeasurable can operate upon us. This will enable the human race to build up an economy for peace, a politics for peace, friendship, brotherhood, education for peace. It will help us to replace the psychology of confrontation by the psychology of cooperation and friendship.

The present educational structure is antagonistic to the above discoveries of modern science. Modern science establishes that we "have to learn about the mind, not to acquire knowledge about the mind.... Learning is possible, when one observes the mind in action". Krishnamurti added, "First of all, one must observe, see with great care the mind and its functioning, listen to what is within and without; out of this arises sensitivity and in sensitivity there arises insight. That insight alone will wipe away sorrow". Vimala Thakar wrote, "We have to educate the cerebral organ to behave in an orderly way. Wherever there is order, there is no tension. Order is perfect relaxation. It is only when there is disorder, chaos, anarchy, there is confusion and therefore, tension. Every confusion is a tension. Not only that orderliness is beauty and elegance; orderliness is total relaxation".[15] She added, "Education in relaxation, education in putting oneself in a state of non-doing, non-motion, relaxation".[15] Education is in the non-dual frame of reference of Mind Continuum. It is awareness of the fact at all moments. Education is possible when one is attentive to the objective and the subjective dimensions of life, which implies that the subject is aware of the wholeness of life. The subject of cognizance, which is engaged in the act of observing, becomes a flame of attentiveness, which becomes the content of consciousness.

In the schools/colleges/universities, we will have to create a new orientation, which is called integral or holistic education. Vimala Thakar wrote:

> *I do hope that all over the world there will be entirely new systems of education, not only new orientations to education, but the very approach to education will have to go through a radical revolution where the acquaintance with the biological and the psychological organism, the rhythm in which the body and the mind moves, will be incorporated in the syllabus of the students.*[16]

As we teach the science of physics and chemistry we will teach the "science of consciousness" in Science faculty, and trans-personal psychology or deep ecology in Art faculty. The burning challenge before mankind is psychological mutation, and we have to resolve the challenge for our own survival and excellence.

Concluding Comment

Vimala Thakar prophetically wrote:

Either mankind learns to end this collective violence, this brutality, this addiction to resolving problems through violence or faces the challenge of racial extinction. There is no other option. Either we die together or we live together.

All the tensions, the conflicts, the wars, the violence, going on in the world, have got seeds and roots in the memory, the cultural belief structure, stored in the individual consciousness. Conflicts and wars exist and will exist as long as we carry the seeds of violence, tensions and conflicts within us. Wars have never solved human problems and they never will. The mental movement is absolutely irrelevant to the exploration of the unconditioned. It is only living in higher hierarchical expression of Consciousness that can resolve all the psychological imbalances.

If a new man, with a poise in his reason and peace in his heart is to emerge, we will have to initiate educating our physical organism and psychological structure, we will have to work for a paradigm shift in the whole system of education. It is going to be an humble and honest effort to look at the facts of life and the problems with which man the world over is confronted today. We have to explore understanding, learning, spontaneity, action, meditation as one harmonious movement.

The principle governing restructuring of education is fundamentally based to assist nature for an evolution of life. The evolution of mankind can only happen when there is understanding and spontaneity; dynamic interplay of intuition and action; and awareness of the transcendent dimension of life. The real goal for each one of us is to awaken and activise in ourselves perfect spiritual consciousness. A true seeker is the one who with his ideal in view, goes on purifying his feeling, thoughts and actions. We have to learn to empty our mind of all worldly thoughts and wait silently for the vision of the universal consciousness. The fundamental research into the manifestation of ever greater potentials and newer dimensions of Consciousness would be our core commitment and action. Consequently we have to work for *restructuring of science, restructuring of education, awareness of human perennial values, and live that understanding every moment.* This is a great new challenge, a new historical dimension, and a thrilling task for educators everywhere in the world.

Albert Einstein, the greatest scientist of all time, wrote that man is here for the sake of other man, for the sake of whole mankind; and indeed for the sake of all human beings. *The fundamental research into the manifestation of ever greater potential and newer dimensions of Consciousness would be our core commitment, action and responsibility.* We are here on this planet to share our findings with all human beings. This is the consummation of human life. It is the greatest vocation in the world.

References

1. Sampooran Singh, Dr (Mrs) Kanwaljit Kaur, Paramjit Singh and AVM Manjit Singh, *Global Education and Human Survival*, Sanbun Publication, March 2002, pp. 35-36, 46; and Sampooran Singh, in *Education for a Global Society*, Delhi: Shipra Publications, 2003, pp. 91-92.
2. Wilder Penfield, in Deepak Chopra, *How to Know God*, London: Rider, 2000, pp. 215, 217.
3. Erwin Schrodinger, *What is Life*? and *Mind and Matter*, Cambridge University Press, 1967, p. 145.
4. Gregory Baterson, in Fritjof Capra, *The Evolution of Consciousness*, New Delhi: National Publishing House, 1983, p. 247.
5. *ibid.*, in Larry Dossey, *Medical Science and Non-Violence*, New Delhi: Gandhi Peace Foundation, 1988, pp. 12-13.
6. J. Krishnamurti, *KFI Bulletin*, 1982, No. 1, p. 40.
7. *Ibid.*, *The Urgency of Change*, London: Victor Gollancz Ltd., 1977, pp. 189-192.
8. R.D. Laing, *The Voice of Experience*, New York: Pantheon, 1982; in Fritjof Capra, *The Turning Point*, London: Wildwood House, 1982, p. 40.
9. *Ibid.*, *The Politics of Experience*, New York: Ballantine, 1978, p. 104; in Fritjof Capra, *ibid.*, p. 420.
10. Vimala Thakar, *Mutation of Mind*, Ahmedabad: New Order Book Co., 1972, pp. 148-49.
11. J. Krishnamurti, in Pupul Jayakar, *Krishnamurti: A Biography*, Harper & Row, 1986, pp. 379-80, 435.
12. Fritjof Capra, *The Tao of Physics*, Bantam Books, 1988 (Ed.), p. 298.
13. J. Krishnamurti, *Krishnamurti to Himself*, London: Victor Gollancz Ltd., 1987, p. 122.
14. Niels Bohr, in Fred Alan Wolf, *Taking the Quantum Leap*, San Francisco: Harper and Row, 1981, pp. 21, 85.
15. Vimala Thakar, *Passion for Life*, Holland: Bookfund Vimala Thakar, Magazine Contact, Issue No 25, pp. 79, 88, 114.
16. *Ibid.*, *Towards Total Transformation*, Ahmedabad: New Order Publishing House, 1970 (Talk at Stanford University, October 28, 1968), p. 79.

Part-II
Symbosis of Science and Spirituality Survival of Mankind

Spirituality is the physics of Consciousness – the wholeness of life and living. It is the awareness of the wholeness of life around you and wholeness within you. Both science and spirituality have non-authoritarian approach.

When pure Consciousness expresses itself in a pure non-dual and non-conceptual frame of reference of Mind Continuum, spirituality emerges. In spiritual science, the emphasis is on the act of perception, the quality of the instruments employed for perception, the quality of Consciousness behind the perception. In spirituality the act of investigation, exploration is conducted on behalf of the whole human race.

Spirituality is understanding the conditioned mind, the matter content of consciousness, and also to perceive the Unconditioned Energy; so it is the science of the wholeness of life. It aims at self-actualization of the hidden potential. Spirituality is fundamental and objective science is a derivative.

When thought-structure coupled with volition encounters a true subject of cognizance (restructured science-education-health), the thought-structure goes to abeyance; and the true subject of cognizance is called "creativity in science". The true subject is aware of the true perception. In this case, restructured science is identical to spirituality. A symbiosis (integration) of science and spirituality is the key to the survival of mankind. It jettisons all chaos, strife, pain, misery and travails, and ushers a new era of freedom, peace and bliss. This is the only vocation of the human race.

9
The Turning Point

Abstract

A new age to understand Universal Consciousness or Life Field in the light of recent discoveries of science has started. There is an energy interaction in the symbolic-dualistic frame (subject object), which takes a few millionth of a second. If this state is observed by restructured science-education, the symbolic-dualistic frame makes a quantum jump to non-dual frame; this happens when there is flow of energy from the Unmanifest, which is represented as Unmanifest → non-dual frame. In technical parlance, the quantum jump is called crossing the "event horizon" or "frequency barrier".

The scientific mode or temper of mind is a relatively new dimension of Consciousness, which takes in its fold the objective and the subjective dimensions of life, and this is called restructured science. Similarly the restructured education deals with sensual and trans-sensual world. When we observe the conditioned psyche with restructured science-education dimension, it regenerates the Intelligence, which resolves all human suffering and bestows order, non-violence and peace.

Man (Homo sapiens) has an evolutionary origin. He emerged at about 2.5 million years ago. Julian Huxley wrote, "Man's evolution is not biological but psychosocial, it operates by the mechanism of cultural tradition, which involves the cumulative self-reproduction and self-variation of mental activities and their products". Man has been engaged, for a million years or more, in making symbols, concepts and images of experience and knowledge (information), which are recorded as memory or cultural belief structure. The memory is metastable, it spontaneously disintegrates and the by-product is called thought.

This article discusses the challenge of thought, resolution of the human paradox, and the turning point. This ends all human suffering and ushers a realm of order, rhythm, harmony, non-violence and peace.

The Challenge of Thought

Modern physics has confirmed that all the concepts we use to describe nature are limited, and that they are not features of reality, but creations of the mind; parts of the map, not of the territory.[1] The space-time matrices is a subjective idea and not an objective fact. Space, time and the physical world of objects and events have no objective reality, apart from the mental concepts that man creates in his mind.

Ilya Prigogine wrote, "When an atom is excited, the electron jumps from one orbit to another. At this very instant the atom emits or absorbs a photon the frequency of which corresponds to the difference between the energies characterizing the electron's motion in each of the two orbits. This difference is calculated in terms of Einstein's formula relating energy to frequency".[2] D.S. Kothari wrote, "The entire process of transition from one state to another is a non-visualisable, unanalysable. *It is indivisible*.... As we think – in any sequence of thoughts – there is always a sudden jump, discontinuous jump from one thought to the next.... Quantum jumps for thought appear to be as much a necessity for the conscious self as these are for the atoms".[3]

Thought is capable of doing discontinuous scanning in one dimension and in one plane of an event. All our thinking (thought flow) an speaking is characterized by linear and sequential structure. The thought is capable of scanning and infinitesimal sample of the world. It is clear that our abstract system of conceptual thinking can never describe or understand the reality completely. We can only expect an approximate representation of reality from such a procedure, and all rational knowledge is therefore limited. Rational knowledge will always offer a partial answer to our problems.[4] Arthur Eddington put it succinctly, "Nature thus provides that knowledge of one-half of the world will ensure ignorance of the other half".[5] We live in a man-made world which is divorced from the reality. Man's life has lost stability and he has begun to suffer.

The realm of rational knowledge is, of course, the realm of science which measures and quantifies, classifies and analyses. Rational knowledge belongs to the realm of the intellect whose function it is to discriminate, divide, compare and categorize. In this way, a world of intellectual distinctions is created, of opposites which can only exist in relation to each other. We have to be constantly aware of the limitations and of the relativity of conceptual knowledge.

Ilya Prigogine suggested the foundation of quantum mechanics is in some way related to the perturbations resulting from the process of observation.... And Heisenberg's uncertainty relation would only express the perturbation created by the measurement process. The act of observation is accompanied by an inevitable disturbance which alters the state of the observed system in an unpredictable manner. What is observed is different from what it was before the act of observation. This is inherent in the nature of things.[6]

Spencer Brown, in a most spectacular passage, wrote, "In order to do so (see or observe), evidently it must first cut itself up into at least one state which sees, and at least one other state which is seen".[7] This subject-object division, due to man's psychosocial evolution, led to energy exchange at the quantum level (via photons), this resulted to distorted (mutilated) subject and distorted object, and this is called subject-object dualism.

Physics talks of subject-object dualism, or cause and effect, or chance, or probability. The dualistic thought, as an instrument, is a media or technology, which was created by man. Thought is matter and deals with modification of matter, which have been used to construct new instruments. It is the symbols, concepts, images. It is expressions and interpretations of the experiential dimension which is taken as real. This is repeated mechanically through verified formulas. Thought is an interpretation, a viewpoint of the experience, it is not an experience, not an actuality. But all of us identify with thought as if beliefs, judgements, conclusions are my identity and my life.[8] Thought moves from "what was" to "what should be"; it is a pendulum movement which misses the "what is", the present.

Thought constructs "I-consciousness", or "ego", or "me". It is a concept and has no factual reality. It is always incomplete and therefore selfish, greedy and so on.

Consciousness is shared by whole mankind. Consciousness is the quantum connection between the numerous layers making up the quantum machine. In other words, we are quantum mechanical "*golems*". Consciousness performs a dual role in the universe. In the world of the quantum, it is both the awareness and the creation of experience. It is the *being* and *knowing* of experience. With the stroke of the twentieth-century quantum eraser, the dividing line between ontology (theory of being) and epistemology (theory of knowing) is rubbed out.[9] The fusion of ontology, epistemology, psychology and science is beginning to emerge.

Mind looking at mind is a similar process. Since atoms with consciousness are looking at atoms with consciousness, we have the same experience of wonder and mystery whenever we look at the universe. We are looking at ourselves. It is this process that we call the universe. Richard Feynman wrote, "The atoms come to my brain, dance a dance, then go out; always new atoms but always doing the same dance, remembering what the dance was yesterday".[10]

We are beginning a new age of awareness, the age of quantum consciousness, the age of the conscious atom. The only instrument available is *observing* or *looking within* ourselves.

Whenever there is a challenge, whenever life is threatened, the sympathetic nervous system – concerned with defence mechanism – comes to the aid of the organism. This sympathetic activity leads to an increase in heart rate and breathing; and more blood is pumped to the muscles so that the animal can either run or fight. The regulation of the sympathetic nervous system and thus of the fight-or-flight mechanism is brought about on the instinctual level. In case of human beings, even after the challenge is over, the mind can go into tension and anxiety from memory of the event alone. So in man, fear is usually related to thinking, not to an actual threat to survival – and therefore the sympathetic nervous system has become unsympathetic to human survival.[11] For the human mind, thought has become the greatest challenge.

Thought invariably follows "self-destructive potential" and never touches the "self-integrative potential". It constructs pain, misery and travails. Every activity of thought leads to increase of entropy, which is directed towards self-annihilation or self-destruction.

A thought is a function of the brain which points to the need of security. It only arises as a response to a challenge – it shows that something remains unresolved, incomplete. So when thought creates these challenges of fear, envy, jealousy, hate and violence, these challenges have to be seen. Only when these are seen, can they be resolved, and when thought has performed its function it comes to an end. A mind that can understand thought is free of thought.[11]

Sri Aurobindo explored the evolutionary crisis and wrote:

> The evolution of mind working upon life has developed an organization and use of matter which can no longer be supported by human capacity without an inner change. The problem is fundamental and in putting its evolutionary Nature is confronting man with a critical choice which must one day be solved in the true sense if the race is to arrive or even to survive.[12]

He suggested that if the human race is to survive, it must work for transcendence of the psychological debris, or inner psychological transformation or mutation of the human psyche. There are many thinkers, such as Teilhard de Chardin, Lecomte du Nouy and Julian Huxley, who also spoke of transformation of psychosocial evolution to spiritual evolution.

Resolution of the Human Paradox

Many eminent scientists – including Albert Einstein, Werner Heisenberg, Erwin Schrodinger, Arthur Eddington, and so on – have stumbled upon another mode of perception (knowing), which is "understanding of life directly, instead of in the abstract, linear terms of representational thinking".[13] Eddington also wrote, "The intimate knowledge (direct, non-dual knowledge) will not submit to codification and analysis; or, rather, when we attempt to analyze it, the intimacy is lost and it is replaced by symbolism".[14] This implies that the frame of reference is oscillating between non-dual and symbolic-dualistic frame. In other words, at the moment of attention there is non-dual frame, if the frame is taken over by inattentivity, the symbolic-dualistic frame appears. J. Krishnamurti wrote:

> At the moment of attention all the conditioning disappears, all the image-

> building comes to an end; it is only when you are not attentive that the whole thing begins.[15]

It appears that the non-dual frame is synergistic, or higher quantum energy potential, or higher frequency than symbolic-dualistic frame.

The nature's law appears to be that if an entity or event at a lower frequency comes in the field of higher frequency, the event at lower frequency explodes and gets annihilated. As the non-dual frame is synergistic to thought, so if thought comes in the field of non-duality, it gets annihilated.

The outer (objective) and the inner (subjective) are not two separate movements, but it is one unitary movement. When science takes in its ambient both the inner and the outer dimensions of life, it is called "restructured science". The restructured science deals with the wholeness of life, so it is synergistic to symbolic-dualistic frame, or thought. If one "observes" or "enquires" or "experimentally interrogates" the conditioned psyche, thought goes to abeyance.

Prigogine stated that, "the behaviour of a reality (is) unrelated to our beliefs, our ambitions, or our hopes".[16] Erwin Schrodinger stated, "a moderately satisfying picture of the world has only been reached at the high price of taking ourselves out of the picture, stepping back into the role of a non-concerned observer".[17] David Bohm concluded that, "to observe nature with a certain kind of objectivity could be called the germ of scientific attitude",[18] or scientific temper.

In the state of observation, the subject is 'non-variable', or 'non-reactional' attentiveness. In this state, one does not jump to evaluate, or to compare. Observation is perception without psychological involvement, without any kind of reaction. Reaction implies evaluation, judgement, condemnation, comparison. Observation is non-evaluatory perception. Observation without the pseudo-observer is the state of wholeness of perception and wholeness of responses.

We have to educate our physical organism and psychological structure through scientific mode of mind. This results in a movement of life from self-knowledge to self-awareness, which is "understanding of life directly". It flows to a quantum jump from symbolic-dualistic frame of mind spectrum to non-dual and non-conceptual frame of reference of Mind Continuum, and from intellect to intuition (insight). So one moves from activity to non-action (no-mind state), from speech to verbal silence, from movement to non-motion gracefully and voluntarily. It energises the Cosmic Intelligence that takes charge of one's being.

The Turning Point

We live today in a globally interconnected world, in which biological, psychological, social and environmental phenomena are all interdependent. We need a change in our thoughts, perceptions and values.[19] We need a quantum jump from symbolic-dualistic world to non-dual and non-conceptual world; from psychosocial evolution to spiritual evolution; from intellect to wisdom; from experience and knowledge to "understanding of life directly". We need restructured education – the holistic education which is training and purifying the instruments of perception. We have to initiate psychological experiments to educate our instruments of perception. And this change is likely to result in a transformation of unprecedented dimensions, a turning point for the planet as a whole. The survival of our whole civilization may depend whether we can bring about such a radical transformation or mutation of our present psyche.

Schrodinger wrote, "No personal God can form part of a world model that has only become accessible at the cost of removing everything personal from it.... I do not find God anywhere in space and time – that is what the honest naturalist tells you".[20] Deepak Chopra wrote:

> To know God personally, you must penetrate a boundary that physicists call "the event horizon", a line that divides reality sharply in half. On this side lies anything that remains within the speed of light; on the other side is anything faster than the speed of light.[21]

"The division of reality sharply in half" does not seem valid.[22] Chopra, on the basis of relativistic physics, stated, "The speed of light is absolute; it is like a wall that no object can crash through. As we approach the wall, time slows down, mass increases, and space becomes curved". He added, "Any place where knowledge stops there is also an event horizon. The brain can't explore beyond where (the limit to which) photons go.... A nervous system is just a machine for sensing photons". A symbolic-dualistic frame of reference deals with subject-object interactions, or photon transfer between subject-object. The quantum jump from symbolic-dualistic frame to non-dual and non-conceptual frame is basically crossing "the event horizon", which implies ending of subject-object dualism and space-time going to abeyance. The non-dual frame is aware of the *fact*, the *truth*, the *present*, the *what is*! The non-dual frame is *open* to the Unmanifest and closed to environments (meaning ending of photon transfer). The non-dual frame is beyond "the event horizon".

Deepak Chopra added, "Quantum physics dips across the border all the time, only it can't stay there very long. When a particle accelerator bombards two atoms, causing a subatomic particle to jump out of its hiding place for a few millionths of a second, the "event horizon" has been crossed". Modern physics just touches the fringes of non-dual frame and then reverts to symbolic-dualistic frame instantly.

We call the "event horizon" as the "frequency barrier". The symbolic-dualistic frame functions at the lowest frequencies of consciousness, and this is written as subject object. The non-dual frame is at the highest frequency of Consciousness. The Unmanifest supplies energy to the non-dual frame for its sustenance, it is written as Unmanifest → Non-Dual frame.

In the symbolic-dualistic frame lies anything that remains within the velocity of light; and on the other side, in the non-dual frame there is anything faster than the velocity of light. The non-dual frame perceives the Reality. In Reality, there is superluminal communion without signal (not communication) between two or more non-dual frames of references of Mind Continuum.

The Einstein-Podolsky-Rosen thought experiment indicates that information can be communicated at *superluminal* (faster-than-light).[23] The superluminal communion can happen between two non-dual frames without a signal. It appears that a new field has been created in the realm of superluminal quantum connectedness.

At the moment of crossing the "frequency barrier", the subject-object energy interaction ends, because the photons are knocked down by the energy of attentivity which is at a higher potential. The object goes to abeyance and dualism ends. The experience and knowledge also stops. All this happens in less than a hundred microseconds. As one approaches the frequency barrier, time slows down, mass increases and space becomes curved; it implies that time-space matrices also go to abeyance. The boundaries of all psychological imbalances, ecological challenges, and so on are true event horizons, so these all go to abeyance. What is left behind is the subject which expresses itself as the *fact*.

The mind-brain-body system cannot explore beyond the velocity of light or the velocity of photons – this is the limitation of the mind-brain-body system. There is no experience and knowledge when photons are knocked down. When experience and knowledge go to abeyance, there is pure perception. In the non-dual frame, there is direct input of the Unmanifest in the brain cells without mediation of the sensual apparatus. The pure perception, or the *fact* is an expression of the Reality. This is the beginning of a new age of consciousness, the age of quantum consciousness. This is the march to peace.

Fred Alan Wolf wrote, "I shall call the quantum wave function a 'qwiff' .... A qwiff pop is a destruction of a qwiff flow .... When the qwiff pops, the qwiff itself vanishes .... The qwiff is a quantum wave function that at best can only describe the probability of the observation and not the actual observation .... Each pop is a sudden disruption, a break in the past, a violation of the law of cause and effect".[24] If we observe the conditioned psyche by scientific mode of mind, we conclude, "It is the symbolic-dualistic frame of reference of mind that constructs the qwiff; or in other words, it is the thought that creates the qwiff. The qwiff describes the probability of the observation, it obeys the law of cause and effect. When symbolic-dualistic frame of reference makes a quantum jump to non-dual frame, it pops; the time-space matrices go to abeyance and the non-dual frame is aware of the '*fact*', the '*pop*'. Pop is synergistic to the qwiff". It appears that the psychodynamics of the mind-brain-body system can adequately explain the qwiff-pop relationship.

If a photon approaches the black hole's immense gravity, the photon is devoured in less than a hundred microseconds. The photon falls into the black hole forever, frozen in time. As light and energy get sucked into a black hole, they disappear from our cosmos. Similarly the high quantum energy potential of attentivity annihilates a photon, which is at a lower quantum energy potential. In case the mind-brain-body system falls to inattentivity, the photon is regenerated (or a new photon gets activised) and the non-dual frame falls to symbolic-dualistic frame of reference. The mind-brain-body system *oscillates* between non-dual realm (attentiveness, intimate knowledge) and symbolic-dualistic field (inattentivity, relative knowledge).

Restructured Science and Restructured Education

Man lives simultaneously in two dimensions – the objective and the subjective. The objective (outer) and the subjective (inner) are not two separate movements, it is a unitary movement, one whole. The classical science deals with the objective dimension and divorces the subjective dimension; hence it leads to conflict and contradiction. We have to restructure science so that it takes in its ambient both the objective and the subjective world. It explores the wholeness of life. Similarly we have to restructure education so that it deals with both the sensual and the trans-sensual dimensions of life.[25, 26, 27]

In the words of Ramana Maharishi, "Illiteracy is ignorance and education is learned ignorance. Both are ignorant of the true Aim".[28] Ken Wilber commenting on above wrote, "The 'true Aim' is the non-dual mode of knowing, while being literate or illiterate concerns the dualistic and symbolic mode of knowing. Ignorance, in other words, is ignorance of the non-dual and non-conceptual mode of knowing, which would instantly reveal the universe to be Mind only".[29]

An Action Plan for Human Survival

An Action Plan was presented in 1994 which has been fovourably received by others active in the United Nations.[30] This was reprinted in 1998.[31] It appears that man, due to his ignorance and arrogance, has not implemented the plan (or a modified plan) for his survival.

We have to restructure science, restructure education and learn to live in the awareness of the perennial values. We have to educate our physical organism and psychological structure through observation by the scientific mode of mind. We have to understand deeply ***Science and "Mind-Brain-Body Interactions"***.[32]

The mutation of conditioned psyche is discussed in two phases, but this is merely an exegetical convenience, a device to make the complete story a little easier to tell:

Phase I: When one observes the conditioned psyche with subject as restructured science-education, it leads to crossing the "event horizon" or "frequency barrier" which results in a quantum jump from symbolic-dualistic frame of mind spectrum to non-dual and non-conceptual frame of reference of Mind Continuum. It bestows perception of the *fact*, the *true*, the *present*, the *non-dual*, the *what is*!

Phase II: When this observation is continued for a long chronological time and is sustained, it blossoms to understanding and learning; to intuition, spontaneity and right action; and the perceptions are born of non-duality and responses are born of spontaneity. This brings man on the fringes of mutation of the mind-brain-body system. It is an inner voyage of self-discovery from self-observation through self-knowledge and self-awareness to Silence, to Intelligence. If man lives his own understanding, it energises the Intelligence, the Wisdom, which resolves all human challenges and bestows a "new global security order", a "non-violent peaceful world". Intelligence is the expression of the highest frequency or the highest quantum energy potential of the Universal Consciousness, the Life Field.

Concluding Comment

Man is an amphibian who lives simultaneously in two dimensions – (a) the *known*, the time-space matrices, the symbolic-dualistic world; and (b) the *Unknown*, the Life Field, the Universal Consciousness, the non-dual world. The *known* offers no guidance, in regard to the much more fundamental problem of the relationship in the *Unknown*. The *Unknown* is synergistic to the *known*.

The psychosocial evolution of man, for one million years or more, has constructed a great variety of symbol-systems, and this has expressed itself as art, science, law, philosophy, art of war, and so on. There has been a fantastic and exponential growth in knowledge. The old chasm between knowledge and wisdom has widened dangerously. Our civilization has entered a new and potentially a dangerous phase of its long and tortuous history. Every activity in the field of symbolic-dualistic world is leading to increase in entropy, which is leading us to self-destruction, or self-annihilation.

The whole mankind is caught in a crisis of perception, a psychic pandemic. There is an urgency to make a quantum jump from intellect to intuition, from knowledge to intelligence, wisdom, from psychosocial evolution to spiritual evolution.

The scientific mode or temper of mind is a relatively new dimension of Consciousness, which takes in its fold the objective and the subjective dimensions of life; it deals with the wholeness of life, and we call it restructured science. Similarly, restructured education deals with sensual and trans-sensual world. When we observe, or enquire, or experimental interrogation the conditioned psyche by restructured science-education mode, it undergoes radical transformation or mutation. The mutated psyche regenerates Intelligence, which takes charge of the *being* and resolves all human challenges and ends all human misery.

All Intelligence or Wisdom is structured in the Unmanifest, the Life

Field, the Universal Consciousness. It is the Life Field that can guarantee human survival and end all human suffering. This ushers a realm of order, non-violence and peace.

References

1. Fritjof Capra, *The Tao of Physics* (Second Edition), Bantam Books, 1983, pp. 147, 277.
2. Ilya Prigogine and Isabelle Stengers, *Order Out of Chaos*, Bantam Books, 1984, pp. 220, 224.
3. D.S. Kothari, *Atom and Self*, in *The Evolution of Consciousness*, New Delhi: National Publishing House, 1983, p. 212.
4. Sampooran Singh, *A Scientific Guide to the Integration of Man, Society and Mankind*, New Delhi: Siddharth Publications, 1987, pp. 158-59.
5. Arthur Eddington, in Ken Wilber, *The Spectrum of Consciousness*, USA: The Theosophical Publishing House, 1982, p. 29.
6. D.S. Kothari, *Some Thoughts on Science and Religion*, The University of Delhi, 31.10.1977 & 03.11.1977, p. 25.
7. Spencer Brown, *Laws of Form*, New York: Julian Press, 1972, pp. 104-05.
8. S.C. Malik, in *The Odyssey of Science, Culture and Consciousness*, New Delhi: Abhinav Publications, 1990, pp. 153-55.
9. Fred Alan Wolf, *Taking the Quantum Leap*, Harper and Row, 1981, pp. 211, 229.
10. Richard Feynman, in Fred Alan Wolf, *ibid*., 1981, p. 228.
11. R.P. Kaushik, *Light of Exploration*, USA: Journey Publication, 1977, pp. 57-63.
12. Sri Aurobindo, in *The Odyssey of Science, Culture and Consciousness*, New Delhi: Abhinav Publications, 1990, p. 41.
13. Alan W. Watts, *The Ways of Zen*, New York: Vintage Books, 1957, p. 10.
14. Arthur Eddington, in Commins and Linscott, *Man and the Universe*,: Washington Square Press, 1969, p. 453.
15. J. Krishnamurti, *Krishnamurti in India, 1970-71*, India: Krishnamurti Foundation, 1971, p. 69.
16. Ilya Prigogine and Isabelle Stengers, *op. cit*., p. 5.
17. Erwin Schrodinger, *What is Life*? and *Mind and Matter*, Cambridge University Press, Ed. 1967, p. 128.
18. David Bohm, *New Scientist*, November 11, 1982, p. 364.
19. Fritjof Capra, *The Turning Point*, London: Wildwood House, 1982, pp. 1-516.
20. Erwin Schrodinger, *op. cit.*, p. 150.
21. Deepak Chopra, *How to Know God*, London: Rider, 2000, pp. 270-71, 271-72.
22. The Universal Consciousness or Life Field expresses Itself in two dominant modes: (a) successive consciousness, where the Transcendent Potential has gone to abeyance and its only linkage is the immanent potential; and further, as an analogy, if we consider the successive consciousness as a drop of water, then the Reality is a fathomless and unbounded ocean, and (b)

Simultaneous Consciousness, or Transcendent Potential, or Life Field, or the Reality, or the Unmanifest. The "event horizon" or the "frequency barrier" is defined as a quantum jump from successive consciousness to Simultaneous Consciousness. It appears that the "event horizon" cannot "divide reality sharply in half", in the words of Deepak Chopra; but it is responsible for the unfoldment of the Reality in a non-dual and non-conceptual frame of reference of Mind Continuum.

23. Gary Zukav, *The Dancing Wu Li Masters*, Fontana/Collins, March 1982, pp. 303-04.
24. Fred Alan Wolf, *op. cit.*, pp. 170, 188.
25. Sampooran Singh, Kanwaljit Kaur and Paramjit Singh, *Global Education and Human Survival: A Scientific Appraisal*, New Delhi: Sanbun Publishers, March 2002, pp. i-xiv, 1-88.
26. Sampooran Singh, *A New Vision of Education and Peace: A Scientific Appraisal*, Chandigarh: Faith Publishers, April 6, 2003, pp. 1-31.
27. Sampooran Singh, Kanwaljit Kaur & Paramjit Singh, Pathways to Peace: *The Wisdom of Listening and the Power of Commitment*, Chandigarh: Faith Publisher, October 2003, pp. 1-42.
28. Ramana Maharshi, in Ken Wilber, *The Spectrum of Consciousness*, USA: The Theosophical Publication House, 1982 (Quest Book), p. 110.
29. Ken Wilber, *ibid.*, p. 110.
30. Sampooran Singh, *An Action Plan for Human Survival*, USA: Psycho Science, Supplement, Fall/Winter, 1994; Reprinted *Psycho Science*, Spring-Summer, 1995, pp. 41-49.
31. Sampooran Singh, Dr (Mrs) Kanwaljit Kaur and Paramjit Singh, *The Masterplan Paradigm for Human Survival and Excellence: The Science of Life*, Chandigarh: Faith Publishers, February 1998, pp. 228-41.
32. Sampooran Singh, *Science and "Mind-Brain-Body Interactions"*, unpublished.

10
March to Peace: The Wisdom of Observation, Listening and the Power of Commitment

Abstract

The restructured science-education is synergistic (higher quantum energy potential) to thought. If the scientific mode of mind observes or listens to a thought (or I-consciousness) the latter gets annihilated. Man has to learn to live in a higher hierarchical expression of consciousness than that of thought, which is living in self-awareness or attentiveness.

Nature invariably expresses Itself in a superluminal (faster-than-light) channel without signals, while the brain-body system, which is free from conditionings (non-dual frame), acts as a material structure for nature's expression. Life is a mysterious wholeness and interrelatedness. It blossoms to intuition (insight), spontaneity and right action. It flowers to understanding and learning. Understanding leads to the perception of the whole. It bestows order, non-violence, peace and survival.

Man (*Homo sapien*) has gathered experience and knowledge (information) over the last one million years or more, and has stored this as memory or thought-structures, in the brain. This has resulted in the conditionings of the human psyche; or subject-object dualism; or symbolic-dualistic frame of reference of mind spectrum. There is an energy interaction between the subject and the object (written as subject↔object). And the energy transfer happens, via photons, in less than a hundred microseconds.[1] This leads to distortion or mutilation of both subject and object. Man is aware of the distorted perception. This constructs pain, misery and all suffering. This is leading man to self-destruction, or self-annihilation. The whole mankind has to explore non-violence, peace and survival.

"What lies behind us and what lies before us are small matters compared to what lies within us". This stimulating observation of R.W. Emerson, has constantly inspired, encouraged and strengthened our resolve to explore **"what**

lies within us". It investigates the healing of the mind-brain-body system. It is an open call to express higher hierarchical level of expression of consciousness eloquently through restructured science-education for a global society and our daily living. It is the integral dimension of Consciousness or Life Field that can resolve all psychological imbalances, ecological challenges, and so on. It is basically an extension of our paper entitled *Pathway to Peace: The Wisdom of Listening and the Power of Commitment.*[2] This is the only guarantee to human peace, survival and excellence.

Dimensions of Consciousness?

The Universal Consciousness or Life Field expresses Itself in two dominant modes: (a) It expresses itself in the symbolic-dualistic frame of reference in the mind spectrum, any event or entity is materialistic and remains within the velocity of light – this is called "successive consciousness"; and (b) It expresses itself in the non-dual and non-conceptual frame of reference of Mind Continuum; it transcends all time-space-causation matrices; any event or entity is encoded in the superluminal (faster-than-light) channel, and everything is interconnected and in communion – and this is called "Simultaneous Consciousness". The successive consciousness is at the lowest frequencies, or in other words, it is at the immanent quantum energy potential of consciousness, or the lowest expression of quantum energy potential of Life Field. The Simultaneous Consciousness is at the highest frequency, or in other words, the transcendent quantum energy potential of Consciousness, or the highest expression of Life Field.[3]

In a recent (1996) scientific conference at Moscow, the Russian scientists proposed, "A life Field, beyond our physical concepts of time, space and the known electromagnetic and gravitational fields". In another conference at Cambridge, it was proposed, "A new scientific approach would include subjective experience as valid information and address all forms of human experience – including intuition, spirituality and mystical encounters".[4]

Scientists have discovered something mysteriously rich in energy (at very high frequency, higher frequency means higher energy) in empty space (the Void, the Universal Consciousness). This has confirmed the perennial wisdom of ancient sages/seers. Now we are on the verge of extending this richness into human dimension.

The Known and the Unknown Dimensions of Man

Man is an amphibian who simultaneously lives in two dimensions – the subjective (inner) and the objective (outer). We have to restructure science itself so that it embraces both the dimensions – this is called the scientific mode of mind. It deals with the wholeness of life. The energy flows from the Unmanifest to non-dual frame of reference, and this is written as "Unmanifest

→ non-dual frame". Similarly education explores the sensual dimension and the trans-sensual dimension. The restructured science-education is at a higher quantum energy potential than thought-structures.

Restructured Science and Education

We have to educate the physical organism, the psychological structure, the instruments of perception, to learn the arts of observation and listening, and to understand the wholeness of life. The essence is: (a) *Restructuring of Science*, which implies that objective science takes in its fold the subjective dimension of man; (b) *Restructuring of Education*, which implies to understand the sensual and trans-sensual dimensions of life; (c) *Awareness of Human Values*, which implies to understand the depth dimension of man; and (d) *to Live in the Light of that Understanding.*[5]

The new vision of education is fundamentally a single dynamic flow of Cosmic Consciousness, Life Field; and the technologies are: (a) *Epistemological Revolution Through Scientific Temper*; (b) *Pilgrimage to the Sacred Temple of Understanding and Learning*; and (c) *Symbiosis of Science and Spirituality*. These technologies are one and the same except a difference of shade. These lead to radical transformation or mutation from non-fact to fact, from intellect to intuition (insight), from conflict to cooperation, and from fragmentary conditioned consciousness to the Wholeness, the Unconditioned Consciousness.[5]

The essence of education is understanding and learning about life directly *here* and *now*. The perennial truth sustained by enquiry or by scientific mode of mind has been discussed by J. Krishnamurti in words:

> In oneself lies the whole world, and if you know how to look and learn, then the door is there and the key is in your hand. Nobody on earth can give you either that key or the door to open, except yourself.[6]

Observation and Listening

Observation: Sullivan put it, "We cannot observe the course of nature without disturbing it",[7] or Andrade "Observation means interference with what we are observing.... observation disturbs reality".[8] George Wald wrote, "The very crux of twentieth-century physics is the recognition that the observer cannot be excluded from his observation; he is an intrinsic participant in them".[9]

Kothari wrote, "In dealing with atomic phenomena we have to recognise that the act of observation is accompanied by an inevitable disturbance which alters the state of observed systems in an unpredictable manner. What is observed is different from what it was before the act of observation. This is inherent in the nature of things. It cannot be eliminated".[10] This is basically due to energy transfer, via photons, between the subject and the object, which in technical parlance is written as Subject↔Object. This transfer of energy happens in less than a millionth of a second. The transfer of energy results in distortion of the subject and the object, and we are aware of the distorted or mutilated subject and object. It implies that the symbolic-dualistic frame of reference of mind spectrum can never observe a true subject and a true object. We have lost the faculty of true observation.

It is most surprising that the ancient seers/sages were aware of this limitation of the art of observation. R.P. Kaushik wrote, "The first stage prior to recognition and naming is a state of non-verbal or thought-free perception (*nirvikalpa jnana* or pure perception). From this pure perception we proceed to thought, recognition and naming (*savikalpa jnana*)".[11]

Listening: The art of listening by symbolic-dualistic frame implies: (a) to listen to the words (object) without translating or interpretation, (b) it recalls the past memory (subject); (c) the subject superpositions itself on the object; (d) this results in distortion of listening; and (e) all this happens in less than a millionth of a second. So listening in symbolic-dualistic frame produces an automatic reaction. Such an interpretation or translation leads to destroying our listening and understanding.

In the non-dual frame of reference of Mind Continuum, listening – (a) it is, first, perceived as a *fact*; (b) there is no recall of memory hence there is no interaction and the perception of fact does not get distorted; (c) we should watch this fact with our whole being, so that we remain at the non-dual level; (d) this bestows true listening; and (e) at the level of the non-dual frame the listening, to be more precise, is called observing.

In principle, the true arts of listening and observation will bring about a tremendous understanding and learning.

Observation by Restructured Science-Education

In the common waking state, or the conditioned psyche, the observer is the past – the experience, knowledge and memory – and the observer is called the pseudo-subject. Physical revolution has never psychologically changed the human state. There is nothing to learn in the psychological field. There is no such thing as psychological evolution. Psychologically, inwardly, there is only the *fact*.

The restructuring of science-education bestows a scientific mode of mind; this is called true-subject as it embraces the wholeness of life. A scientific mind takes up "the role of a non-concerned observer (true-subject)", in the words of Schrodinger.[12] It is "unrelated to our beliefs, our ambitions, our hopes", in the words of Prigogine.[13] It must not construct a pseudo-observer, or a pseudo-subject. It must see in the present and reject the past remembrances. The values in restructured science-education are subjective truth coupled with the objective truth, integrity and honesty regarding the wholeness of life.

The time-hallowed discrimination between subject and object has been jettisoned. Schrodinger prophetically wrote:

> The world is given to me only once, not one existing and one perceived. Subject and object are only one. The barrier between them cannot be

> said to have broken down as a result of recent experience in the physical sciences, for this barrier does not exist.... The world is given but once. Nothing is reflected. The original and the mirror-image are identical. The world extended in space and time is but our representation.[14]

Prigogine wrote, "It is physics that presupposes an observer situated within the observed world. Our dialogue with nature will be successful only if it is carried on from within nature".[15] Geoffrey Chew in a prophetic message says:

> Our current struggle with (certain aspects of advanced physics) may thus be only a foretaste of a completely new form of human intellectual endeavor, one that will not only lie outside physics but will not even be describable as 'scientific'.[16]

Max Planck, the father of quantum mechanics, summed up the quest of science in terms:

> Science .... means unresting endeavor and continually progressing development toward an aim which the poetic intuition may apprehend, but which the intellect can never fully grasp.[17]

Physics of Consciousness?

(a) The Committee on Science and Technology, USA, submitted in June 1981 a report to the House of Representatives, USA, that further research on "***The physics of consciousness***" should be undertaken, and concluded that recent experiments:

> .... suggest that there exists an 'interconnectiveness' of the human mind with other minds and with matter; .... that the human mind may be able to obtain the information independent of geography and time.[18]

(b) Fritjof Capra wrote, "All their theories of natural phenomena, including the "laws" they describe, are creations of the human mind; properties of our conceptual map of reality, rather than of reality itself".[19]

(c) Fred Alan Wolf wrote, "Quantum physicists discovered that every act of observation made of an atom by a physicist disturbed the atom.... Our activity introduces discontinuities in whatever we are observing.... To Niels Bohr, 'discontinuity was a fundamental truth'."[20]

(d) When Consciousness with higher frequency encounters consciousness with lower frequencies, the latter gets annihilated. In other words, when subject at the higher quantum energy potential of Consciousness bombards an object with lower quantum energy potential of consciousness, the object gets exploded (cold fission), so the object goes to abeyance.

The restructured science-education deals with the wholeness of life. The non-variable attentivity, or reaction-free attentivity, or objectless-attention takes over as the true-subject. The subject sees that the movement of thought-structures is independent of my volition, it is independent of the movement of the 'ego', the "me", the "I-consciousness". It implies that the concept of the ego is independent of the momentum of thought. As the concept of the ego is at a lower quantum energy potential than the momentum of thought, so it is knocked down, or the I-consciousness or ego goes to abeyance. The mental movement has a momentum independent of the ego.[21] As an analogy, we may compare it to a Particle Accelerator.[22]

The relationship that the thought has an existence independent of "me" is established, and the vanity that it is my thought coupled with volitions ends and its place is taken by humility. This suggests that there is nothing like an individual mind, there is nothing like an individual I-consciousness. Thought is the product of collective human effort through centuries. This gives a sense of freedom.

(e) If the observation is continued, which implies to see the momentum of thought in utter freedom. The photons meet an impregnable wall – called "event horizon"[23] or "frequency barrier". If the photon (thought) comes within the field of restructured science-education, it is bombarded by the high energy potential and photons annihilate themselves, or in other words, the object goes to abeyance. With the abeyance of the object, time-space-causation and psychological time vanish. There is cessation of image-construction, all dualism ends. And all psychological imbalances go to abeyance. We are left with the perception of the *fact* in a non-dual and non-conceptual frame of reference of Mind Continuum. Basically it is the symbolic-dualistic state that makes a quantum jump to the non-dual state.

When non-dual frame is stabilized over a long chronological time, the Intelligence is regenerated, which ends all human suffering and bestows non-violence and peace.

There is no "object out there", there is only "subject in here". The true-subject expresses itself as the *fact*, the *true*, the *present*, the *non-dual*, the *what is*! If observation is further continued, the subject also vanishes. What is left behind is pure Consciousness bathing the brain cells; this blossoms to intuition (insight), spontaneity and right action; it flowers to understanding and learning; it heals the diseased mind-brain-body system; it resolves all psychological imbalances, ecological problems and other challenges; and it ushers a new era of order, harmony, non-violence and peace.

(f) Ervin Laszlo wrote that there are "only ripples upon ripples, waves upon waves, interpenetrating and propagating in a seamless sea".[24] This indicates a oneness or interconnectedness of the universe. The superluminal (faster-than-light) channel without signals indicates interconnectiveness of Minds, or to be more precise, there is only one Mind which governs the whole cosmos and beyond. The one Mind is the Life Field, the Universal Consciousness.

Understanding and Learning

The symbolic-dualistic frame is repetitive and mechanical, so it has no capacity for understanding and learning. It is only capable of minor invention, which is basically a modification of the known, it constructs new gadgets with modification of the known. The non-dual frame bestows understanding and learning, it is capable of new discoveries.

So long as the mind is agitated, it can have no understanding. Understanding comes when the mind is very quiet, equipoised, balanced. Only the truth can liberate the mind from its own ideation. Learning flows from understanding. The non-dual frame bestows intuition (insight), spontaneity and right action. All intuition, spontaneous action, understanding, learning, all perennial knowledge and values are embedded in the Unmanifest, the Reality, the Life Field, the Universal Consciousness.

Many eminent scientists – including Albert Einstein, Werner Heisenberg, Erwin Schrodinger, Arthur Eddington, and so on – have stumbled upon another mode of perception (knowing), which is "understanding of life directly, instead of in the abstract, linear terms of representational thinking".[25] This implies "understanding the wholeness of life". The symbolic-dualistic frame of reference of mind spectrum does not have the capacity for understanding. Understanding expresses itself in the non-dual and non-conceptual frame of reference of Mind Continuum. *Understanding is a movement in the non-duality, a movement of the whole being.* Understanding comes only when the mind is very quiet, tranquil, silenced.

Understanding leads to a perception of the whole. Perception has nothing to do with knowledge (information). *The perception are born of non-duality and responses are born of spontaneity.* It awakens the awareness, the intelligence, which arises out of observation, self-knowing and compassion.

David Bohm, in a lecture in USA, in April 1977, said:

> The ultimate perception does not originate in the brain or any material structure, although a material structure is necessary to manifest it. The subtle mechanism of knowing the truth does not originate in the brain.[26]

It implies that ultimate perception happens in a trans-sensual (non-dual) dimension, which is timeless (beyond brain). As all knowledge is structured in Consciousness, ultimate perception is a happening when Universal Consciousness bathes the brain cells. Perception bestows understanding the wholeness of life, which flowers to intuition, spontaneity and right action; it is a happening in the non-dual and non-conceptual frame of reference of Mind Continuum.[27]

If you understand what is the known, you will perceive the *fact* in the non-dual frame. The non-dual is the extraordinary silence, or in other words, the creative emptiness in which alone reality can enter. Reality can only come to that which is *being*, which understands *what is*! The fact is *what is*!, or in other words, the mirror-image of reality.

One has to learn to sustain the fact. The very fact of being aware of "*what is*" is truth. It is truth that liberates and not your striving to be free. The observation by scientific mode of mind ends all psychological suffering, which implies awareness of the fact, the truth. *In ending there is renewal, while in continuity there is death, there is decay.* Human survival is embedded in ending of human suffering, because in ending there is renewal of order, non-violence and peace.

Nature invariably expresses itself as fact in a non-dual frame of reference of Mind Continuum. It encodes a message on the superluminal channel. Jack Sarfatti, in 1975, proposed faster-than-light connections. He elevated the EPR effect to a first principle of quantum mechanics. He called this theory the "superluminal transfer of negentropy without signals".[28] (Negentropy is another term for order.) Love, or Truth, or Beauty, or Reality is superluminal transfer of negentropy.

Nature expresses itself on the "superluminal transfer of negentropy without signals" as *fact*, and then facts are understood and learning happens. The time is now ripe when we realize that in nature everything happens simultaneously and everything is interrelated. It is an integrated and self-emerging system, a totality that exists. Life is a mysterious wholeness. It is non-fragmentary, undivisible whole. It is homogeneous, undifferentiated and dimensionless. It is timeless, eternal and infinite. Life is immeasurable and unnamable. Ervin Laszlo discussed the emerging concepts of the cosmos in words:

The parts of this totality are not radically separate: they are integral elements of the whole. Every "thing" in the cosmos is a coherent "event" and every event is the summed result of all the other events.... The cosmos is a system of interacting waves: it is a sea where the effect of the smallest perturbation propagates throughout the entire field.... the new physics and the new cosmology do not recognize discrete things and independent events in the universe, only ripples upon ripples, waves upon waves, interpenetrating and propagating in a seamless sea.

Love is not different from truth. Love is that state in which the thought process, as time, has completely ceased. Truth is not cumulative. It is from moment to moment. Eternity is in the moment, the *now*.[29] Beauty is reality, now, not tomorrow.

Intuition

All knowledge is structured in consciousness. This recognizes that consciousness alone can be totally self-referral, so the pure mind or non-dual mind can perceive the consciousness. When non-dual state of mind is presented as an object to pure consciousness, then it reveals itself. In the non-dual mind, the intelligence is born of the inner silence, of the inner space. The intelligence is uncluttered by the movement of thought. The dimension of intelligence, of unconditioned energy, of Fundamental Vibration, has no past, so it is free of all conditioning and divisive loyalties, etc. It bestows insight. Insight implies something which you perceive without any logic, without any motive. It is called immediate perception. *The perception are born of non-duality, of silence, and responses are born of spontaneity*. Perception has nothing to with knowledge. Intelligence begins to use the human brain and help the human being to look at the world in quite a different way.[30]

Spontaneity

A non-dual mind perceives the truth spontaneously. It implies that there is no psychological time between expression of Life Field in the brain cells and its perception in the brain. Further there is no psychological time between pure perception and action. So the expression of Life Field in the brain cells, the perception and action are one dynamic flow. The whole Consciousness is a seamless, fluid web of process structures, it is one dynamic flow, and cannot be disassembled or split into simpler and simpler parts.

Commitment

The "*power of commitment*" is a perennial value. It has nothing to do with the sensual or objective world. It arises from the dimension of the trans-sensual world, or non-dual state of mind, or subjective realm of man. It is not a commitment to a particular part, but the wholeness of life. The commitment is not related to the sensual world. It flows from understanding and learning.

Silence is a dimension of Consciousness. The Unconditioned energy bathes the brain cells. It bestows the power of commitment which self-emerges in a non-dual frame of reference of Mind Continuum. Man can grow into that dimension and live from there.[31]

Scientific Discoveries

Kothari wrote, "A scientific theory is always provisional. It is a guide to further development, to its own replacement. A theory or explanation is always tentative, waiting to be replaced by an improved, more comprehensive theory".[32]

The new theorems, unless trivial, are, of course first "discovered" and then proved. Discovery is a flash of insight. It happens when there is a quantum jump from symbolic-dualistic frame of reference of mind spectrum to non-dual and non-conceptual frame of reference of Mind Continuum. Kothari added, "A rational account of a discovery, the proof, follows and not precedes the discovery".

Discovery is a happening in the timeless moment. All great discoveries have arisen from the realm of the timelessness, so scientific enquiry is not a cerebral activity and gathering information, but a quest of the Timelessness, the Divinity, the Life Field. A selfless, unfaltering devotion to truth and use of knowledge for man's uplift – higher hierarchical expressions of Consciousness – are the only vocation of a scientist.

Spirituality

Spirituality is understanding the conditioned mind, the matter content of consciousness, and also to perceive the Unconditioned Energy; so it is the science of the wholeness of life. It aims at self-actualisation of the hidden human potential. Spirituality is the awareness of the wholeness around you and wholeness within you. Spirituality explores what the truth is, what the *fact* is, and what the Reality is behind the truth. Spirituality teaches us to live with the *fact* and transcend the *non-fact*. An integration of science and spirituality is the key to the survival of mankind. It jettisons all pain, misery and travails, and ushers a new era of freedom, peace and bliss. This is the only vocation of the human race.[33] There is an urgency for symbiosis of science and spirituality.

Peace

It is a total way of living. As long as there is inherent violence in the human psyche, there can never be sustainable peace in the world. Peace requires the mutation of individual psyche. It is only holistic education that helps man to grow from knowledge to understanding, from confrontation to cooperation, from rationality to intuition, from intellect to intelligence.

We are on the crossroads of destiny; we must take the road from information gathering and experience to understanding and learning, from violence to non-violence, to motivation of sharing rather than dominating, complementarity and not competition, harmony and not discord; and in the final analysis education alone that can bring this about.[34]

Love is not different from truth. Love is peace. Peace is that state in which the thought process, as time, has completely ceased. Peace is a state of being. Peace is our birthright.

The Action Plan (Paradigm Shift) for Human Peace and Survival

The Action Plan is discussed in two phases, but this is merely an exegetical convenience, a device to make the complex story of the purification of mind-brain-body system a little easier to tell:

Phase 1: When we observe the conditioned psyche with true-subject as restructured science-education, the I-consciousness is first knocked off as it is at a lower quantum energy potential than thought. One observes that the thought movement is independent of my volition, is independent of the movement of the ego. If this observation is continued, one observes that photons cannot cross the "event horizon", or "frequency barrier", or "time barrier" and these get annihilated; or in other words, thought goes to abeyance and there is perception of the *fact*, the *true*, the *present*, the *non-dual*, the *what is*![21]

Phase 2: If we continue observing or listening to the *fact*, without justification or condemnation or motivation, we observe there is only the fact and nothing else. The fact itself is the operator, not you operating on the fact. If you leave the fact alone, it has no time. The fact is free from fear. The fact grows, which implies that there is release of tremendous energy[35] and the fact moves to higher quantum energy potential. This is expressed as Pure Consciousness → Fact. It helps the fact to make a quantum jump in the vertical dimension of Consciousness. When the fact is left alone, it is operating and you do not have to do a thing about it, it matures and blossoms. Immediacy of perception dissolves the fact, and one revels in Pure Consciousness or Intelligence. Only in Intelligence there is complete security. Intelligence sees the false as the false and ends the false. Mutation happens in the timelessness. To understand Consciousness or an enquiry into the subject is the perception of that immeasurable state.

Schrodinger wrote that, "Consciousness is never expressed in the plural, only in the singular".[36] Similarly, the over-all number of Mind Continuum is just one and the whole mankind, including all other existences, shares it. The observation and listening happens in a non-dual frame of reference of mind continuum.[37]

The Action plan ends all human suffering and ushers an era of order, harmony, non-violence and peace. We have to implement this plan (or a modified one after receiving responses from world scientists) now and here.

Concluding Comment

We are today at a historic juncture in the long and tortuous history of the human race on Planet Earth. We stand at a bifurcation where one road is leading to Buddha (Intelligence, Wisdom) and the other road to bomb (self-destruction and self-annihilation). We have to explore the nature and quality of man's mind and life. We have to undertake an inner voyage of self-discovery from self-observation through self-knowledge and self-awareness to Silence, to Life Field. This flows to Intelligence and Wisdom. It is the Unmanifest, or Life Field, or Universal Consciousness that can guarantee peace and human survival.

We urgently need "Symbiosis of Science and Spirituality".[38] If man escapes his responsibility he will meet the same fate as the dinosaurs; and he will do so at his own peril, and that of India and mankind.

References

1. Sampooran Singh, *The Turning Point: The Ending of Human Suffering; The March to Order, Non-violence and Peace*, unpublished.
2. Sampooran Singh, Dr (Mrs) Kanwaljit Kaur and Paramjit Singh, *Pathways to Peace: The Wisdom of Listening and the Power of Commitment*, Chandigarh: Faith Publishers, October 2003, pp. 1-42.
3. Sampooran Singh, *The Pilgrimage to the Third Millennium*, Chandigarh: Faith Publishers, January 1995, p. 19.
4. Joan Walton, *Scientists Tune into Consciousness*, New Science, Winter, 1996
5. Sampooran Singh, Dr (Mrs) Kanwaljit Kaur and Paramjit Singh, *op. cit.*, October 2003, p. 23.
6. J. Krishnamurti, *You Are the World*, KFI, 1981, p. 158.

7. J.W.N. Sullivan, *The Limitations of Science*, New York: Mentor Books, 1949, p. 140.
8. E.N. da C. Andrade, *An Approach to Modern Physics*, New York: Doubleday Anchor Books, 1957, p. 255.
9. George Wald, in *Synthesis of Science and Religion*, Mumbai: The Bhaktivedanta Institute, June 1988, p. 19.
10. D.S. Kothari, *Some Thoughts on Science and Religion*, The University of Delhi, October 31 and November 3, 1977.
11. R.P. Kaushik, *Towards a New Consciousness*, Delhi: Darshan Yoga Sansthan, 1974, p. 63.
12. Erwin Schrodinger, *What is Life*? and *Mind and Matter*, London: Cambridge University Press, Ed. 1967, p. 128.
13. Ilya Prigogine and Isabelle Stengers, *Order Out of Chaos*, Bantam Books, 1984, p. 5.
14. Erwin Schrodinger, *op. cit.*, pp. 137, 146.
15. Ilya Prigogine, *op. cit.*, p. 218.
16. Geoffrey Chew, in Gary Zukav, *op. cit.*, p. 331.
17. Max Planck, *The Philosophy of Physics*, New York: Norton, 1936, p. 83.
18. *Report to the House of Representatives*, USA: Committee on Science and Technology, June 1981, in R.G. Jahn, February 1982, *Proceedings of the IEEE*, Vol. 70, No. 2, p. 165; in Sampooran Singh, et. al., *The Masterplan Paradigm for Human Survival and Excellence*, Chandigarh: Faith Publishers, February 1998, Preface, p. xiii.
19. Fritjof Capra, *The Tao of Physics*, Bantam Books (Second Edition), 1988, p. 277.
20. Fred Alan Wolf, *Taking the Quantum Leap*, Harper & Row, 1981, pp. 2, 21, 85.
21. Vimala Thakar, *Through Silence to Meditation*, Ahmedabad: Vimal Prakashan Trust, 1998, pp. 22-27; and Silence, Holand Bookfund, Vimala Thakar, 1989, p. 6.
22. As an analogy, we may compare it to a "Particle Accelerator", where if a particle comes within the field of higher quantum energy potential, it gets annihilated.
23. Deepak Chopra, *How to Know God*, London: Rider, 2000, pp. 270-71.
24. Ervin Laszlo, *Emerging Concepts of the Cosmos*, *The Scientific and Medical Network Newsletter No. 49*, August 1992, p. 28; in *The Pilgrimage to the Third Millennium*, Chandigarh: Faith Publishers, January 1995, p. 31.
25. Alan W. Watts, *The Ways of Zen*, New York: Vintage Books, 1957, p. 10.
26. David Bohm, in Gary Zukav, *The Dancing Wu Li Masters*, Fontana/Collins, 1982, p. 327.
27. Sampooran Singh, *A Review on Symbiosis of Science and Spirituality*, Dagshai Cantt, The Atam Science Trust, October 7, 2002, pp. 5-6.
28. Jack Sarfatti, in Gary Zukav, *op. cit.*, pp. 310-311.
29. J. Krishnamurti, *The First and Last Freedom*, London: Victor Gollancz Ltd., 1977, pp. 287, 286.
30. Sampooran Singh, *The Human Paradox*, Chandigarh: Faith Publishers, 1988, pp. 12-13, 11, 19.

31. Sampooran Singh, Dr (Mrs) Kanwaljit Kaur and Paramjit Singh, *op. cit.*, p. 35.
32. D.S. Kothari, *The Perception of Truth in Science and Philosophy*, in *The Odyssey of Science, Culture and Consciousness*, New Delhi: Abhinav Publications, 1990, pp. 53, 49.
33. Sampooran Singh, *A Review on Symbiosis of Science and Spirituality*, *op. cit.*, pp. 16, 17.
34. *ibid.*, p. 19.
35. J. Krishnamurti, *Krishnamurti on Education*, Orient Longman, KFI, 1974, p. 90.
36. Erwin Schrodinger, *op. cit.*, pp. 94, 95.
37. Sampooran Singh, *Nature's Evolutionary Impulse*, Vedanta: Ramakrishna Vedanta Center, U.K., November-December 2003, No. 314, pp. 260-65 and January-February 2004, No. 315, pp. 13-19.
38. Sampooran Singh, Dr (Mrs) Kanwaljit Kaur and Paramjit Singh, *Spirituality for the Scientific Age* and *Towards a Culture of Harmony and Peace*, unpublished.

11
Science and "Mind-Brain-Body Interactions"

Abstract

This article explores the "mind-brain-body interactions" (energy exchange) in the light of recent discoveries in science. It discusses the symbolic-dualistic frame (time) of mind spectrum, and the non-dual frame (timelessness) of Mind Continuum. It investigates the holistic transformation (mutation) in man, and the purpose of nature in man. It is basically man's dialogue with nature in the non-dual frame, or from within nature.

Physics talks of subject-object dualism, or cause and effect, or chance, or probability, these are valid only in symbolic-dualistic frame of reference. Nature invariably expresses itself in the non-dual frame rejecting chance, probability, cause and effect, error, and so on. Life has a set purpose in man and in universe, a glorious pattern and a rigid quantum logic.

Nature expresses itself as understanding and learning, and these flow to intuition, spontaneity and right action. Nature responds to arts of observation (seeing) and listening (hearing), which are non-subjective and non-authoritarian. Nature deals with the wholeness of life, so pain, misery, conflict are transcended. Nature expresses itself as order, harmony, non-violence and peace.

The symbolic-dualistic frame of reference is a closed system – closed to the Absolute, Life Field, and open to the environment, matter. The non-dual frame of reference is an open system – open to the Absolute and closed to the environments or matter.

In a non-dual frame, there is a direct input of the Unmanifest in the brain cells without mediation of the sensual apparatus. It is a state of no-mind, it is a state of pure perception where the past does not interfere, so there is no interaction with the environments or matter.

All wisdom or intelligence is structured in the Unmanifest, the Life Field. All discoveries in science, sculpture, music, paintings are manifestations of the Universal Consciousness.

We have to understand and learn to make a quantum jump from symbolic-dualistic frame to non-dual and non-conceptual frame of reference of Mind

Continuum. We have to explore the conditioned psyche through observation and listening. It is an inner voyage of self-discovery from self-observation through self-knowledge and self-awareness to Silence, to Life Field. It is the Unmanifest, or Life Field, or Universal Consciousness that can guarantee human survival and peace.

Modern physics has confirmed that all the concepts we use to describe nature are limited, that they are not features of reality, but creations of the mind; parts of the map, not of the territory. The time-space matrices in quick succession are properties of our conceptual map of reality, or mind-brain-body system, rather than of reality.[1] There are no frontiers to the human mind except what we erect ourselves. Man's lack of knowledge about himself is appalling.

There is no explanation, no clue whatsoever, as to *how* the interaction between mind and matter occurs. We define interaction as energy exchange. Paradoxically, nothing is more common place than the *reality* of this interaction. My writing this sentence, and your reading it, is itself an expression of mind-brain-body interaction, and yet nothing is more intractable, nothing a greater mystery. Apart from this mystery, how does this interaction at the level of symbolic-dualistic frame of reference expresses itself as feelings, motives, desires, violence, fear and so on.

The only real problem of philosophic thought, or religious thought, or scientific thought is mind-brain-body problem. Karl Popper wrote, "Western philosophy consists mainly of world pictures which are variations of the theme of body-mind dualism and the problems of method connected with them".[2] Max Born has said convincingly "that today theoretical physics is actual philosophy".[3]

The brain is a seat of intense chemical and electrical activity, of myriad of shifting patterns of electric currents. The brain or atom belongs to the external, objective world (public) of time-space and matter-energy. Self or mind belongs to the internal subjective world (private). As the brain is *within* natural science and the mind *outside* natural science, the problem of mind-brain interaction becomes the riddle of riddles – the greatest mystery of all. What we call symbolic-dualistic mind (with a lower case 'm') is a conditioned energy, moving through the body, monitored by brain, operating through the senses. What we call Mind (with upper case 'M') is the non-dual and non-conceptual frame of reference of Mind Continuum. Candance Pert wrote, "White blood cells are bits of brain floating around in the body".[4] This led to the concept that mind-brain-body is one system. It can be investigated objectively. Mind is subjective. The brain-body can be considered as an object to the mind and studied objectively. There is no problem more mysterious than the mind-brain-body interaction. Modern science has not made it less intractable. On the other hand it has added a new urgency and also a new poignancy.

Man is an amphibian who lives simultaneously in two worlds – (a) the *known*, the symbolic-dualistic frame of reference of mind spectrum; the time-space-causation matrices; it expresses itself as pain, misery and travails; it leads to self-destruction or self-annihilation; and (b) the *Unknown*, the non-dual frame of reference of Mind Continuum; the Universal Consciousness or Life Field expresses itself in the brain.

The field of classical or conventional science has dealt with the world of outer experience and knowledge (information), and has neglected the whole world of human inner experience, where all our sense of values, meanings, goals, and so on come from. The awareness of profound harmony, order, non-violence and peace are totally lacking.

This article presents mind-brain-body interactions in the light of physics. It discusses the symbolic-dualistic frame of reference in the mind spectrum and the non-dual frame of reference in the Mind Continuum. This highlights the historic transformation *in* man and the purpose of nature *in* man. It is man's experimental dialogue with nature.

This article also discusses a new vision of *Reality* that will require a fundamental change in our thoughts, perceptions and values. We have to initiate an enquiry, or exploration, or experimental interrogation of "restructured science and restructured education" and "mind-brain-body interactions". This leads to "understanding of life directly", learning and living in the light of one's own understanding. This flows to radical transformation or mutation of the conditioned psyche. It results in I-consciousness and thought-structures going to abeyance. This bestows order, non-violence and peace.

Mind-Brain-Body Interactions (Symbolic-Dualistic Mind)

(a) Swami Vivekananda, in his historic address in the Chicago World Parliament of Religions in 1893, said:

> Is man a tiny boat in a tempest, raised one moment on the foamy crest of a billow and dashed down into a yawning chasm the next, a powerless, helpless wreck in an ever-raging, ever-rushing, uncompromising current of cause and effect? .... The heart sinks at the idea, yet this is the law of Nature.

A Vedic sage stood up before the world and proclaimed the "great truth"; and Swami Vivekananda put it in his words:

> I have found the Ancient One (Universal Consciousness, Life Field), who is beyond all darkness, all delusions; knowing Him alone, you shall be saved from death over again.[5]

He added, "Every soul is potentially divine. The goal is to manifest this divinity within.... This is the whole of religion. Doctrines, or dogmas, or rituals, or books, or temples, or forms, are but secondary details".[6]

(b) Ruth Nanda Ashen, in prophetic words, wrote:

> The new order, including those of modern physics, is a unifying order, but it is not causality; it is purpose, and not the purpose of the universe

> and of man, but the purpose *in* the universe and *in* man.... we need a calculus of potentiality rather than of probability, a dialectic of polarity, one in which unity and diversity are redefined as simultaneous and necessary poles of the same essence.... Our response must be new.[7]

Man is not a package of pathology, but the package of potentials of consciousness. The quantum jump implies higher quantum energy potential of consciousness so that our response is from higher potential of consciousness.

(c) We Live in a World of Shadows: Man is confronted with an exceptional crisis because it is in the field of ideation, conceptualization, matter-constructs, sensate values, and so on. The crisis has arisen that at least during the last 2,500 years, or more, man thought that he is living in the actual world (pure perception or conscious of the actual territory), but in fact, he has been living in a map world or symbolic world, or mental construction, or world of shadowgraphs (distorted perception). Arthur Eddington, in a most spectacular passage, summarized thus, "*The frank realization that physical science is concerned with a world of shadows is one of the most significant of recent advances*".[8] In more precise words, man jettisoned the Reality, the Truth, and started living in the symbols, the concepts, the false. And this living constructed pain, misery, conflict, and so on.

(d) Psychosocial Evolution: Julian Huxley wrote, "Man's evolution is not biological but psychosocial, it operates by the mechanism of cultural tradition". The trend of evolution at the human level is quality and fulfilment. He added:

> Once greater fulfilment is recognised as man's ultimate or dominant aim, we shall need a science of human possibilities to help guide the long course of psychosocial evolution that lies ahead.[9]

(e) Modern physics explores an object, it studies constructs of the symbolic-dualistic frame of reference, it studies the known and suggests a modified known. Physics talks of local action, or cause and effect, or randomness, or chance, or probability, or error, but this is true of symbolic-dualistic frame of reference. In the non-dual frame, there is no room for cause and effect, no probability, no error in the cosmic order. Life has a set purpose *in* man and *in* universe, a glorious pattern and a rigid quantum logic. Life's laws are immutable and eternal.

(f) The symbolic-dualistic frame is called 'successive consciousness', it is at the lowest frequencies of consciousness. The non-dual frame is called 'simultaneous Consciousness', it is at the highest frequency of consciousness. It is the awareness or attentivity that is responsible for input of energy which results in a quantum jump from successive consciousness to simultaneous consciousness. It is the Unmanifest that expresses itself at different quantum energy potentials. We have to educate ourselves so that man lives in a higher level of awareness. Higher level of consciousness leads to radical transformation or mutation of the psyche. The physics of the whole process is a deep mystery.

(g) Subject-Object Dualism: The modern science tells us that an observer cannot observe without altering what he sees. The very crux of twentieth-century physics is the recognition that the observer cannot be excluded from his observation; he is an intrinsic participant between them.

(h) Wave-Particle Complementarity: We can demonstrate that light is particle-like with the photo-electric effect, that it is wave-like with the double-slit experiment, and that it is both particle-like and wave-like with Compton scattering. This is basically due to our *interaction* with it. Complementarity is the concept developed by Niels Bohr to explain the wave-particle duality of light. Einstein, using the photoelectric effect, 'proved' that light is particle-like, and Young, using the phenomena of interference, 'proved' that light is wave-like. Both of these complementary aspects of light (particle and wave) are necessary to understand the nature of light.... In short, what we experience is not external reality, but our interaction with it. Wave-like behaviour and particle-like behaviour are properties of our interactions.

The mind has an in-built mechanism to make a quantum jump from the 'dimension of measure' to the 'dimension of value' and both are complementary to each other. The 'dimension of measure' belong to the symbolic-dualistic frame of reference, and the 'dimension of value' to the non-dual frame of reference of Mind Continuum.

The wave-aspect is an expression at a higher quantum energy potential (or higher frequency) than that of particle-aspect. This deep understanding would radically alter the level of expression of consciousness of the observer, reorientating it towards a perception of the 'unbroken wholeness' of which everything is a form.[10]

(i) Causal Determinism: Werner Heisenberg stated the celebrated Principle of Uncertainty or Principle of Indeterminancy in simple language as "it is impossible to give a well-defined meaning to both the position and the motion of a particle (velocity) at the same time". This has put an end to causal determinism in physics – and thereby in philosophy. A deeper exploration of mind-brain-body system suggests that causal determinism (Law of Cause and Effect) is only applicable to symbolic-dualistic frame of reference of mind spectrum, but this law is transcended when the mind makes a quantum jump to non-dual frame of reference of Mind Continuum. As all natural processes are in the realm of non-duality, so these are beyond the causal determinism.

(j) The Second Law of Thermodynamics: It states that disorder can never spontaneously give rise to order. More precisely, this law legislates nature's trading account in a quantity called entropy, which is roughly, the degree of disorder in a physical system.[11] Entropy governs the relative field of mind-brain-body system. Every activity of the mind-brain-body system gives rise to higher entropy, which implies that the degree of disorder rises.

The increase of entropy expresses the increase of probability, of disorder. The rise in entropy is leading the mind-brain-body system to self-destruction or self-annihilation.

There is no disorder in the absolute rest frame, so entropy does not enter there. The relative frame is an infinitesimal part of the absolute field; it is at the lowest frequencies of consciousness.

(k) Discontinuous World: Human activity, in the symbolic-dualistic frame of reference, introduces discontinuities in whatever we are observing. Niels Bohr believed that discontinuity was a fundamental truth; or in other words, it was fundamentally a discontinuous and quantum jumping world.[12] The Bohr electron follows a discontinuous path of quantum jumps. These continuities are fundamental to the new physics of the twentieth century. These discontinuities are basically due to energy exchange between conditioned consciousness and environments (matter).

We see discontinuities in the non-dual frame of reference of Mind Continuum. There is flow of energy (one way flow) from the Unmanifest to the immanent quantum energy potential of the system. The discontinuities in the non-dual system are basically due to the expression of the Unmanifest at different quantum energy potentials.

(l) Quantum Jump/Leap: It implies: (a) the tiny but explosive jump that a particle of matter undergoes in moving from one level of consciousness to another level of consciousness; and (b) it means taking a risk, going off into an unchartered territory with no guide.

When an atom makes a transition, or a jump from one stationary state to another, the actual process cannot be visualised or even imagined. There are no intermediate stages. The entire process of transition from one state to another is a non-visualisable, unanalysable unity. It is indivisible. It is a quantum jump.[13]

There is an analogous situation for thought processes. Quantum jumps for thought appear to be as much a necessity for a conscious self as those are for the atom.

The quantum jump in an atom has been experimentally observed. A trapped ion blinks on and off and each blink is a quantum jump. It is a striking illustration that things occur discontinuously in nature. The ultimate unit doing the dance by itself. The atom is in one state and moves to another, these are indivisible energy levels. An atom will jump to one or the other of the upper levels, absorbing one photon. Then a short time later, it spontaneously emits a photon and returns to the ground state.[14]

The Non-Dual Frame of Reference of Mind Continuum

During the last five decades, Nature, out of its benediction, has been releasing its secrets, bit by bit, in the minds of many eminent scientists, who engaged themselves in persistent and rigorous "experimental interrogation" of the conditioned psyche. This has led to many discoveries or dominant understanding of man, and are as under:

(a) Many eminent scientists – including Einstein, Schrodinger, Heisenberg, Eddington, and so on – have stumbled upon another mode of knowing. They called it intimate, or non-dual, or direct, or intuitive mode of knowing. They led to "understanding of life directly instead of in the abstract linear terms of representational thinking". The intuitive perception is in the non-dual mode of mind, or direct expression of the true subjective dimension of life. David Bohm discussed:

> The ultimate perception does not originate in the brain or any material structure, although a material structure is necessary to manifest it. The subtle mechanism of knowing the truth does not originate in the brain.[15]

Intuition is a direct input of the Unmanifest in the brain cells of the non-dual frame of reference of Mind Continuum without mediation of the sensual apparatus; or in other words, there is no interaction between the Unmanifest and the mind-brain-body system.

(b) Man simultaneously lives in two dimensions – objective and subjective, or symbolic-dualistic mode and non-dual realm, or the *outer* and the *inner*. The outer and the inner are not two separate movements, but one dynamic whole. It deals with the Wholeness of Life. This leads us to: (i) *Restructuring of Science:* It implies that science must take in its fold both the objective (outer) and the subjective (inner) dimensions; (ii) *Restructuring of Education:* it investigates both the sensual and the trans-sensual dimensions of life; (iii) *Awareness of Human Values:* the perennial values flow from the trans-sensual dimension of life; and (iv) *Living in the Light of One's Own Understanding.* Awareness happens when you observe yourself without any identification, without any comparison, without any condemnation.

(c) There is an urgency for a quantum jump from symbolic-dualistic frame to non-dual frame, from conflict and confrontation to harmony and cooperation with nature, from disorder to order, from violence to non-violence and so on. The symbolic-dualistic frame is at the lowest frequencies of consciousness, while the non-dual frame is at the highest frequency of Consciousness. It implies that for a quantum jump to happen the mind must remain attentive. As long as one is attentive, he revels in non-dual realm; and the moment he becomes inattentive, he falls to symbolic-dualistic frame of reference.

(d) Understanding, learning and action are a cyclic process, which results in a spiral going upwards in a vertical dimension; it is a growth in the vertical dimension. The higher the understanding, the higher the expression of the Consciousness in the human mind-brain-body system.

(e) Life consists of two dimensions – (i) the *closed* system, often called conditioned psyche, the relative field; it is closed to the Absolute and open to the environments; and (ii) an *open* system, often called absolute rest frame; it is open to the Absolute, the Life Field and closed to the environments. The closed system leads to disorder and higher form of complexity of thought-structures, or cultural belief structure. The open system leads to higher and higher expression of Consciousness. The open system is a pointer to order, harmony, non-violence and peace. Living in a closed system is antagonistic to the laws of life.

(f) The man-made world of civilization and culture is a construct of the symbolic-dualistic frame of reference of mind spectrum. It is a tiny bit of the wholeness of life. As an analogy we say that the man-made construct is equivalent to a drop of water, while the Wholeness of Life is compared to a fathomless and unbounded ocean.

(g) All intelligence, wisdom, including knowledge, is structured in Consciousness. All discoveries flow from the Unmanifest. A *Mind* with a scientific temper is tuned to perceive the Wholeness of Life, it bestows a perception of intuition, spontaneity, harmony, human values, non-violence, truth, love and compassion; so it is in communion with the Universal Consciousness, the Life Field.

(h) Our experimental dialogue with nature will be successful only if it is carried on from within nature.[16] It implies that dialogue can only happen from a non-dual and non-conceptual frame of reference of the Mind Continuum. Such a dialogue bestows an experimental verification in the mind of the experimenter. The purpose of nature *in* man is to perceive the highest quantum energy potential of Consciousness and to live in order, non-violence and peace.

A Vision on Transformation in Man

The radical transformation or mutation in man happens when there is "man's experimental dialogue (interrogation) with nature", and this is possible through the arts of observation and listening. It bestows perception of the *fact*, the *true*, the *present*, the *non-dual*, the *what is*! Only the truth can liberate the mind from its own ideation; to see the truth, the mind must realize the fact that as long as it is agitated, it can have no understanding. The key to transformation is understanding and learning about the wholeness of life.

(a) Herman Weyl, in a most spectacular passage, wrote:

> Scientists would be wrong to ignore the fact that theoretical construction is not the only approach to the phenomenon of life; another way, that of understanding from within (interpretation), is open to us.... This inner awareness of myself is the basis for the understanding of my fellowmen.[17]

Ilya Prigogine expressed that he completely agrees with Herman Weyl. The subjective world expresses itself in a non-dual frame of reference of Mind Continuum as intuition (insight), spontaneity and right action. It also expresses itself as understanding of life directly and learning. It manifests as spontaenous activity and irreversibility. The objective and the subjective world are now closing together, instead of rupture and opposition, there is a growing coherence of our wisdom or intelligence between man and nature, and between man and man.

(b) Voltaire wrote about human destiny, ".... everything is governed by immutable laws.... everything is prearranged.... everything is a necessary effect .... we are both equally foolish, both toys in the hands of destiny".[18] All wisdom or intelligence is structured in pure consciousness. If the subject is restructured science (the wholeness of life), all dichotomy between experience and perception, reason and intuition will vanish and we perceive the Oneness of Life.

(c) Maurice H. Wilkins, Nobel Prize in Medicine, 1962, wrote:

> The world view engendered by science is in many ways a very undesirable one because it focuses attention on the material aspects of life and on technical solutions to human problems. I think this is very bad for human attitudes.... It is also a question of how science is developed through educational institutions and through institutionalized science. I agree with Einstein that the sort of scientific education we have now has produced a narrow-minded way of thinking amongst scientists, so that they give no proper attention to the moral and psychological dimensions.[19]

We will have to develop a new science through educational institutions which covers the wholeness of life. We need to transform the whole structure of science and education because we have touched the quantum level, where Unmanifest expresses itself as perfect order, non-violence and peace. And in restructuring of science and education and understanding observation and listening, our survival is embedded.

(d) Bohm wrote, "The new physics requires a new "instrument of thought" (a non-dual thought).... which would radically alter the consciousness of the observer, reorientating it toward a perception of the "unbroken wholeness" of which everything is a form.[20] Eugene Wigner wrote:

> I think it is necessary to develop a new tool to understand consciousness just as we had to develop quantum theory to understand chemistry. It was necessary to have a new physics book for electricity and magnetism, and similarly I think new concepts and a new basis will be necessary for physics to understand a phenomenon such as life – that is, if we ever understand it at all or if we ever incorporate it into physics.[21]

Sudarshan stated, "Creativity is a case of the suspension or transcendence of causality and not a violation of it. Poetry, religion and art deal with creativity".[22] Wigner wrote, "We have two kinds of reality, namely the reality of the external world, with which physics and its laws are concerned, and the reality of consciousness. Science, driven by the spirit of enquiry and guided by reason, should concern itself with both realities".[23] Werner Heisenberg wrote:

> A rationalistic play with words and concepts is of little assistance here (spiritual truths); the most important preconditions are honesty and directness. But since ethics is the basis for the communal life of men, and ethics can only be derived from that fundamental human attitude which I have called the spiritual pattern of the community.[24]

Edwin Hubble wrote:

> Science is necessarily restricted to one aspect of the universe – the objective world of phenomena. It deals with probable knowledge only, its methods are empirical, its philosophy is pragmatic.... The world of pure values, that world which science cannot enter, has no concern whatsoever with probable knowledge. There finality – eternal, ultimate truth – is earnestly sought.[25]

Max Planck wrote:

> It is a fact that there is a point, one single point in the immeasurable expanse of mind and matter, where science and therefore every causal method of research is inapplicable, not only on practical grounds, but also on logical grounds, and will always remain inapplicable. This is the point of [our] individual awareness.[26]

The above statements are basically man's dialogue with nature in the non-dual frame of reference, or from within nature.

(e) In a recent (1996) scientific conference at Moscow, the Russian scientists proposed, "A Life Field, beyond our physical concepts of time, space and the known electromagnetic and gravitational fields". In another conference at Cambridge, it was proposed, "A new scientific approach would include subjective experience as valid information and address all forms of human experience – including intuition, spirituality and mystical encounters".[27]

(f) The subject-object interaction (energy transfer via photons) determine the properties and results in the emergence of causation. Werner Heisenberg wrote:

> Natural science does not simply describe and explain nature: it is part of the interplay between nature and ourselves.[28] ....What we observe is not nature itself, but nature exposed to our method of questioning.[29] ....When we speak of a picture of nature provided by contemporary exact science, we do not actually mean any longer a picture of nature, but rather a picture of our relation to nature.[30]

(g) The rational knowledge is sterile to transform our thinking. Einstein says:

> The whole of science is nothing more than refinement of every day thinking.
>
> Science can denature plutonium but science cannot denature evil in the heart of man.... The splitting of the atom has changed everything except our mode of thinking and thus we drift towards unparalleled catastrophe.[31] ....Related truths can be achieved by process of reasoning.

Einstein refers to intuitive faculty of mind and says:

> Physics is essentially an intuitive and concrete science. Mathematics is only a means for expressing the laws that govern phenomena.[32]

He added, "I cannot conceive a genuine scientist without profound faith", and faith comes from intuition. Scientific truth is time dependent and logic dependent. A truth is timeless, independent of symbolic-dualistic frame of reference of mind spectrum. All newness, all originality, all creativeness of science and arts lie in the intuitive faculty of the mind.

(h) Einstein stated, "Our new idea is simple: to build a physics valid for all coordinate systems", or frames of reference of the mind spectrum. He wrote, "The chief motivating force behind all technical effort ought to be supreme concern for the destiny of man". It seems so urgent and important that we understand the wholeness of human life, or the systems view of human life, or the actualization of the highest integrative potential of man, or the physics of consciousness.

(i) In 1932 after a World Disarmament Conference in Geneva, Albert Einstein, in an unusual bout of anger, exclaimed:

> .... they (politicians and statesmen) have cheated us; they have fooled us. Hundreds of millions of people in Europe and in America; billions of people have been and are being cheated, traded and tricked out of their lives, health and well-being.[33]

These remarks are equally valid today. In principle, man is deceiving himself. He has become a puppet, a mechanical golem. He has lost sensitivity, he has lowered his level of perception, he has started living in a synthetic world which is a construct of his mind, and man's world is totally divorced from Reality.

Exploration of the Known and the Unknown Dimensions of Man

Table 1: Comparative Exploration of the Known and the Unknown Dimensions of Consciousness

The Known Dimension of Consciousness Symbolic-Dualistic Frame	*The Unknown Dimension of Consciousness Non-dual and Non-conceptual Frame*
1. Rational knowledge is a system of abstract concepts and symbols.	Intuitive knowledge gives holistic truths.
2. Successive Consciousness (characterized by the linear and sequential structures).	Simultaneous Consciousness (the expression is simultaneous).
3. Studies experience and knowledge (information) partially).	Explores perception in its totality.
4. Repetitive and mechanical processes, no newness or originality.	Every action is new and original.
5. Decreases sensitivity and brings dullness.	Enhances perceptive sensitivity and makes mind alert.
6. No understanding and no learning.	Understanding and learning the wholeness of life.
7. Recorded as memory	No recording. Every perception is new.
8. Logical reasoning (Derivative)	Intuition (Fundamental).
9. Psychological time	Timelessness.
10. Self-knowledge (dynamics of conditioned psyche)	Self-awareness of the wholeness.
11. Science without values, Education without values.	Restructured science and restructured education bestows perennial values.
12. Divisive in nature and adds to inward suffering. Constructs pain, misery and conflict.	A benediction which heals inward suffering.
13. Leads to self-destructive potential of mind, to self-annihilation.	Guides to non-violence and peace.
14. Divorces spirituality	Leads to self-integrative potential. Path to survival and excellence.
15. Mind cannot grow inwardly, or flower psychologically.	Symbiosis of science, religion and spirituality. Mind grows in a vertical dimension.
16. No capacity to lead to basic discoveries.	All intelligence or Wisdom is structured in the Unmanifest or Universal Consciousness.
17. Objective science gives relationships between symbols, concepts, images and so on. It cannot give us nature of reality.	The restructured science and restructured education bestows a profound understanding of reality and relationship between man and nature and between man and man.
18. The symbolic-dualistic frame is *closed* to the Absolute and *open* to the environments. Objective science expresses itself as cause and effect, chance, probability, and so on. Every activity is antagonistic and in opposition to the laws of nature, so it adds to conditionings.	The non-dual frame is *open* to the Absolute and *closed* to the environment, so it rejects cause and effect, chance and probability. It leads to truth, love and compassion, order and peace.

Mankind urgently needs a quantum jump from symbolic-dualistic frame to non-dual and non-conceptual frame; or from lowest frequencies to the highest frequency of Consciousness; or a symbiosis of "restructured science, and spirituality". One has to understand and live in Now and not in the past. It is the feeling of Now, and not knowledge of Now.

Experimental Dialogue with Nature

(a) Restructuring of Science: Many eminent scientists – Erwin Schrodinger, Karl Pribram, Albert Einstein, Hideki Yukawa, Niels Bohr, E.P. Wigner, Roger Sperry, J.C. Ecceles, Pere Teilhard de Chardin, and so on – state that the objective (outer) and the subjective (inner) dimensions of man are a single dynamic flow, a unitary movement.[34] It is imperative to restructure science so that it takes in its ambient the objective and the subjective dimensions of man. Similarly education must treat the sensual and the trans-sensual dimensions. The trans-sensual dimension of life flows to human values. So the restructured 'science-education' treats the wholeness of life, and hence it is synergistic (higher quantum energy potential of consciousness) to thought structures with their volitions. This is often called the scientific mode or temper of mind, or true subject.

(b) Observation for Transcendence of 'I-consciousness: The arts of observation (seeing) and listening (hearing) are basically non-subjective and non-authoritarian approach. These depend on the understanding of the wholeness of life. It implies a quantum jump from symbolic dualistic knowledge to non-dual perception, from sensual world to understanding of life directly; from knowledge (information) of parts to understanding the wholeness of life.

The boundary, or the wall, between symbolic-dualistic frame of mind spectrum, and non-dual and non-conceptual frame of reference of Mind Continuum is called "the event horizon"[35], or "the frequency barrier". [As an analogy, the parallels in the field of matter-energy are "sound barrier" in jet travel, and "gravitational barrier" in space travel]. To cross the "frequency barrier" between symbolic-dualistic frame to non-dual frame, we require the thought-structure and I-consciousness to go to abeyance, and then the energy is supplied by the Unmanifest, so the non-dual frame is synergistic (higher quantum energy potential, or higher frequency) to the symbolic-dualistic frame. Crossing the 'frequency barrier' requires an inner voyage from time to timelessness. It bestows higher perceptive sensitivity to the mind-brain-body system. The frequency barrier divides: (i) the successive consciousness where anything or event remains within the speed of light, and (ii) simultaneous Consciousness where anything is faster than the speed of light. The inner voyage for this quantum jump to happen is essential for human survival and excellence.

(c) The restructured science-education deals with the wholeness of life. The non-variable attentivity, or reaction-free attentivity, or objectless-attention takes over as the true-subject. The subject sees that the movement of thought-structures is independent of my volition, it is independent of the movement of the 'ego', the "me", the "I-consciousness". It implies that the concept of the ego is independent of the momentum of thought. As the concept of the ego is at a lower quantum energy potential than the momentum of thought, so it is knocked down, or the I-consciousness or ego goes to abeyance. The mental movement has a momentum independent of the ego.[36]

The relationship that the thought has an existence independent of "me" is established, and the vanity that it is my thought coupled with volitions ends and its place is taken by humility. This suggests that there is nothing like an individual mind, there is nothing like an individual I-consciousness. Thought is the product of collective human effort through centuries. This gives a sense of freedom.

(d) Transcendence of Thought-Structures: If the observation is continued, which implies to see the momentum of thought in utter freedom. The photons meet an impregnable wall – called "event horizon"[37] or "frequency barrier". If the photon (thought) comes within the field of restructured science-education, it is bombarded by the high energy potential and photons annihilate themselves, or in other words, the object goes to abeyance. With the abeyance of the object, time-space-causation and psychological time vanish. There is cessation of image-construction, all dualism ends. And all psychological imbalances go to abeyance. We are left with the perception of the *fact* in a non-dual and non-conceptual frame of reference of Mind Continuum. Basically it is the symbolic-dualistic state that makes a quantum jump to the non-dual state.

(e) Observation of the Fact: Between any two consecutive facts, there is the *Unmanifest mathematical space* – may be called *Hilbert space*[38] or *Unmanifest emptiness*. It implies that between any two consecutive facts, there is discontinuity and this continuity is the Unmanifest, so it oscillates between the Unmanifest and the fact [This is analogous to space between two thoughts, and atomic objects].

If we continue observing or listening to the *fact*, without justification or condemnation or motivation, we observe that there is only the fact and nothing else. The fact itself is the operator, not you operating on the fact. If you leave the fact alone, it has no time. The fact is free from fear. The fact grows, which implies that there is release of tremendous energy[39] and the fact moves to higher quantum energy potential. This is expressed as Pure Consciousness → Fact. It helps the fact to make a quantum jump in the vertical dimension of Consciousness. When the fact is left alone, it is operating and you do not have to do a thing about it, it matures and blossoms. Immediacy of perception dissolves the fact, and one revels in Pure Consciousness or Intelligence. Only in Intelligence there is complete security. Intelligence sees the false as the false and ends the false. Mutation happens in the timelessness. To understand Consciousness or an enquiry into the subject is the perception of that immeasurable state.

There is no "object out there", there is only "subject in here". The true-subject expresses itself as the *fact*, the *true*, the *present*, the *non-dual*, the *what is*! If observation is further continued, the subject also vanishes. What is left behind is pure Consciousness bathing the brain cells; this blossoms to intuition (insight), spontaneity and right action; it flowers to understanding and learning; it heals the diseased mind-brain-body system; it resolves all psychological imbalances, ecological problems and other challenges; and it ushers a new era of order, harmony, non-violence and peace.

(f) State of Awareness: When non-dual frame is stabilized over a long chronological time, the Intelligence is regenerated, which ends all human suffering and bestows non-violence and peace. This is often called the state of the emptiness of consciousness, which bestows intimate feel of aliveness. In that aliveness and alertness there is the flavour of awareness (meditation), which is not attention, or attentivity. It is a state of non-communicable Truth. It is a non-subjective and non-objective state. It is an expression of harmony, coherence and order. It is non-cerebral, non-psychological exploration of the Reality.[40]

Meditation is a new approach to life, in which we observe the objective challenge, the subjective reactions and the causes of the subjective reactions simultaneously, without identifying oneself to any of them. This results in freedom. Then the moment of reaction will not carry me over with it, but I will be ahead of the reactions. By the time the reactions reach the true subject, there is awareness of the objective challenge, then the reactions get dissipated. I will not be a victim of my reactions, but I see them as I see the objective challenge. Freedom from the momentum of reaction in relationship is real freedom. Natural response is a response out of the understanding of the wholeness of the challenge, reactions and the roots of reactions. Meditation is the all inclusive attention while moving in life.[41]

(g) The Oneness of Universe: Ervin Laszlo wrote that there are "only ripples upon ripples, waves upon waves, interpenetrating and propagating in a seamless sea".[42] This indicates a oneness or interconnectedness of the universe. The superluminal (faster-than-light) channel without signals indicates interconnectiveness of Minds, or to be more precise, there is only one Mind which governs the whole cosmos and beyond. The one Mind is the Life Field, the Universal Consciousness.

The Art of Living

(a) J. Krishnamurti, in answer to Fritjof Capra's question, stated, "First you are a human being; then you are a scientist. As a human being, you have to go beyond thinking to solve the human dilemma. As a scientist, you can use your analytic mind, which is wonderful and marvelous, but in a more general sense, as a human being, you need to make contact with the metaphysical state of consciousness".[43] The classical science, in the process of showering us with material benefits, has all but erased transcendence from our reality map. Krishnamurti's advice to every scientist of the world that we our reality map. Krishnamurti's advice to every scientist of the world that we must regenerate this reality map and then do science. Intuition (insight) is fundamental and reason is a derivative.

(b) Albert Einstein is reported to have told poet Saint-John Perse that "poets and scientists are akin in having sudden illuminations, and that discovery and inventions are logical, intellectual, and analytical only *after* imagination makes its leap forward".[44] The sudden illumination is intuition or insight. D.S. Kothari also discussed intuition that, "The new theorems, unless trivial, are of course first discovered and then proved.... It is not a step-by-step process, but a sort of quantum jump .... A rational account of a discovery, the proof, follows and not precedes the discovery". Fundamental discoveries which go deep into the nature of things are the result of uncommon intuition and faith, not cold logic.[45] We urgently need a quantum jump from reason to intuition. This will be rewarding to the scientists and also to mankind.

(c) Maurice H. Wilkins, Nobel Prize in Medicine, 1962, wrote, "The open-minded inquiry of the scientist is not something peculiar to science itself but is a characteristic of the good way of living for human beings in general. It is equivalent to the religious concept of love, where you are always giving attention to new developments and new possibilities. To say that the essence of science is that you are always inquiring and open-minded is to say that you are in fact living a virtuous life".[46] B.D. Josephson, Nobel Prize in Physics, 1973, wrote, "Our nervous systems are designed to allow us to interact not only the material level of existence but also with the spiritual levels.... This would be a part of the scientific training of the future; scientists working in this field would try to develop their nervous systems so that they could explore these regions".[47]

(d) Karan Singh stated, "We stand at a juncture wherein we are in supreme peril – not a peril from another species, not a peril from outer space, but a peril from ourselves. From deep within the human psyche, there has developed this terrible power that threatens not only our generation but generations yet to be born, and not only our race, but all species of life on this planet".[48]

(e) Ervin Laszlo wrote, "Thanks to the convergence of science with intuitive insight, we now possess the critical mass of essential concepts and ideas to fulfil the dream. We can be more than conscious parts in an evolving totality; we can be parts that truly fulfil their potentials and become conscious of the totality – and come to live in harmony with it".[49]

Our Survival is at Stake?

(a) Conditioning of the Psyche: During the last three thousand years, or more, many eminent scientists, philosophers-mystics, educationists, politicians have echoed that our civilisation based on the institutionalized religions, objective science and educational structures have led the mankind to "world of shadows" or "to an abyss" – called conditionings, thought-structures, I-consciousness – and this is leading to self-destruction or self-annihilation. Erwin Schrodinger, Nobel laureate in Physics, 1931, wrote that for more than two thousand five hundred years, "We were not aware of it (the subjective dimension); we thought that we are dealing with the world itself".[50] George Wald, Nobel Prize in Medicine, 1967, wrote:

> We have achieved the heady eminence of being able to exterminate ourselves, taking much of the rest of life on earth with us – the first self-extinction of a species of which we know.... The conscious drama will go on elsewhere, ignorant of our demise, caring nothing; but we will have no further part in it.... By now the problem is not transcendence, but survival.[51]

Jacques Monod, Nobel Prize winning biologist, wrote:

> .... the nineteenth century scientism saw leading infallibly upward to an empyrean noon for mankind, whereas what we see opening before us today is an abyss of darkness.[52]

President Vaclav Havel, Czech Republic, stated that today it seems that, "as if something (conditionings) were crumbling, decaying and exhausting itself, while something (spirituality) else, still indistinct, were arising from the rubble". He added, "Our inner self continues to have a life of its own.... Transcendence as the only real alternative to extinction".[53] In another speech, he stated that "all humanity is the experience of transcendence in the broadest of the word.... modern man has lost: his transcendental anchor"[54] If we continue our present way of living, it appears that man will meet the same fate as dinosaurs – who dominated this planet for millions of years – who disappeared from the face of the earth, leaving only fossils.

(b) Resolution of the Human Paradox: Nature, out of its benediction, has bestowed man two faculties: first, *known*, called self-knowing, which is gathering knowledge (information) and experience; and second, *Unknown*, called *inner self-awareness. The human being seems to be born with the faculty of self-awareness and understanding.* That understanding, which is not knowing, has a perceptive sensitivity. So the activity of knowing is not the act of understanding. Understanding and learning come when cerebral activity goes to abeyance voluntarily. Intelligence leads to self-awareness, to perception, to understanding and to action.[55]

We urgently need a quantum jump from psychosocial evolution to spiritual evolution, from intellect to intuition (insight), from experience and knowledge (information) to perception, from fragmentary and dualistic approach to "understanding of life directly". This involves a major conceptual shift of jettisoning rupture, opposition, conflict to growing coherence of our relationship between man and nature and between man and man. This surely involves the expression of a new, higher mode of consciousness that cuts across the barriers of race, religion, nationality and ideology, and knits the entire human race into a unified whole.

The time is now ripe when we realize that when the Unmanifest expresses itself creativity happens. The reality of the external world is at the lowest frequencies, and reality of the Consciousness (inner world) is at the highest frequency, so, in principle, both are one and there is no dichotomy. Further intuition (insight) is fundamental and reason is a derivative. Man has to develop the wisdom to direct the evolutionary process towards his fulfilment rather than towards his destruction. The pursuit of restructured science and restructured education requires "to develop a new tool to understand consciousness" in the words of Eugene Wigner and the new tool is to live in non-dual and non-conceptual frame of reference where there is creativity, intuition, understanding and learning. We must develop physics (and other sciences) which are blended with creativity, intuition and understanding. It is science of the Wholeness of Life, or Oneness of Life. An Action Plan or Paradigm Shift has been discussed in earlier papers.[56] The Plan will blossom to human survival and excellence.

Scientists have discovered something mysteriously rich in energy (at very high frequency, higher frequency means higher energy) in empty space (the Void, the Universal Consciousness). This has confirmed the perennial wisdom of ancient sages/seers. Now we are on the verge of extending this richness into human dimension. The Life Field introduces a quality of the absolute into relativities and confers on perishable things a texture of eternity.

Spirituality

We find ourselves in a universe that breeds life and possesses the very particular properties that make that possible. Life is a multi-dimensional phenomenon and human beings are expressions of that phenomenon are themselves multi-dimensional organisms. Life is self-generated, self-sustained, self-controlled, self-regulated phenomenon. Life seems to be an energy of Intelligence, that has been unfolding, unwinding, uncovering, manifesting itself for billions of years and yet has remained ever virgin and ever fresh. We call Life Eternal, Infinite, Immutable.

Spirituality is the science of life and living. Its dimensions are homogeneous, undifferentiated and non-dimensional, so there is no spontaneous organic inter-relatedness of all that exists. Spirituality is homogeneous wholeness. In spiritual science, the emphasis is upon the act of perception, the quality of the instruments employed for perception, the quality of consciousness behind the perception. In spirituality we proceed from the whole and with the awareness of the whole we observe the parts.

Spirituality is an experimental science and as precise as mathematics. It is verifiable by another explorer; but this *wisdom* or *truth* cannot be passed by one experimenter to another (non-communicable truth). So every body is all alone to undertake this exploration. Man is only a conduit for expression of

Spirituality rejects the man-made social cosmos; it rejects subject and object split, which is called subject-object dualism; it rejects all processes of evaluations and comparisons by reason or intellect; it rejects all grafting of conceptual reality on the perceptual reality; and it rejects all pain, misery and conflict. It imbibes the immutable laws of nature, the true expression of nature, so it goads man to live in pure perception – the *fact*, the *true*, the *present*, the *non-dual*, the *what is*! Spirituality is exploring nature *here* and *now*, where the brain-body is used as a mirror to reflect nature's expression. In principle, we explore nature's expression or manifestation at different hierarchical levels of expressions of Consciousness.

The whole mankind is unknowingly caught in an "inner structural crisis" at symbolic-dualistic frame of reference of mind spectrum. Spirituality has the potential to rectify the inner crisis, and understand and live according to the immutable laws of nature. This is a pointer to persuade the human beings to grow into the maturity of love, friendship and peace.

Concluding Comment

According to one estimate, twenty thousand scientists around the globe are currently engaged in the exploration of consciousness. Never before, so many subjective scientists, in addition to philosophers-mystics, were investigating consciousness.

Karan Singh quoted George Wald and stated, "The great disciplines of science and religion are in fact linked because they are both explorations into truth. Science has tended to go outwards and look into the very structure of matter, into the heart of the endless universe outside, while religion has turned the spotlight within and has sought to plumb the depths of the *atman*, or the inner spirit".[57] There is a growing coherence and oneness between pure religion and restructured scientific education. There is harmony between the two disciplines.

If we continue the present march of objective science and pursuit of our psychological imbalances, we are going to convert the Planet Earth into a charred-out cinder circling the sun into eternity? The path for our survival is as difficult and dangerous as a razor's edge, but there is no other path, no soft option for man today, in the nuclear age. We have to make the crucial transition to a higher consciousness through symbiosis of restructured science and restructured education and spirituality. The answer to this question will decide the future of the human race.

We are today at a historic juncture in the long and tortuous history of the human race on Planet Earth. We stand at a bifurcation where one road is leading to Buddha (Intelligence, Wisdom) and the other road to bomb (self-destruction and self-annihilation). We have to explore the nature and quality of man's mind and life. We have to undertake an inner voyage of self-discovery from self-observation through self-knowledge and self-awareness to Silence, to Life Field. This flows to Intelligence and Wisdom.

It is Universal Consciousness or Life Field that can resolve all human challenges and ushers an era of order, rhythm, non-violence and peace.

India was once the "Light of Asia". She has now to fulfil her historic role of being the Light of the World. India has been called upon *now* and *here* to save Life in the world from extinction. Will she wake up from deep slumber and "*Lead Kindly Light Amid the Encircling Gloom*". We share the privilege and the responsibility.

References

1. Fritjof Capra, *The Tao of Physics*, Bantam Books (Second Edition), 1988, pp. 147, 277.
2. Karl Popper, in D.S. Kothari, *Some Thoughts on Science and Religion*, New Delhi: Shri Raj Krishan Jain Memorial Lecture at the University of Delhi, October 31 & November 3, 1977, pp. 18-19.
3. Max Born, in D.S. Kothari, *ibid.*, p. 15.
4. Candance Pert, in Fritjof Capra, *The Web of Life*, Flamingo, 1997, p. 277.
5. Swami Vivekananda, *Complete Works*, Vol. 1 (eleventh edition), p. 11.
6. *Ibid.*, p. 124.
7. Ruth Nanda Ashen, in Jonas Salk, *Man Unfolding*, New Delhi: Allied Publishers Private Limited, 1972, pp. xi-xix.
8. Arthur Eddington, *The Nature of the Physical World*, Cambridge University Press, 1928, Introduction; in Erwin Schrodinger, *What is Life? & Mind and Matter*, Cambridge University Press, 1967, p. 130.
9. Julian Huxley, *Evolution After Darwin*, Vol. 1, p. 21.
10. Sampooran Singh, *A Renewed Look at the Wave-Particle Complementarity, The Survival of Mankind and Quantum Jump from Violence to Peace*, New Delhi: Books India International, October 1999, pp. 82-87; Bulletin of the Theosophy Science Study Group, India, June-August 1988, Vol. 26, Nos 3 & 4, pp. 35-37.
11. Paul Davies, *Superforce*, New York: Simon and Schuster, 1984, p. 224.
12. Fed Alan Wolf, *Taking the Quantum Leap*, San Francisco: Harper and Row, 1981, pp. 21, 85.
13. D.S. Kothari, *Atom and Self*, in *The Evolution of Consciousness*, New Delhi: National Publishing House, 1983, p. 212.
14. New York Times, *Interplay of Light and Atoms*, in The Invincible, Vol. IV, No. 6, November-October 1987, pp. 6-10.
15. David Bohm, in Gary Zukav, *The Dancing Wu Li Masters*, Fontana/Collins, 1982, p. 327.
16. Ilya Prigogine and Isabelle Stengers, *Order Out of Chaos*, Bantam Books, 1984, p. 218.
17. Herman Weyl, in Ilya Prigogine, *ibid.*, p. 311.
18. Voltaire, in *ibid.*, pp. 257-258.
19. Maurice H. Wilkins, in *Synthesis of Science and Religion*, Mumbai: The Bhaktivedanta Institute, June 1988, pp. 36, 35.
20. David Bohm, *op. cit.*, p. 326.

21. Eugene Wigner, in *Synthesis of Science and Religion, ibid.*, p. 261.
22. E.C.G. Sudarshan, in *Synthesis of Science and Religion, ibid.*, p. 243.
23. Eugene Wigner, *ibid.*, p. 242.
24. Werner Heisenberg, *The Physicist's Conception of Nature, ibid.*, p. 3.
25. Edwin Hubble, *The Nature of Science and Other Lectures, ibid.*, p. 119.
26. Max Planck, *Where is Science Going?, ibid.*, p. 183.
27. Joan Walton, *Scientists Tune into Consciousness*, Winter 1996; in Sampooran Singh, et al., *Global Education and Human Survival*, New Delhi: Sanbun Publishers, March 2002, pp. 72-73.
28. Werner Heisenberg, *Physics and Philosophy*, London: Allen and Unwin, 1963, p. 75.
29. *Ibid.*, p. 57.
30. *Ibid.*, *The Representation of Nature in Contemporary Physics*, Daedalus, Vol. 87, No. 3, 1958, p. 107.
31. Albert Einstein, in Ken Keyes Jr, *The Hundredth Monkey*, Vision Books, Ed. February 1984.
32. *Ibid., from Letters*, Maurice Solovine, Gauthier-Villars: Paris, 1956.
33. *Ibid.*, in *Bridging Science and Spirituality*, Mount Abu: Bramha Kumaris World Spiritual University, 1988, pp. 67-68.
34. Sampooran Singh, Dr (Mrs) Kanwaljit Kaur, Paramjit Singh, and Amarjit Singh, *March to Peace: The Wisdom of Observation, Listening and Power of Commitment*, Chandigarh: Faith Publishers, January 20, 2004, p. 2.
35. Deepak Chopra, *How to Know God*, London: Rider, 2000, pp. 270-71.
36. Vimala Thakar, *Through Silence to Meditation*, Ahmedabad: Vimal Prakashan Trust, 1998, pp. 22-27; and Silence, Holand Bookfund, Vimala Thakar, 1989, p. 6.
37. Deepak Chopra, *How to Know God*, London: Rider, 2000, pp. 270-71.
38. D.S. Kothari, Some *Thoughts on Science and Religion*, The University of Delhi, 31st Oct & 3rd November 1977, p. 25.
39. J. Krishnamurti, *Krishnamurti on Education*, Orient Longman, KFI, 1974, p. 90.
40. Vimala Thakar, *Ego*, Mumbai: Vimal Parivar, November 1993, pp. 20-27.
41. *Ibid., Mutation of Mind*, Ahmedabad: New Order Book Co., 1972, pp. 122, 150, 151, 170.
42. Ervin Laszlo, *Emerging Concepts of the Cosmos*, *The Scientific and Medical Network Newsletter No. 49*, August 1992, p. 28; in *The Pilgrimage to the Third Millennium*, Chandigarh: Faith Publishers, January 1995, p. 31.
43. Fritjof Capra, *Synthesis of Science and Religion, op. cit.*, p. 273.
44. Albert Einstein, *ibid.*, p. 51.
45. D.S. Kothari, *The Perception of Truth in Science and Philosophy*, in *The Odyssey of Science, Culture and Consciousness*, 1990, pp. 48-49.
46. Maurice H. Wilkins, *Synthesis of Science and Religion, op. cit.*, p. 29.
47. B.D. Josephson, *ibid.*, p. 44.
48. Karan Singh, *ibid.*, p. 23.
49. Ervin Laszlo, in *The Odyssey of Science, Culture and Consciousness*, New Delhi: Abhinav Publications, 1990, p. 238.
50. Erwin Schrodinger, *What is Life? & Mind and Matter*, Cambridge University Press, 1967, p. 130.

51. George Wald, in Kishore Gandhi, *The Evolution of Consciousness*, New Delhi: National Publishing House, 1983, pp. 6, 13, 14.
52. Jacques Monod, *Chance and Necessity*, New York: Vintage Books, 1972, p. 171; in *Synthesis of Science and Religion*, *op. cit.*, p. 127.
53. President Vaclav Havel, *Speech, Philadelphia*, July 4, 1994.
54. *ibid.*, *Speech*, *Stanford University*, September 29, 1994.
55. Sampooran Singh, *The Human Survival* (*SS*: *11*), Chandigarh: Faith Publishers, February 1993, p. 105.
56. *ibid.*, *A Review on Symbiosis of Science and Spirituality*, Dagshai Cantt: The Atam Science Trust, 7.10.2002, pp. 29-30; & Sampooran Singh, et al., *Pathways to Peace*: *The Wisdom of Listening and the Power of Commitment*, October 2003, p. 36.
57. Karan Singh, in *Synthesis of Science and Religion*, *op. cit.*, pp. 22, 28.

12

Inter-Faith Harmony and Social Cohesion

Abstract

The twentieth century has witnessed unparalleled destruction and unimagined progress, the cruelest mass killings in human history and the most outstanding breakthroughs in human welfare. It is now quite clear that humanity is transiting into a new kind of global society. What is missing is the wisdom and compassion to transcend destruction. Knowledge proliferates, but wisdom languishes. Human survival involves the growth of a creative and compassionate global consciousness.

The humanist science and technology (restructured science-education-health) has brought mankind on the control panel of the transformation or mutation of the human psyche. This has the capacity to save mankind from annihilation. We have to undertake a pilgrimage from knowledge to understanding, from thinking to awareness, from rationality to intuition, from quietness to silence, and from silence to Life Field. And if man fails in his responsibility, he does so at his own peril, of India and of mankind.

Man (*Homo sapiens*) is an amphibian who lives simultaneously in two dimensions – (a) the *known*, the man-made social cosmos (world) of time-space-causation matrices which is a construct of the mind-brain-body system and is not a feature of reality; it is at the lowest quantum energy potential of consciousness called the "*successive consciousness*"; and (b) the *Unknown*, the highest quantum energy potential of Consciousness, or supreme Intelligence, the Life Field, the *Brahman*, the *Self*; it is called the "*Simultaneous Consciousness*", as it expresses Itself simultaneously in all existences. Man is only a transient being living in mental consciousness (successive consciousness) but with the possibility of acquiring a new consciousness, the Truth-Consciousness, and capable of living a life perfectly harmonious, free of all social conflicts, free of all violence and aggression, free of all psychological imbalances and ecological crisis.

Man has been trying to end conflict, violence, aggression, war in the psyche for the last 10,000 years or so, but all intellectual approaches, thought-structures, beliefs, concepts have proved to be sterile in transforming the human psyche; rather these have added to human conditionings. We have brought the human race to an explosive situation and our life is poised on a razor's edge. The discovery of this truth led me to an *inner urge* to explore the "*Science with Creativity*", or "*Symbiosis of Science and Spirituality*". I even felt that "it was being explored and put in rational language through me, rather than by me". It is the record, in words, of the "celestial music and inner harmony".

Modern physics has confirmed that "all their theories of natural phenomena, including the 'laws' they describe are creations of the human mind; properties of our conceptual map of reality, rather than of reality itself – parts of the map, not of the territory".[1] As a symbol we may compare the "*known*" (man-made social cosmos) to a drop of water and the "*Unknown*" (Life Field) to a fathomless and boundless ocean. The known is in the field of time, so it is limited, finite, transient and ephemeral. The Unknown is in the realm of the timelessness, so it is Eternal, Infinite and Immutable.

This article explores harmony and challenges of society. In the words of Ilya Prigogine, who won the Nobel Prize in 1977, "There is a need for new relations between man and nature, and between man and man .... (today man) carries an ethical responsibility".[2] As all chaos, strife, violence are embedded in the conditioned psyche, so this article explores all the frames of reference of the conditioned psyche and the Unconditioned Energy. It is now quite clear that the humanity is transiting into a new kind of global society – from psychosocial evolution to spiritual evolution. In other words, we need a quantum jump from successive consciousness to the Truth Consciousness which rejects disharmony and societal challenges, and bestows harmony and ending of all psychological imbalances.

Despite the astounding scientific and technological developments during the 20th century, humanity remains fractured and fragmented, torn by strife and chaos, by turmoil and tension. The last century witnessed unparalleled destruction, the cruelest mass killings in human history, and outstanding breakthroughs in the field of materialism adding to human comfort. We find ourselves poised at a crucial crossroads in the long and tortuous history of the human race on Planet Earth.

Religion

Every highly developed religion of the world has two dimensions – (a) historically conditioned socio-economic-political expression of religion which is local, parochial, transient, mutable, and deals with social customs; and (b) the universal core of religion which is eternal, infinite, immutable and non-local. The authenticity of the truths of the universal core of religion lies in their being perceived by spiritual experimenters and their being capable of re-verification by others.

The universal core of religion is a quest for discovering order in consciousness, or the inner world, or subjective dimension of man, or the search of truth that sustains man and all existence. Pure religion is also known as spirituality. This suggests that both pure science and pure religion are in search of truth, and are not antagonistic but complementary disciplines of life. The symbiosis of science and spirituality will produce fully integrated human beings, and thus help to evolve a complete human civilisation for which the world is ripe and waiting.

The conditioned mind-brain-body system constructs the time-space matrices in quick succession. The succession leads to the concept of causation. What we call mind (with a lower case 'm') is a conditioned energy, moving through the body, monitored by brain, operating through the senses. It is a movement of the structured framework of patterns and the dynamics of behaviour patterns are shared by all human beings. Mind is scattered vibrations. The time-space matrices represent the relative field of consciousness. It constructs pain, misery, suffering and travails. All knowledge, concepts, ideals, intellectual approaches are sterile in transforming the conditioned psyche, or to bring in radical revolution in man.

What we call Mind (with upper case 'M') is the non-dual frame of reference of Mind Continuum. The Universal Consciousness expresses. Itself in a non-dual and non-conceptual frame as intuition, spontaneity, love and compassion, order and harmony. It is the first-hand personal acquaintance of the quantum jump of the conditioned energy consciousness to Universal Consciousness; or in other words, a quantum jump from fragmentary perception to the perception of the Wholeness.

Science

In one of the first scientific symposia on "*Altered States of Consciousness*", it was discussed that "in high-level creativity, the mind can do weeks of work in a short time"; Houston said that "Albert Einstein claimed he thought not only with words but visually and with his muscles, an indication that somehow an *other* state of consciousness was making direct inputs into his rational thinking"; many speakers recommended the establishment of 'inner space labs'; Elkes said, "new ways must be found to accurately correlate physically measurable changes – brainwaves, chemical and harmonal fluctuations – with what is happening inside"; and Houston remarked, "We are perhaps on the verge of a golden age of brain and mind research".[3] Schumacher wrote, "The period of great convergence .... the language of spiritual wisdom can now be understood also as the language of sanity"; Bryce-Smith wrote, "We face an evolutionary challenge that requires a moral response: we shall survive only if we can show that we deserve to survive"; Bernard Lovell talked of the coming together of science and theology as "the only hope for the world in its present perilous state".[4]

The nature of consciousness was discussed in a scientific conference at Moscow. The Russian scientists proposed "the idea of a Life Field beyond our physical concepts of time-space and the known electro-magnetic and gravitational fields". In another conference at Cambridge University, it was discussed that "a new scientific approach would include subjective experience as valid information and address all forms of human experience – including intuition, spirituality and mystical encounters".[5] In another article, we discussed the "Science with Creativity is an experiment of creating 'New Man' – a human being who lives in harmony with the inner and the outer, with himself and his environments, and who is free from all ideologies and conditionings that now divide humanity".[6]

The States of Mind

(a) The Scientific Mode of Mind: About 1600 Kepler and Galileo simultaneously and independently formulated the principle that the laws of nature are to be discovered by measurement. All propositions were to be confined to that which was objectively measurable and verifiable.

A scientific mind takes up "the role of a non-concerned observer", in the words of Schrodinger.[7] It is unrelated to our beliefs, our ambitions, our hopes", in the words of Prigogine.[8] It must not construct a pseudo-observer, it must see the present and reject the past remembrances or future projections. Charles Sherrington wrote, "The material world has only been constructed at the price of taking the self, that is, mind, out of it, removing it; mind is not part of it; obviously, therefore, it can neither act on it nor be acted on by any of its parts".[9] Erwin Schrodinger wrote, "Mind has erected the objective outside world of the natural philosopher out of its own stuff. Mind could not cope with this gigantic task otherwise than by the simplifying device of excluding itself – withdrawing from its conceptual creation. Hence the latter does not contain its creator".[10] Schrodinger quoted C.G. Jung[11] and B. Spinoza[12] who hold the same opinion. Many eminent scientists – including Albert Einstein, Werner Heisenberg, Erwin Schrodinger, Arthur Eddington, and so on – based on a rigorous, consistent and persistent methodology of objective scientific research, have stumbled upon another mode of perception (knowing), which is "understanding of life directly, instead of in the abstract, linear terms of representational thinking".[13] The new mode of perception bestows "direct understanding of life", non-comparative perception, non-evaluatory perception. It is to observe non-subjectively. This suggests that the scientific art of observation gives a quantum jump from symbolic-dualistic frame to non-symbolic, non-dualistic and non-conceptual frame of reference of Mind Continuum. The non-dual frame has the capacity of understanding of life directly", which is in the realm of timelessness.[14]

The scientific mode of mind bestows on us "objective values", which implies that it leads to an invention (not discovery) of the modified use of the existing material) and this leads to progress of technology or progress of materialism.

(b) Restructuring of Science: Karl Pribram wrote, "Brain science must deal with the awareness of awareness. It could no longer afford to shut out that part of the world which we call subjective".[15] Roger Sperry "spoke of the importance of a neglected area of science, namely the study of human subjective experience".[16] John Eccles also emphasized "to deal scientifically with the neglected area of inner conscious awareness".[17]

Roger Sperry wrote, "When the brain is whole, the unified consciousness of the left and the right hemispheres adds up to more than the individual properties of the separate hemispheres". So these studies raised the issue of consciousness in a new way. He adds, "When parts come together in a new whole, this new whole exhibits features – emergent properties – that can't be predicted as a rule from the parts, and cannot be explained in terms of the parts".[18]

The role of consciousness suggests to consider the objective and the subjective as one whole, or in other words, the outer (objective) and the inner (subjective) are not two separate movements but a unitary movement, one whole. The objective science must restructure itself to take in its ambient both the objective and the subjective dimensions of life. The energy exchange between the subject and the object takes less than a hundred microsecond. The restructured science deals with the wholeness of life.

Education

Indian thought has always upheld the view that all values proceed not from matter, but from the spirit (inner quantum energy potential of man). All ethical values are essentially spiritual values, they proceed from the trans-sensual dimension of human personality. Recently, many eminent scientists – Werner Heisenberg, Edwin Hubble, Max Planck, G.F. Chew, Bertrand Russell, Lincoln Barnett, Max Born and so on – have concluded that we cannot impart value education at reason or intellect level.[19] Value is the implicate order of thought, the subjective nature of thought, the true nature of thought. Value education is to cultivate capacity for understanding of relationship which is life in its wholeness. All ethical or spiritual values are embedded in the trans-sensual dimension of life. Education in the true sense is helping the individual to be mature and free, to flow greatly in love and goodness. The highest function of education is to bring about an integrated individual who is dealing with life as a whole. The action plan or paradigm shift has been recently discussed.[20]

It seems that developing an alternative way of living, an alternative culture, a spiritual culture, radically different from the materialistic, consumerist culture, is the content of new education – we may call it self-education. Divinisation of humanity is the purpose of education.[21]

Health

All diseases are maladjustment of the human mind-brain-body system, or conditioned human psyche, or crisis of perception. Health is a complete physical, mental, psychological and social well-being; it is a dynamic state which is equipoised and balanced. A living in the *present*, the *now* heals; and living in the *past* and *duality* suppresses man's perceptive sensitivity and enhances disease. The curative functions of medicines have to be supplemented with understanding and living the "symbolic-dualistic frame of reference of mind spectrum (successive consciousness)" and the "non-dual and non-conceptual frame of reference of Mind Continuum (Simultaneous Consciousness)".

There is an urgency to restructure medical education which takes in its ambient the psychodynamics of the mind-brain-body system, or in other words, the wholeness of life. We have to jettison the illusion of the *past*, and to live in the *present*, the *now*. We have to restructure the biomedical model and adopt the holistic ecological perception. The restructured medical education takes in its ambient the wholeness of life.

It is imperative that medical profession and other scientists explore the healing of the psyche or ending psychic pandemic. The secret of healing is not embedded in our mind-brain-body system, but in higher awareness bestowed on us by Cosmic Consciousness or Life Field. And if man escapes his responsibility, he will do so at his own peril and that of India and mankind.[22]

Spirituality

It is the science of life and living. Life is a holistic phenomenon. It consists of two dimensions – the indirect processes are psychological imbalances and ecological crisis; and direct processes which are non-cerebral and the non-psychophysical and these bestow true perception of pure Consciousness.. The energy of thought, of feelings, of sound are indirect processes; and are of different frequencies; and these express themselves as experience, knowledge, verbalizing, so these are partial, limited, finite, transient, ephemeral.

All indirect energies emerge out of emptiness. It is direct expression of Life Field in a pure mind. It is an inner unconditional relaxation, which is tension-free state of the nervous system and a pressure-free state of the chemical system of our being. The emptiness, the silence gives the feel of aliveness, alertness and aloneness. It is a consummation of holistic growth, which results in a different texture of our being, a different quality of our relationship, a different culture altogether. The exploration is on behalf of the whole human race, on behalf of the whole human family.

Mind and Consciousness

Consciousness is the primary datum of human existence. In the world of the quantum, it is both the awareness and the creation of experience. It is the *being* and the knowing of experience. With the stroke of the twentieth-century quantum eraser, the dividing line between ontology (theory of being) and epistemology (theory of knowing) is rubbed out.[23] The fusion of ontology, epistemology and psychology is beginning to emerge.

Wilder Penfield, the renowned neurosurgeon, concluded that the fact that the mind can survive brain trauma and function under anesthesia points very strongly to the separate existence of the mind. He concluded that, "it is the mind which experiences and it is the brain which records the experience ....

our minds and our brains are not at all the same".[24] Brain is a part of the objective world, it can be investigated objectively. Mind is subjective, it is a vibrational self-conscious energy. The control of the mind over the body is an incontrovertible fact of personal experience. He again wrote, "To suppose that the consciousness or the mind has location is a failure to understand neurophysiology".[25] The mind appears to be everywhere. It is observing on the scale of atoms and molecules, neurons, cells, tissues, muscles, bones, organs – in other words, it is observing on all scales of physical existence. The time required for a signal and its perception is about five thousandths of a second.[26] Wolf wrote, "We are beginning a new age of awareness, the age of quantum consciousness, the age of the conscious atom. By looking within ourselves, we may be able to solve the problems facing us on the final frontier – the frontier of the human spirit".[27]

Paul Davies concluded, "The laws which enable the universe to come into being spontaneously seem themselves to be the product of exceedingly ingenious design. If physics is the product of design, the universe must have a purpose, and the evidence of modern physics suggests strongly to me that the purpose includes us".[28]

Mutation or Radical Transformation of Psyche

(a) Ending of I-consciousness: We transfer what is happening at the biological level and transfer it to the psychological level; and this we call ego, or I-consciousness, so the 'I' is not an entity, but a concept. I-consciousness has no factual reality. It has a conceptual reality. It has an ideational reality. We graft the subjective on the perceptual level and transform it to conceptual level.

The I-consciousness is at the successive consciousness. The subject of cognizance or true-subject is at the immanent quantum energy potential. When I-consciousness (thought coupled with various volitions) encounters the true-subject, the I-consciousness is knocked off because it is at the lowest quantum energy potential. This suggests that (i) the movement of thought-time is independent of I-consciousness; (ii) it is the movement of the humanity, and (iii) both the I-consciousness and thought-structure have a conceptual reality.[29,30]

(b) Ending of Thought-Time: Every partial experience and fragmentary knowledge is stored in the human psyche, and it is called the cultural belief structure. It is meta-stable and disintegrates to thought structure. When thought-time comes in the realm of true-subject, the photons meet an 'impregnable wall', or "event-horizon", or "frequency-barrier", and are dissipated, which implies that thought-time goes to abeyance and what is left behind is the *fact*, the *true*, the *present*, the *what is*![31] When thought-time subsides, it is a discontinuity in the mental stream, and in philosophical language, it is called "no-mind" state, which comes gracefully and voluntarily. There is no fear, no acquisitiveness, no desire for ownership and possession, no desire to dominate over others. There are no psychological imbalances, no ecological crisis, no human challenges. A new human race, with a new culture and a new dynamics, a new foundation for human relationship will be there. Suddenly there is an inner stillness. This state is highly alert and aware, but no thinking.

(c) Ending of the Fact: If we continue observing or listening to the *fact*, without justification, or condemnation, or motivation, we observe that there is only the *fact* and nothing else. The fact itself is the operator and not you operating on the fact. If you leave the fact alone, it has no time. The fact is free from fear. The fact grows, which implies that there is release of tremendous energy and the fact moves to higher quantum energy potential. This is expressed as Pure Consciousness → Fact.

(d) Intuition: The "understanding of life directly, instead of in the abstract, linear terms of representational thinking" bestows intuition, spontaneity, sensitivity and creativity. These are metaphysical (quantum logic) insight, falling outside the realm of duality, or the intellectual construction. *Understanding, learning and action are a cyclic process, which results in a spiral going upwards in a vertical dimension; it is the growth in vertical dimension. The higher the understanding, the higher the expression of consciousness in the human mind-brain-body system.* The higher sensitivity or creativity leads to a perception of pure consciousness.

(e) Intelligence: Intelligence perceives life in its wholeness, and responds to it out of its own wholeness. The perception through intelligence, through sensitivity is a perception in non-duality. In love, beauty there is instantaneous perception. The perception and the response have no time lag between them.

The communion with the Divine, the Eternal, the Infinite is a happening; it is not the result of my doing. The uncovering of the secret is not a human achievement. It is something that happens, when the human being puts all his conditionings into relaxation, and remains before the mystery of Life in utter humility, in utter nakedness of his psyche. The consummation of human growth is the manifestation of the Divine, the Self-Unfoldment of man. Life is consummation of human growth. Then the mystery unfurls itself, uncovers itself. It uncovers itself into us.

All great discoveries have arisen from the realm of the timelessness, so scientific enquiry is not a cerebral activity and gathering information, but a quest of the timelessness. It also suggests that observation (seeing) is direct, listening (hearing) is direct, understanding and learning of life is direct, all action arising from understanding and learning is direct; but thinking is indirect. All direct relationship is in communion with the immutable laws of nature; and indirect relationship is antagonistic to the laws of nature.

Life – A Manifestation of Harmony

Arthur Eddington wrote, "One hundred billion stars make a galaxy; one hundred billion galaxies make a universe. Our own galaxy, the Milky Way, contains about one hundred billion stars". It is a vast thing. It takes light, travelling at 300,000 kilometers per second, about 100,000 years to cross the Milky Way. And yet that Milky Way is just a tiny spot in the universe we know. There is a remarkable harmony, order, and unity of nature. Some physicists are deeply impressed by the way that the laws of nature hang together that they feel compelled to believe that there is something behind it all; in a pithy phrase of Fred Hoyle, "The universe is a put-up job".

Many physicists find unexpected harmony and coherence of nature. The Second Law of Thermodynamics states that disorder can never spontaneously give rise to order. More precise, this law legislates nature's trading account in a quantity called entropy, which is roughly the degree of disorder in a physical system. When ordered energy becomes disordered, the entropy rises. The second law forbids the entropy of a complete system from falling.[32]

Karan Singh stated, "Scientific breakthroughs and technological ingenuity have given us the capacity to overcome all these challenges. What is missing is the wisdom and compassion to do so. Knowledge proliferates, but wisdom languishes. It is this yawning chasm that will need to be bridged if we are ever to reverse the present trend towards disaster, and it is here that education in the broadest sense of the term (and mutation of the conditioned psyche) assumes such a vital position".[33] We have not cared to restructure our education.

Man saw the fury of nature (or disorder) on December 26, 2004 which through tidal waves generated by earthquake led to a disaster (Asian quake disaster toll 1,25,000 on 01.01.2005). It is basically a release of bottled up energy, which must have been stored over thousands and thousands of years. My *inner* says that if mankind understands, learns and lives the conscious Presence, the Life Field, then such energy of violence would not have been stored; and perhaps such disasters would have been averted, or the suffering would have been mitigated. The answer to this paradox is understanding and living *Science with Creativity* which implies *Restructuring of Science, Education and Health*, and this blossoms to *Symbiosis of Science and Spirituality*. Obviously it is 'you' and 'I', 'we' and 'they', and all mankind is responsible for this disaster and we have to take urgent steps to ameliorate the suffering.

Life is not a struggle for existence. It is a manifest of the inner harmony. It leads to harmonious relationship between man and nature, man and man, man and other animate and inanimate existences. They form a harmonious wholeness. The cosmic life is in harmony and human beings are born to manifest that harmony on the human level.

Society is social-economic-political structure. A harmonious living ends all societal entropy, all societal challenges. There is ending of all violence and aggression, free of all conflict and tension, free of all psychological imbalances, ecological crisis and other challenges.

Paul Davies wrote, "We cannot fail to be struck by the ingenuity of the natural world? Nature is astonishingly clever in the way it gets things to work". Nature's benediction flows at all moments. We have to learn and live the immutable laws of nature.

Concluding Comment

(a) The humanist science and technology (restructured science-education-health) has brought mankind on the control panel of the transformation or mutation of the human psyche. This has the capacity to save mankind from annihilation. We want a collective psychological experiment where every human being is an equal partner. Such an experiment flows to new relation between man and nature, between man and man, and awareness of human values. The transformation happens in the *individual* and he becomes a nucleus to transformation in the *collective*.

(b) We have to undertake a pilgrimage within, which blossoms to "New Human Being". The new human being understands, learns and lives in the Truth Consciousness – he lives in the harmony of the inner and the outer, with himself and his environments, and who is free from all ideologies and conditionings that now divide humanity. He must be free from political activity, (institutionalized) religious conformity and tradition – which means free from the weight of time, free from the burden of the past, free from all the action of will. This only is the social, psychological, and even the political revolution.

(c) We have to undertake a pilgrimage from knowledge to understanding, from thinking to awareness, from rationality to intuition, from quietness to silence, and from silence to Life Field.

(d) Brain is a part of the objective world, it can be investigated objectively. Mind is subjective, it is a vibrational self-conscious energy. Wilder Penfield concluded separate existence of the mind, so our minds and our brains are not at all the same. It is the mind which experiences and it is the brain which records the experience.

(e) If a thought coupled with volition encounters the true-subject, the I-consciousness is knocked off because it is at the lowest quantum energy potential. This suggests that (i) the movement of thought-time is independent of I-consciousness; (ii) it is the movement of the humanity, and (iii) both the I-consciousness and thought-structure have a conceptual reality.

(f) When thought-time comes in the realm of true-subject, the photons meet an 'impregnable wall', or "event-horizon", or "frequency-barrier", and are dissipated, which implies that thought-time goes to abeyance and what is left behind is the *fact*, the *true*, the *present*, the *what is*!

(g) Many physicists are deeply impressed by the unity and order, by harmony and coherence of nature. The exploration suggests complete harmony between the objective and the subjective, and ending of all social-economic-political challenges. The laws which enable the universe to come into being spontaneously seems themselves to be the product of ingenious design. The evidence of modern physics suggests that, man being a part of nature, must understand, learn and live in order and harmony.

(h) The arts of observation, listening, understanding, learning, action, and silence in the no-mind state, are direct; while thinking is indirect. Nature invariably expresses Itself as *direct*. If man understands and lives in communion with the laws of nature, there is emergence of a "New Man" – a human being who lives in harmony with the inner (subjective) and the outer (objective), with himself and his environment, and who is free from all human challenges, societal entropy, ideologies and conditionings that now divide humanity. A new balanced human being, who has no imbalances (no impurities), but who has a grandeur of inner equipoise and majesty of outer sensual balance, emerges on this Planet Earth.

It presents the first scientific model of "Symbiosis of Science and Spirituality". The words of the "Union of Concerned Scientists (April 1993)"[34] *are true today that "This ethic must motivate a great movement, convincing reluctant leaders and reluctant Governments and reluctant peoples themselves to effect the needed changes". And if man fails in his responsibility, he does so at his own peril, of India and of mankind.*

References

1. Fritjof Capra, *The Tao of Physics* (*Second Edition*), Bantam Books, 1988, pp. 277, 147.
2. Ilya Prigogine and Isabelle Stengers, *Order Out of Chaos*, Bantam Books, 1984, p. 312.
3. Constance Holden, *Science*, 9 March 1973, Vol. 179, No. 4077, pp. 982-83.
4. Ian Breach, *New Scientist*, 26 May 1977, Vol. 74, No. 1053, pp. 478-79.
5. Joan Walton, *New Science*, Winter 1996.
6. Sampooran Singh, *Science with Creativity : A Vision on Human Survival and Excellence*, unpublished.
7. Erwin Schrodinger, *What is Life*? and *Mind and Matter*, London: Cambridge University Press, Ed. 1967, p. 128.
8. Ilya Prigogine, *op. cit.*, p. 5.
9. Charles Sherrington, in Erwin Schrodinger, *op. cit.*, p. 128.
10. Erwin Schrodinger, *ibid.*, p. 131.
11. C.G. Jung, in Erwin Schrodinger, *ibid.*, p. 129.
12. B. Spinoza, *ibid.*, p. 131.
13. Alan W. Watts, *The Ways of Zen*, New York: Vintage Books, 1967, p. 10.
14. Sampooran Singh, *Relationship and Society*, unpublished.
15. Karl Pribram, in Ken Wilber, *The Holographic Paradigm and Other Paradoxes*, Boulder: Shambhala, 1982, p. 17.
16. Roger Sperry, in *Synthesis of Science and Religion*, Mumbai: The Bhaktivedanta Institute, June 1988, p. 199.
17. John Eccles, in *Synthesis of Science and Religion*, *ibid.*, p. 199.
18. Roger Sperry, *Bulletin of the Theosophy Science Study Group*, India, February

& April 1988, p. 23 and June & August 1988, p. 27.

19. Sampooran Singh, Dr (Mrs) Kanwaljit Kaur and Paramjit Singh, *Global Education and Human Survival: A Scientific Appraisal*, New Delhi: Sanbun Publishers, March 2002, pp. 46, 35-36.
20. Sampooran Singh, *Science and Value Education*, Dilip, October/December 2004, pp. 34-41.
21. Vimala Thakar, *Ego*, Bombay: Vimal Parivar, 1995, pp. 31, 40.
22. *Ibid., The Healing of Conditioned Mind-Brain-Body System: The Ending of Psychic Pandemic*, unpublished.
23. Fred Alan Wolf, *Taking the Quantum Leap*, Harper and Row, 1981, p. 229.
24. Wilder Penfield, in Deepak Chopra, *How to Know God*, Rider, 2000, pp. 217, 214.
25. *Ibid., The Mystery of the Mind*, Princeton: Princeton University Press, 1975, p. 109; in Fred Alan Wolf, *op. cit.*, p. 244.
26. Fred Alan Wolf, *op. cit.*, p. 236.
27. *Ibid.*, p. 248.
28. Paul Davies, *Superforce*, New York: Simon and Schuster, 1984, pp. 243, 232.
29. Vimala Thakar, *Through Silence to Meditation*, Ahmedabad: Vimal Prakashan Trust, 1998, pp. 22-27.
30. *Ibid., Silence*, Holland Bookfund, 1989, p. 6; and *ibid., The Invincible*, July-August 1987, Vol. IV, No. 4, p. 14.
31. Deepak Chopra, *How to Know God*, *op. cit.*, pp. 270-71.
32. Fred Hoyle, in Paul Davies, *Superforce*, 1984, pp. 223-24.
33. Karan Singh, in *Education for a Global Society*, Delhi: Shipra Publications, 2003, p. 2.
34. *World Scientists' Warning to Humanity*, USA: Issued by Union of Concerned Scientists, April 1993.

13

Exploring Consciousness and Mutation of Human Psyche

Abstract

Human consciousness is a multidimensional activity – the conditioned energy (symbolic-dualistic frame of reference of mind spectrum) and Unconditioned Energy (non-dual and non-conceptual frame of reference of Mind Continuum). The conditioned energy gathers psychological imbalances, ecological challenges, which express as nervous tension and chemical pressures.

Mutation is a sudden, abrupt change, happening in the totality of the being – it is basically a quantum jump from symbolic-dualistic frame to non-dual and non-conceptual frame. The Intelligence is energised which ends all psychological imbalances, ecological crisis and all other challenges, and one revels in order, harmony, non-violence, peace and bliss.

Human consciousness is a multi-dimensional activity and it functions in three dominant modes: (a) biological, (b) conditioned energy as experience and knowledge (information), the *known*, the manifest, the psychosocial, and (c) Unconditioned Energy, the *Unknown*, the Unmanifest. The biological autonomous processes – seeing or hearing, smelling or breathing – are movements in the realm of non-duality; and these are functioning in an interrelated, intertwined, choiceless, complex and yet homogeneous way as expressions of supreme Intelligence which surrounds and interpenetrates us in Nature. The *known* processes gather experience and knowledge, store these in the brain, and act as ego, the "me". This is mechanistic, repetitive and dualistic process. The *Unknown*, the choiceless energy of Unconditioned Consciousness or Awareness, which is an energy born of the emptiness of Silence, which is born by the non-action of conditioned consciousness, and this functions without any volition, or ego, or I-consciousness. The movement of organic Intelligence in the complexity of cosmic Life is a movement of non-duality and conducted in non-duality. It is an energy that permeates the whole being, vibrates in the whole being, that becomes the source of perception and response and has no relation to the willing, the volition, the choice. So Awareness is choiceless.

The Consciousness, the Life Field expresses itself in two dimensions. The *known* is at the lowest frequency of consciousness, or lowest quantum energy potential of Consciousness. The *Unknown* is at the highest frequency, or highest quantum energy potential. Each higher stage is synergistic to its junior components; it includes them but is more than them. When the higher stage bombards the lower stage, it dissipates the lower stage which goes to abeyance. The higher illuminates the lower, and not *vice versa* – this lays foundation of a creative principle. The mutation or radical transformation of the human psyche *happens* when it is bombarded by higher consciousness, so mutation is governed by Consciousness, or by the Unmanifest. This article explores Consciousness and mutation of the conditioned psyche.

We are all human beings, we are the rest of mankind, we are the rest of humanity. Your brain is not yours, it has evolved through thousand and thousand years. The memory or thought-structure is shared by mankind. Thinking is a playback of thought coupled with feeling and response of the memory – cultural, racial, religious, communal – which is called willing. Willing flows to reaction. Thinking is common to all of us, so thinking is the movement of all mankind, it is not individual thinking. You are concerned with global thinking. You are concerned with the world, with all humanity. We live in the past and the mind can project itself into future. We are conditioned. We have constructed an aggressive and violent society. There is global disorder, the threat of wars, poverty, fear, and so on – we are responsible for that. We live in pain, misery and conflict. Conflict wears out the brain, makes the mind old. We have gathered psychological imbalances and ecological challenges. Life is movement in relationship – man with nature, man and man, man and other organisms, and man and environments. The thought structure cannot have right relationship with others.

The answer to this paradox is embedded in psychological search, seeking out inwardly and finding out for oneself a tremendous area that is beyond all thought. It is to transcend thought or ego, be aware of the *fact* in the non-dual realm, the "understanding of life directly", learning and to explore the all-inclusive dimension of Consciousness or Life Field. It bestows understanding, intuition (insight), spontaneity, right action and meditation. That action is born of intelligence – intelligence, love and compassion, all go together. This is one harmonious movement of the dimension of Consciousness or Life Field. This is really a sacred thing and the beauty is in itself and not beyond it.

Knowledge

Human consciousness is a multidimensional activity. In the common waking state, it expresses itself at the lowest hierarchical expression of consciousness. At this level, the mind cuts itself into the subject and the object, and the subject gathers experience and knowledge (information) and stores it in the brain as frequency and amplitude information – called memory.

Knowledge is always an indirect activity, so knowledge is invariably fragmentary, divisive and dualistic.

The conditioned I-consciousness has voluntarily accepted the role of a pseudo-observer, so it reacts to a challenge. Observing and reacting are incompatible. Observing, evaluating and assessing are incompatible.

The thought-energy acts as per its motives, desires, acquisitiveness, psychological needs. Thought constructs habit patterns, so it reacts as personal preferences, or prejudices. Thought accepts ideologies. The thought-energy is controlled, shaped, refined, sophisticated by civilisation, by culture.

Our daily living is nothing but a chain of reactions to the compulsions that surround us – socially, culturally, economically and politically. I accept the authority of the past, of the compulsions and pressures. So I live in the past with pain and sorrow.

The measurements, the symbols representing the measurements, are the substances (matter) of the past. There is no such thing as psychological evolution. There is nothing to learn in the psychological field. Psychologically, inwardly, there is only the *fact*, which is in the non-dual realm. To be a light to oneself requires great freedom, not a conditioned brain.

Thinking is the movement of all mankind, it is not individual thinking. Experience and knowledge stored in the brain, as memory, is thinking.

Observation

The first stage prior to recognition and naming, identifying, judging or the association of ideas and emotions, is a state of non-verbal or thought-free perception. To observe non-subjectively, to observe without any evaluatory comparison or judgement. The occurrence of seeing or hearing, smelling or breathing is a movement in the realm of non-duality. There is no subject-object division and no movement born of the tension of duality. There is no time gap between observation and action; it is a kind of simultaneity which seems to be the principle of Cosmic Life.

Sullivan wrote, "We cannot observe the course of nature without disturbing it". Andrade stated, "Observation means interference with what we are observing .... observation disturbs reality". George Wald stated, "The very crux of twentieth-century physics is the recognition that the observer cannot be excluded from his observation; he is an intrinsic participant in them". The observer and the observed, at this level, are one and the same.

When I observe my own psyche, I see that I am a part of the psychological load and I participate in its activity at every moment, so I am the disorder, the violence. In this observation, the subject and the object get integrated, and the ego, the I-consciousness goes to abeyance. This brings about a humility. If you observe the racial thought, it flowers, blossoms and withers away, and intelligence expresses itself. Humility is the prerequisite of learning; humility is the soil in which intelligence flowers. Intelligence is the unconditioned energy of the whole being.

Human consciousness is a multidimensional activity; it implies that it can express itself at different quantum energy potentials of consciousness. It can express itself to inquire into the meaning of Life, the nature of the Reality, the Truth sustaining facts. Each quantum energy potential of Consciousness can knock down a structure at a *lower* quantum energy potential, so it bestows unpolluted, uncontaminated and non-evaluatory perception.

Human interference introduces discontinuities, it is fundamentally a discontinuous and quantum jump world. Niels Bohr stated that discontinuity is a fundamental truth. The discontinuity happens when the energy from the Unmanifest flows to the manifest raising its quantum energy potential, or the immutable law of nature – Unmanifest → manifest – expresses itself.

Attentivity

Observation without an observer and an observed is attentivity. The whole being gets tuned into a state of attentivity, where there is neither subject, nor object. It is a subject-free and object-free attentivity. It is not an activity on the plane of duality. It is just sensitivity of the whole being that becomes attentivity.

The restructured science sees both the objective and the subjective dimensions of life, or the wholeness of life. The Subject of Cognizance is non-variable attentivity, or reaction-free attentivity, or objectless attention, it is also called true-subject, or a mind with a scientific temper. It is tuned to perceive the *fact*, the *true*, the *present*, the *what is*!, the *non-duality*. It bestows a perception of "understanding of life directly", or in other words, it bestows intuition (insight), spontaneity and right action. The pure perception bestows non-violence, harmony, human values, truth, love and compassion, so it is in communion with the Consciousness, the Life Field.

Awareness

When attentivity is established over a long chronological time, it makes another quantum jump and it is called awareness. It is awareness of the Wholeness, the Oneness of Life.

Awareness energises the supreme Intelligence. The movement of organic Intelligence in the complexity of cosmic Life, is a movement born of non-duality and conducted in non-duality. There is no association with volition. It is non-voluntary. Awareness brings mutation or transformation of the whole being. Mutation is a radical change in the quality of the mutant, in the substance of the mutant. It is not a peripheral change. It is a radical change in the whole of the mutant. Mutation leads to non-verbal communion between what is outside me and what is within me.

Understanding, Learning, Action and Meditation

Understanding is not in the symbolic-dualistic frame of reference, but it is embedded in the non-dual frame. It is timeless perception. Understanding comes only when the mind is very quiet. Perception has nothing to do with knowledge (information). The perceptions are born of non-duality and responses are born of spontaneity. Spontaneity bestows right action. It is insight alone that can decondition the brain.

Understanding, learning, action and meditation are one harmonious movement. These belong to the realm of non-duality. If this is clearly understood, there is no wastage of energy. To learn there must be total freedom. This one movement is also called meditation. Meditation, or the transcendence of the conditioned mind, is not the result of physical or mental activity. It is in the relaxation, it is in the non-movement, it is in the non-action of the conditioned, that the non-conditioned begins to operate.

Meditation: The roots of every action, every thought and every emotion, are in the subconscious. And that subconscious regulates, directs and controls the nature of the conscious action. The scientific mode of mind, all-inclusive attention, observes the objective challenge, the nature of subjective reactions and the causes of the subjective reactions simultaneously, and such an awareness is always ahead of the reactions. When such reactions face the awareness, they lose their momentum and I am not carried by the reactions. The reaction exhausts its momentum and arrives at the simple, innocent silence. It results in freedom from the momentum of reactions.

A response which is free from the momentum of reaction is a natural response. So natural response is a response which results out of understanding of the wholeness of challenge, reactions and the roots of reactions. The natural response is in the realm of non-duality. Natural response emerges when reaction comes face to face to the light of attention, so we expose the reaction to our attention, to our sensitivity, to our alertness.

The simultaneous awareness of the total movement is meditation. It is a new approach to life, in which we observe the objective challenge, the subjective reactions and the causes of the subjective reactions simultaneously without identifying oneself to any of them.[1]

Brain: My brain is the brain of humanity. Every cell in the human brain holds memory of the million years of man. So memory or thought-time is of the humanity. Mutation is a psychological revolution of the whole memory content. It implies to transcend or decrease the matter content of consciousness. Mutation is growing into a new dimension of consciousness, or in other words, a new quality of consciousness with its pristine glory.

Krishnamurti and David Bohm discussed the quality of brain and mind and their interactions in words:

> I have a million years of experience, and it has given me a certain capacity. And I realize at the end of it all there is no relationship between me and truth.... Is it a shock to discover that your brain, and your mind, your knowledge, are valueless? All your examinations, all your struggles, all the things that you have gathered through years and years, centuries, are absolutely worthless?.... What you have done or not done is

> absolutely of no value.... The ground says, whatever you have done 'on earth' has no meaning.... It has given me a tremendous shock to discover the truth, that all the churches, prayers, books, have absolutely no meaning – except how we can build a better society, and so on.... So there is only one thing, and that is to discover that *all* that I have done is useless – ashes![2]

It is a shock to discover that whatever man has gathered (the known, thought, memory) in centuries is worthless, ashes. It is evident that humanity took a wrong turn. This has to be rectified by an inner journey of self-discovery from self-observation through self-knowledge to Silence. Man has to be a light unto himself. Truth is a process of self-discovery. No one can give the truth to another. Only the truth can liberate the mind from its own ideation.

Krishnamurti wrote, "Every night the brain has been breaking and entering into something immense .... It is a state in which the mind, the brain was no longer in operation – only the source was in operation".[3] It implies that during deep sleep, the Unmanifest, the Immense, the Source takes over the brain. When one awakes after deep sleep, the brain, the mind reverts back to successive consciousness – the old memory, the time.

Living with the Fact: Understanding of life directly is a perception of the *fact*, the *true*, the *present*, the *what is*!, the non-dual perception. When you listen deeply, or you see instantly or directly, the seed has been sown. And that fact, if it is left alone, is operating; you do not have to do a thing about it. The fact itself is the operator, not you operating on the fact. The fact grows and matures. If you leave it alone, it has no time. When the fact is left completely it blossoms instantly. Let the fact operate, let it live, as the seed that has been planted in the womb, is living. Let the fact live.[4]

Thought has no place in the investigation of the inner because thought is a part, limited, fragmentary, divisive and dualistic. When you see that the subject is the object, there is direct perception of the fact. Krishnamurti prophetically wrote, "India has lived all these centuries on ideas. She has to move from a life based on ideas, which are non-facts, to seeing the fact only". He added, "Something had been destroyed in India, but a presence waited in the land .... The benediction of the timeless is waiting, the ground is ready. Can we create something that is of this benediction".[5] My brain is the brain of humanity. The brain says, in words of Krishnamurti, "I am humanity, and what humanity has done, I have done? And I see that it does not produce benediction and therefore all that has been done, has no meaning".

The Unknown energy is a cosmos, the known energy is a chaos. We have to transform the known to the Unknown through higher expression of the Unknown. It is the Consciousness that can transform the chaos to cosmos. To live a life rooted in attention and awareness adds a new dimension to the quality of daily perception and living at every moment; it is living in communion with the immutable law of nature.

Silence

It is the total renunciation of the movement of conditioned mind, so the mind goes to non-action. The discontinuity of the conditioned ego takes one to a different dimension altogether, where knowing or experience may not happen at all; but perception, understanding, learning and action – a new dynamics of behaviour – may get activised. The non-action of the I-consciousness – or in other words, *silence* of the conditioned mind – is a different dimension, where the highest frequency of Consciousness bathes the brain cells, so there is no identification, no fear, no violence, no psychological imbalances, no ecological challenges, and so on. Silence is totally unconditioned, the emptiness is unconditioned. It is a nudity of Consciousness.

The motionlessness of speech, the motionlessness of the thought structure, the motionlessness of the subject (the ego, the self), then that relaxation is the emptiness of consciousness. Thought-freeness of consciousness shall release Awareness. If Awareness is stabilised over a long chronological time, it blossoms to perception of Spirituality, the Wholeness of Life.

It is an inner voyage of self-discovery from self-observation through attentivity to Silence. It is a journey from human consciousness (successive consciousness) to Simultaneous Consciousness, and living in the dimension of Silence. It is an enquiry on behalf of the human race for an alternative way of living, an alternative culture, an alternative dynamics of human relationships.

The living and moving Silence settled down in the neurochemical system is called state of meditation. Meditation implies to see and live in the wholeness of life. Meditation is an alternative way of living, living in the Emptiness.

Spirituality

It is the physics of consciousness. It is an all-inclusive awareness to every aspect of Life. Life in its wholeness is holy and sacred. Reality is measure-free and concept-free. The emptiness (Life Field) which fills the room is immeasurable. When you are aware of the *fact*, the truth behind the fact, you learn which blossoms to the transformation of the psyche.

Mutation

All experiments over the past uncountable aeons regarding relationship of the mind to acquisition, accumulation and preservation of wealth; relationship to knowledge, experience and authority have not changed; or in other words, man has miserably failed in radical transformation of his psyche.

Man urgently needs mutation of mind. Some of the characteristics of mutation are as under:

(a) Mutation is a sudden, abrupt change, happening in the totality of a thing. It is decrease in the matter content of consciousness, it enhances perceptive sensitivity of consciousness, which will flow into higher hierarchical level of expression of quantum energy potential of consciousness. The mutation of the human mind is unconditional total decrease of matter content in the whole human consciousness.

(b) Mutation will destroy everything that the mind has built the enclosure around itself in the name of defence, of security in the psychological field. This will bestow freedom which leads to ending of ego, the "I-ness", the "me".

(c) Mutation implies a quantum jump from symbolic-dualistic frame of reference of mind spectrum to non-dual and non-conceptual frame of reference of Mind Continuum. It bestows awareness of the fact, which is the wholeness of life, the fact. This regenerates intelligence.

(d) Intelligence bestows intuition (insight), spontaneity and right action. There is freedom from tension, conflict and contradiction. One lives in the timelessness. Every act is fulfilling.

(e) Man emerges into a new dimension of life, in which direct perception of rality will be the channel for his consciousness to function.

(f) Man oscillates between different quantum energy potentials of Consciousness, the Life Field. If man is aware of the highest quantum energy potential or maximum frequency of Consciousness, he is said to live in the spiritual realm, the Absolute Realm. If he lives in the lowest quantum energy potential of consciousness, he is said to live in the relative field of consciousness or successive consciousness, or sensual world.

(g) Perception happens in a non-dual frame of reference. Perception is the foundation for mutation. Perception and mutation, if sustained for a long chronological time, flower and blossom to love. Love is state of being in which one has arrived at the humility to act without the ego. In this state, one does not want to use life as a means to an end, but is satisfied in living life as an end in itself.

(h) Perception is direct communion with reality, with truth, or spontaneous movement in truth and reality. Perception is timelessness. Truth or reality or life is a moving thing, a living thing, which is fresh every moment, which is new every moment. And to move with it, to get into the rhythm of that free movement of life, which is freedom, which is mutation, which cannot be seen or perceived at through any action of the mind.

(i) The non-identification with the subjective world, non-identification with the momentum of the mind, results in silence, which blossoms to intelligence. The intelligence is the untapped, unexplored and unactivised dimension of Consciousness, and this is the foundation of all real religions. The intelligence guides the socio-economic-political relationships and heals all psychological imbalances, ecological challenges, and so on.

Concluding Comment

Life is a multi-dimensional phenomenon and human beings who are expressions of that phenomenon are themselves multi-dimensional organisms. Life is self-generated, self-sustained, self-perpetuating and self-regulated phenomenon. Life is Infinite, Eternal and Immutable. Life is an expression of Supreme Intelligence.

Life, or Universal Consciousness expresses itself in three dominant modes: (a) biological – seeing or hearing, smelling or breathing, waking state or deep sleep state, and so on; (b) the *known*, the time-space matrices, the manifest, the conditioned energy; and (c) the *Unknown*, the Timelessness, the Unmanifest, the Unconditioned Energy. As a symbol, we may compare the known to a drop of water and the *Unknown* to a fathomless and boundless ocean. The known is limited, finite and transient, and the Unknown is immutable, infinite and eternal.

We have brought the human race to an explosive situation and our life is poised on razor's edge. From deep within the human psyche (known) there has developed a terrible power (conditioned psyche) that threatens the extinction of all life-supporting systems in a foreseeable period of time. The conditioned psyche is leading the human race to self-annihilation or self-destruction.

Life oscillates between different quantum energy potentials of Consciousness. When higher energy state encounters lower energy state, the latter goes to abeyance. Human being seems to be born with two dominant faculties – faculty of self-awareness and understanding of life directly. Self-awareness is in the field of duality, time, the manifest, and understanding is in the realm of non-duality, timelessness, the Unmanifest. It is the Unmanifest that governs the wholeness of life.

Life is the Divinity, the Wholeness. To understand life directly, we are going to undertake a journey from fragmentation to homogeneity, from partialness to Wholeness, from the movement of knowing to the action of understanding, from the slumber of knowledge and experience to the state of being Awake or Awareness. This involves a major conceptual shift to complementarity in place of competition, convergence in place of conflict, holism in place of hedonism. We must heal the split within the human psyche; we must gather the diverse fragments of human consciousness and weld them into a glowing whole; we must effect the transition to a holistic world-view to replace the present fractured and fragmented consciousness of the human race.

It is the birth-right of every human being to grow into that dimensional transformation, an entirely new way of living. This will guarantee human survival and excellence.

References

1. Vimala Thakar, *Mutation of Mind*, Ahmedabad: New Order Book Co., 1972, pp. 122, 150-151, 164, 171.
2. J. Krishnamurti & David Bohm, *The Ending of Time*, San Francisco: Harper and Row, 1985, pp. 105, 107, 110.
3. J. Krishnamurti, in Pupul Jayakar, *Krishnamurti: A Biography*, San Francisco: Harper and Row, 1986, pp. 391, 392.
4. J. Krishnamurti, in *Within the Mind: On J. Krishnamurti*, KFI, 1982, pp. 148-149.
5. Pupul Jayakar, *Krishnamurti: A Biography*, *op. cit.*, pp. 379, 380, 389.

14
Exploring Consciousness? : One

Abstract

Man is an amphibian who lives simultaneously in two dimensions – (a) the known, the time-space-causation matrices; and (b) the Unknown, the timeless. As a symbol, we may compare the known to a drop of water, and the Unknown to a fathomless and boundless ocean. The known is in the field of time, so it is limited, transient. The Unknown is in the realm of timeless, so it is eternal, infinite and immutable.

The known is at the lowest quantum energy potential of consciousness, while the Unknown is at the highest quantum energy potential of Consciousness. When consciousness with lower frequency encounters the higher frequency, the lower frequencies gets annihilated.

We want a revolution, psychological revolution, in which the whole human race is an equal partner. This revolution regenerates Intelligence, which resolves all human challenges.

It is not difficult to travel back in time to the earliest human attempts at observation. In a newborn baby, you are witnessing the early human observer. The child sees through its wholeness. But in addition it sees the particular when it grasps a finger held before its eyes, or it puts its thumb in its mouth. It is becoming aware of its hand and mouth, and brings subtle division between itself and the outside world – this distinction is often called the separation of the "*out there*" from the "*in here*", or in other words, the "*subject-object distinction*". The child was becoming conscious, which means awareness of the concept "I am". The "in here" experience was "I", the *subject*, the "out there" experience was the *object*.

Man has an evolutionary origin. The *Homo sapiens* appeared at about 2.5 million years ago. Perhaps the concepts of "in here" and "out there" arose in a rudimentary form with the beginning of stone age – "I (subject) go for search of food (object)". The evolution of the nervous system hastened this growth of awareness. It appears that man alone has awareness of the Self, as the *subject* of experience, along with awareness of the not-self, as the *object* of experience, of both the *within* and the *without* of nature. The *within* is called the subjective and the *without* is called the objective dimension of man. Man is unique in developing and nourishing his subjective dimension of life.

This article discusses the Universal Consciousness, the Life Field, the *Brahman*, the One Unmanifest-manifested. It also discusses science especially its role to investigate the *known* (human psyche) and the Unknown (the Life Field), the Reality. It also discusses spirituality for the scientific age, or in other words, "Symbiosis of Modern Science and Spirituality". It flows to radical transformation or mutation of the human psyche, which energises the Intelligence. The Intelligence blooms and blossoms to ending of all human challenges and man revels in nonviolence and peace.

Mysterious Consciousness?

Consciousness: George Wald wrote, "The simple fact is that consciousness gives us no physical signals.... We have no way of identifying either the presence, or absence of consciousness, let alone its contents". He added:

> One can put together everything that we have learned, and add to it everything that workers in this area hope to learn; and none of it comes anywhere near, or even aims in the direction of *what it means to see*.... Seeing – the event in consciousness – seems to lie in another universe, unapproachable by science.... Science offers no way of testing for the mere presence or absence of consciousness, let alone its content.... Science has no way of dealing with consciousness, not even to identify its presence or absence.... Since consciousness presents us with no identifying signals, nothing measurable, no apparent energy expenditure, it would be a comfort for a scientist to deny that it exists, or to declare it irrelevant.... Science tells us nothing about consciousness, nor does it promise ever to do so.... Science does not only discover, it creates, makes real what it asserts to be real.... science becomes a Tao, a Way: the way to the boundaries.[1]

There is no problem more mysterious than the mind-brain-body interaction. Modern science has not made it less intractable. On the other hand, it has added a new urgency and also a new poignancy.

Hideki Yukawa concluded, "Unless something rises into the consciousness it cannot become an object of rational thought, or rational consideratin".[2] Eugene Wigner stated, "The measurement is not completed until its result enters our consciousness".[3] Roger Sperry wrote, "Current concepts of the mind-brain relation.... gives full recognition to the primacy of inner conscious awareness asa causal reality".[4]

In a recent scientific conference at Moscow, the Russian scientists proposed, "A Life Field, beyond our physical concepts of time, space and the known electromagnetic and gravitational fields". In another conference at Cambridge, it was proposed, "A new scientific approach would include subjective experience as valid information and address all forms of human experience – including intuition, spirituality and mystical encounters".[5]

Mystery of Man: Arthur Eddington hinted at the emergence of the mystery of man from the study of the mystery of physical nature:

> Knowledge is only an empty shell – a form of symbols. *It is knowledge of structural form, and not knowledge of content.* All through the physical world runs that unknown content, *which must surely be the stuff of our consciousness.*[6]

Teilhard de Chardin concludes:

> Finally, it breaks down completely with man, in whom the existence of a within can no longer be evaded, because it is the object of a direct intuition and the substance of all knowledge.
>
> *The true physics is that which will, one day, achieve the inclusion of man in his wholeness in a coherent picture of the world.*[7]

Lincoln Barnett wrote:

> Niels Bohr puts it, "We are both spectators and actors in the great drama of existence". *Man is thus his own greatest mystery.... Least of all does he understand his noblest and most mysterious faculty: the ability to transcend himself and perceive himself in the act of perception.*[8]

The great neurologist, Charles Sherrington, wrote:

> Today, Nature looms larger than ever and includes more fully than ever ourselves. It is, if you will, a machine, but it is a partly mentalised machine and in virtue of including ourselves, it is a machine with human qualities of mind. It is a running stream of energy – mental and physical – and unlike man-made machines, it is actuated by emotions, fears, and hopes, dislikes and love.[9]

Richard Feynman wrote:

> Living things, masses of atoms, DNA, protein .... Dancing a pattern ever more intricate.... Out of the cradle onto the dry land.... Here it is standing.... Atoms with consciousness.... Matter with curiosity.... Stands at the sea .... Wonders at wondering.... I.... A universe of atoms.... An atom in the universe.
>
> That is what it means when one discovers how long it takes for the atoms of the brain to be replaced by other atoms, to note that the thing which I call my individuality is only a pattern or dance. The atoms come into my brain, dance a dance, then go out; always new atoms but always doing the same dance, remembering what the dance was yesterday.[10]

Man is a miniaturized cosmos. Since atoms with consciousness are looking at atoms with consciousness, we have the experience of wonder and mystery whenever we look at the universe. We are looking at ourselves; mind looking at mind. It is this process which we call the universe (time-space matrices).

Consciousness performs a dual role in the universe. In the world of the quantum, it is both the awareness and the creation of experience. It is being and the knowing of experience. Awareness implies that the Consciousness is expressing itself at the highest quantum energy potential, or at the highest frequency (higher frequency means higher energy); it is also called the Unmanifest. Experience implies that the 'successive consciousness' expresses itself at the lowest quantum energy potential as time-space matrices, or matter – it is called the manifest. Consciousness may be written as Unmanifest manifest. With the stroke of the twentieth-century quantum eraser, the dividing line between ontology (theory of being) and epistemology (theory of knowing) is rubbed out. The fusion of ontology, epistemology, psychology, philosophy, and physics is beginning to emerge.

Wave-Particle Duality: We can demonstrate that light is particle-like with the photoelectric effect, that it is wave-like with the double-slit experiment, and it is both particle-like and wave-like with Compton scattering. Physicists have demonstrated that light is both particle-like and wave-like, depending on how they looked at. Both of these complementary aspects of light (particle and wave) are necessary to understand the nature of light. Particle-like behaviour and wave-like behaviour are properties of our interactions. In 1923, Compton showed that X-rays also behaved as particle-like and wave-like, and this phenomenon is called Compton scattering.

The evidence gathered in the development of quantum mechanics indicates that subatomic "particles" constantly appear to be making decisions and they seem to know instantaneously what decisions are made elsewhere, and elsewhere may be as far away as another galaxy! This suggests that the photon in the double-slit experiments somehow "know" whether or not both slits are open and they act accordingly. Henry Stapp writes:

> The central mystery of quantum theory is 'How does information get around so quick?' How does the particle know that there are two slits? How does the information about what is happening everywhere else get collected to determine what is likely to happen here?[11].

There is no definite answer to this question. Some physicists, like Walker, speculate that photons may be conscious!

> Consciousness may be associated with all quantum mechanical processes.... since everything that occurs is ultimately the result of one or more quantum mechanical events, the universe is 'inhabited' by an almost unlimited number of rather discrete conscious, usually nonthinking entities that are responsible for the detailed working of the universe.[12]

Richard Feynman wrote:

> How does the particle find the right path? .... all your instincts on cause and effect go haywire when you say that the particle decides to take the path that is going to give the minimum action. Does it "smell" the neighboring paths to find out whether or not they have more action?[13]

The One Unmanifest-Manifested

There is One Reality, the Unmanifest manifested. The nature of the Unmanifest is 'Deep-Silence', 'Inactivity (Non-Vibration)', 'Passivity', 'Immutability', 'Eternal'; and also Activity (Vibration), dynamism and illimitable potency to manifest as many. Activity (Active Reality) expresses which is self-hidden in the Silence (Passive Reality); both are different states of the Supreme Reality, the obverse and reverse affirmations; both are Pure Existence and conscious Force, or Self-Existent Bliss and Supreme Intelligence, and are One. It is the Active Reality which manifests as the perfect freedom and omnipotence of an eternal divine activity in innumerable galaxies as inanimate world and plant and animal kingdoms.[14]

The Active Reality is a powerful, highly directional, monochromatic (same frequency) and coherent (the wave is in one phase) beam of energy; or a state of highest frequency (higher frequency means higher energy), or high quantum energy potential. It is homogeneous, differentiated and dimensionless. It is Omnipresent, Omnipotent and Omniscient. At the level of man, the intense energy of insight expresses itself as direct perception of the fact, the Reality. The insight acts upon the brain matter and transmutes it, so the brain matter transforms and attains a higher order through insight, or higher quantum energy potential of Consciousness. The insight or the understanding permits the powerful subtlest energies to penetrate the known and transform the known. The subtlest energy (higher potentials) knocks the thought out of commission, not by thinking, not by reasoning, but rather a direct transformation due to its contact with the subtlest energy. The insight is the supreme Intelligence.

The Universal Consciousness (highest quantum energy potential) heals, because it has the dimensions of creativity and spontaneity, nonviolence, beauty and harmony, love and compassion. Man has to choose between healing and annihilation, between the *Buddha* (Wisdom) and the *bomb* (annihilation). Obviously the decision is to work honestly, sincerely and truthfully for mutation of one's own psyche.[15]

Science

The twentieth century was dominated by science and technology. Science and technology have made great breakthroughs in medicine, agriculture, communication, travel and industry. Globalisation at the physical level is taking place as the result of the accelerative progress of science and technology and the thrust of psychosocial evolution, but global consciousness is not keeping pace with the globalisation. Globalisation at the material level is not a higher stage in the evolution of man. Technologically, man has progressed incredibly. Yet he remains psychologically immature, and he remains as has been for thousands of years; fighting, greedy, envious, and burdened with great sorrow, pain and travails. He has failed so far to overcome the age-old problems of hunger, poverty, disease, unemployment, ignorance, and so on.

Science is the search for truth. Objective science aims at nothing but making adequate statements about its object. The scientist only imposes two things, namely truth and sincerity, imposes them upon himself and upon other scientists. Science never imposes anything, science states. The aim of science is to study nature and human experience objectively. Objectivity and precision, in the external world are the pillars of science. All its answers are tentative and breed further questions.

Knowing is an intellectual activity, a cerebral activity. We live in the conceptual, ideational world – it is man-made socio-political-economic world. The whole mankind is conditioned. Conditionings are not the content of bondage, but accepting the authority of conditioning is the beginning of bondage. Knowledge is relevant to the man-made world. Knowledge and experience is relevant to the collective relationships so far as they are concerned with time and space. Acceptance of the authority of knowledge (information) and experience as a measurement, to measure the whole of life, is the beginning of misery and sorrow. The notions, the myths, the illusions, that have been handed down generation after generation, are the psychological inheritance. Our present education, present society and our way of living adds to these inheritance, conditioning.

Le Compte du Nouy, the known French scientist, explained the aim of science in words:

> Science describes facts, objects and phenomena minutely, and tries to join them by what we call laws, so as to be able to predict events in the future.[16]

All human observations are relative to the system of reference or the scale of observation. This points to both the limitations of sensorial apparatus as well as the limitation of the instrument which interprets the sense-impacts. Charles-Eugene Guye, Swiss physicist, said, "The scale of observation creates the phenomenon".[17]

The conditioned human psyche uses the sensorial apparatus and constructs the time-space matrices in quick succession. The subject (past memory) is never aware of the *fact*, the *true*, the *present*, the *what is*?, but invariably aware of the *non-fact*, the *false*, the *past*, the *what should be*! Both the subject and the object are continuities of change. The object is a distorted part. So the observation of the non-fact leads to a distorted perception, which is called the crisis of perception, the psychic pandemic.[18] Such scientific observation leads us to a world of illusion, a system of symbols, science expresses average and probable results, which enhances the conditioning of the psyche.

Swami Ranganathananda wrote:

> And if science is to progress further, it has to choose for investigation this field of the mystery of man which towers over its erstwhile study, namely, the mystery of the external universe. This is a vast field of study

> – the field of man's self-awareness, the field of his consciousness, his ego, his being the *subject* and not the object. Science will find here a vaster and more fascinating and rewarding field of study than in external nature.[19]

Spencer Brown, in a most spectacular passage, wrote:

> But *in order* to do (see) so, evidently it must first cut itself up into at least one state which sees, and at least one other state which is seen. In this severed and mutilated condition, whatever it sees is *only partially* itself.[20]

Eddington put it succinctly, "Nature thus provides that knowledge of one-half of the world will ensure ignorance of the other half". And Haldane muttered that "the universe is not only queerer than we suppose, it is queerer than we can suppose". The split of the mind into the seer (subject) and the seen (object) introduced dualism. The error of dualism forms the root of intellection and is therefore next to impossible to uproot by intellection.

About 1600, Kepler and Galileo simultaneously and independently formulated the principle that the laws of nature are to be discovered by measurement. All propositions were to be confined to that which was objectively measurable and verifiable. The very crux of twentieth-century physics is the recognition that the observer cannot be excluded from his observation; he is an intrinsic participant in them. Every act of observation of an atom by a physicist disturbs the atom; observation adds or subtracts a photon from the atom. As Sullivan put it, "We cannot observe the course of nature without disturbing it".[21] Andrade also stated, "Observation means interference with what we are observing.... observation disturbs reality".[22] Physicists concluded that *objective measurement and verification could no longer be the mark of absolute reality, because the measured object could never be completely separated from the measuring subject – the measured and the measurer, the verified and the verifier, at this level, are one and the same.* The subject cannot tinker with the object, because subject and object are ultimately one and the same thing.

Erwin Schrodinger put it bluntly:

> Subject and object are only one. The barrier between them cannot be said to have broken down as a result of recent experience in the physical sciences, *for this barrier does not exist.*[23]

Werner Heisenberg's conclusions are very clear:

> From the very start we are involved in the argument between nature and man in which science plays only a part, so that the common division of the world into subject and object, inner world and outer world, body and soul, is no longer adequate and leads us into difficulties.[24]

Erwin Schrodinger heartily concurs, and states simply, "These shortcomings can hardly be avoided except by abandoning dualism".[25]

Our activity introduces discontinuities in whatever we are observing. These discontinuities are fundamental to the new physics of twentieth-century.

To Bohr, discontinuity was a fundamental truth.[26] In an atom, there is a quantum jump from one orbit to another, and the quantum jump is a discontinuity, a timeless realm. So in an atom there is an evidence of time → timeless → time, or continuity → discontinuity → continuity.

Many eminent scientists – Einstein, Schrodinger, Heisenberg, Eddington, and so on – stumbled upon the intuitive or intimate mode of knowing. Eddington wrote:

> We have two modes of knowledge which I call symbolic knowledge and intimate knowledge .... The intimate knowledge will not submit to *codification and analysis*; or, rather, when we attempt to analyze it the *intimacy is lost* and it is replaced by symbolism.[27]

Alan W. Watts stated, "unconventional knowledge (is) the understanding of life directly; instead of in the abstract, linear terms of representational thinking".[28] If we are to know the Reality, we have to "understand life directly".

Physicists have showed a new meaning for the word *order*. This new order, the basis of the new physics was not found in the particles of matter. Rather it was found in the minds of the physicists. The order of the universe may be the order of our own minds.

Deepak Chopra wrote, "To know God personally, you must penetrate a boundary that physicists call '*the event horizon*', a line that divides reality sharply in half. On this side lies anything that remains within the speed of light; on the other side is anything faster than the speed of light".[29] We call the boundary, or the wall, between symbolic-dualistic frame and non-dual and non-conceptual frame of reference of mind spectrum as 'the frequency barrier' or 'the event horizon'. We assume the mass of thought as electronic rest mass, m_0 [$m_0 = 9.1091 \times 10^{-31}$] and c (velocity of light) is 3×10^8 meters/sec. Based on this assumption, the energy of the frequency barrier comes out to be 8×10^{-14} joules, or 0.5×10^6 electron volt. It means that the frequency barrier represents half-a-million electron volts (0.5 MeV). On this basis, the energy comes out to 8×10^{-14} joules, or 0.5×10^6 electron volts. The frequency of vibration comes out to be 10^{20} vibrations/sec. This has not been measured, but we take it as a first approximation. It gives an order of magnitude. Perhaps, it is the energy level of the "frequency barrier", which makes it difficult to make a quantum jump from time to timelessness. It is limited, an intellectual approach, and requires further exploration.[30]

Whenever any species is in danger of extinction, some evolutionary impulse/urge within the race itself pushes to a new mode of consciousness which would enable it to overcome the danger. This is known as self-unfoldment of consciousness. This survival mechanism has in fact been largely responsible for evolution so far.

Ervin Laszlo discussed the emerging concepts of the cosmos in words:

> The parts of this totality are not radically separate: they are integral elements of the whole. Every "thing" in the cosmos is a coherent "event"

> and every event is the summed result of all the other events.... The cosmos is a system of interacting waves: it is a sea where the effect of the smallest perturbation propagates throughout the entire field.... the new physics and the new cosmology do not recognize discrete things and independent events in the universe, only ripples upon ripples, waves upon waves, interpenetrating and propagating in a seamless sea.[31]

Every thing in the cosmos is a coherent event and every event is the summed result of all other events. The observable world appears to have emerged from the "quantum vacuum (potentialities)" that constitutes the "ground energy state" of the universe. The ground energy state of the cosmos is the Eternity.[32]

Mankind needs a new "instrument of thought", in the words of David Bohm;[33] "the poetic intuition .... which the intellect can never fully grasp", in the words of Max Planck;[34] a completely new form of human intellectual endeavour", in the words of G.F. Chew.[35] These physicists point out that the symbolic-dualistic frame of reference or relative frame of reference of the mind spectrum cannot understand the 'intuition' or 'new form of human intellectual endeavour', but it is the non-dual frame of reference of mind spectrum that has the capacity to perceive the truth.

David Bohm said, "Man is potentially infinite".[36] The potential of the human race is infinite as life is infinite. It appears that the potential contained in the human beings to peel off layers after layers of untruth and reach the core of truth is infinite. *Spirituality is defined to maximize the potential of the human being and to live in that highest potential.*[37]

The Urgency of Self-Discovery

(a) Awareness is the movement of Intelligence. It is the response of the whole being to the communion with the fact.

(b) To live is to be related, and for getting related the awareness of '*what one is*' is absolutely necessary.

(c) An inquirer has the humility to face the global problems in his own life, in his own being; in his relationship with others to face them there.

(d) It is the understanding of life directly that resolves the inheritance.

(e) The awareness of the infinite and eternity makes you extremely sensitive in your relation to the moment *here* and *now*. The awareness of the whole puts you in correct perspective of the particular. We have to deal with the *present*, the *now*, the *here*.

Human Survival

The single planetary civilization to which we all belong confronts us with global challenges. We stand helpless before them because our civilization has essentially globalized only the surface of our lives. But our inner self continues to have a life of its own.[38] Dag Hammarskjold declared, "I see no hope for permanent world peace. We have tried and failed miserably. Unless the world has a spiritual rebirth, civilization is doomed".[39] Andre Malraux wrote, "The third millennium will be spiritual or there will be no third millennium".[40] The answer to this paradox is "rooted in self-transcendence, or transcendence as the only real alternative to extinction". George Wald stated, "By now the problem is not transcendence, but survival. If we are to survive, there must be a fundamental change in direction, not only organizational, but in our ways of thinking. We have to think different thoughts and to want different things".[41] Jacques Monod wrote, "What we see opening before us today is an abyss of darkness.[42]

Spirituality

Man alone has the capacity to understand and decode the deep mysteries of nature. He seems to be born with the faculty of self-awareness and understanding. *Spirituality is understanding the conditioned mind, the matter content of the consciousness, and also to perceive the unconditioned energy; so it is the science of the wholeness of life. It aims at self-actualization of the hidden human potential. Spirituality is fundamental and objective science is a derivative. An integration of science and spirituality is the key to the survival of mankind.* It jettisons all pain, misery and travails, and ushers a new era of freedom, peace and bliss.

Spirituality is a science that deals with the wholeness of life – the inorganic world, the inanimate and animate organisms and the individual man. It begins with the awareness of the whole. In spirituality, the subject is awareness of the whole and it bestows pure perception. In spiritual science, the emphasis is upon the act of perception, the quality of the instruments employed for perception, and the quality of the consciousness behind the perception.

Both science and spirituality have a non-authoritarian approach. *Spirituality is defined to maximize the potential of the human being and to live in that highest potential.* It heals the diseased mind-brain-body system. Healing implies to minimize the matter content (debris) of consciousness, which enhances the perceptive sensitivity of the system. It energises the supreme Intelligence (Wisdom) which ends all human psychological imbalances, ecological challenges, and other human problems.

Spirituality teaches us the art of living. It bestows intuition (insight), creativity, spontaneity and right action. It implies to have understanding of life directly. It teaches us to live without any psychological debris. It is living according to the laws of nature. It is man's only vocation.

Symbiosis of Science and Spirituality

Man is an amphibian, who lives simultaneously in two dimensions – (a) the *known*, the time-space matrices, which is at the lowest frequency of consciousness, and (b) *Unknown*, the timelessness, the highest frequency of Consciousness. As an analogy, we may compare the *known* to a drop of water and the *Unknown* to a fathomless and boundless ocean.

The leading discoveries of dynamic interplay of contemporary science and spirituality are as under:

(a) Modern science has confirmed that Consciousness is the primary datum of human existence.

(b) In a recent (1996) scientific conference at Moscow, the Russian scientists proposed, "A Life Field, beyond our physical concepts of time, space and the known electromagnetic and gravitational fields". In another conference at Cambridge, it was proposed, "A new scientific approach would include subjective experience as valid information and address all forms of human experience – including intuition, spirituality and mystical encounters".[43]

(c) Many eminent scientists – including Einstein, Heisenberg, Schrodinger, Eddington – have stumbled upon another mode of knowing, which has been called intimate, or direct, or non-dual knowledge. The non-dual knowledge aims at the understanding of life directly, instead of in the abstract, linear terms of representational thinking. The non-dual and non-conceptual mode of knowing is called the "Epistemological Revolution". It bestows intuition (insight), learning and right action. The intuitive mode is fundamental and intellect is a derivative. It bestows a perception of the wholeness of life. It leads to a quantum jump from the symbolic-dualistic frame of reference of mind spectrum to non-dual frame of reference of the Mind Continuum.

(d) Modern science has shown that the objective (outer) and the subjective (inner) are not two separate movements, but one, a unitary movement, a whole. The outer and the inner are interdependent and inseparable. It is evident that the objective science must restructure itself to take in its fold the subjective dimension of man, and this is called restructuring of science, or scientific mode of mind. The scientific mode sees the wholeness of life, so the subject is synergistic to thought. If thought comes in the field of the true subject, it is bombarded and it gets annihilated (cold fission), or in other words, thought goes to abeyance.

(e) The restructured science deals with the wholeness of life. The non-variable attentivity or reaction-free attentivity or objectless attention takes over as the true-subject. The subject sees that the movement of thought coupled with volitions is independent of the "me". It implies that the concept of the ego is independent of thought momentum. As the concept of the ego is at a lower quantum energy potential than the momentum of thought, so it is knocked down, or the I-consciousness or ego goes to abeyance. The mental movement has a momentum independent of I-consciousness.[44] The vanity that it is my thought coupled with volition ends and its place is taken by humility.

(f) The true subject watches the movement of thought coupled with volition. It is aware simultaneously of the objective challenge, the subjective reactions and the causes of the reactions, then it results in freedom.[45] The subject is ahead of the reaction, so the subject sees the reaction, as it sees an objective challenge. The thought with its volition goes to abeyance. This blossoms to pure perception. A simple perception can result in an objective approach to all the subjective world. There is awareness of the racial consciousness. The pure consciousness sees itself. *The law of nature is that true subject is synergistic to thought structure; and thought structure is synergistic to I-consciousness, ego*.

(g) Understanding, learning and action are a cyclic process, which results in a spiral going upwards in a vertical dimension; it is growth in the vertical dimension. The higher the understanding, the higher the expression of consciousness in the mind-brain-body system. The higher sensitivity or creativity leads to a perception of the pure consciousness.

Concluding Comment

It is necessary to incorporate the science of mind (or consciousness) in the curricula of all Universities, Colleges and Institutions. Every human being has to get introduced to psychodynamics and limitations of the mind-brain-body system; and that will flower and blossom to new inter-relationship between mind-brain-body system, or the *outer* and the *inner* of man.

The human life from womb to tomb is a school, whose significance is to understand and learn from yourself; in other words, to be ***light*** to yourself. The word light signifies higher frequency of consciousness, so darkness is associated with lower frequency of consciousness. Light is associated with non-dual and non-conceptual frame of reference of Mind Continuum, while darkness is associated with symbolic-dualistic frame of reference of mind spectrum. Light is intuition (insight) while darkness is sensual knowledge (information) and experience.

Spirituality explores what the truth is, what the *fact* is, and what the Reality is behind the truth. It teaches us to live with the *fact* and transcend the *non-fact*. Only the truth can liberate the mind from its own ideation. The truth leads to mutation of the psyche.

We have to work for *restructuring of science*, *restructuring of education, awareness of human perennial values*, and *live that understanding every moment*. It is symbiosis of modern science and spirituality that has the capacity to resolve our psychological imbalances, ecological problems and other human challenges. It re-establishes Order, Harmony, Non-Violence and Peace in life and the world. This establishes the manifestation of ever greater potentials and newer dimensions of Consciousness or Life Field.

When Consciousness with higher frequency encounters consciousness with lower frequency, the latter gets annihilated. In other words, when subject at the higher quantum energy potential of Consciousness bombards an object with lower quantum energy potential of consciousness, the object gets exploded (cold fission), so the object goes to abeyance.

The human survival and excellence is embedded in the higher quantum energy potential of Consciousness. Our core commitment and vocation is to make a quantum jump from intellect to intuition (insight), from symbolic-dualistic frame to non-dual frame of reference of Mind, from *non-fact* to *fact*, from action to non-action (total relaxation); and then live in the light of that perception and understanding. It is to understand and live according to the immutable laws of nature.

The restructured science and the restructured education are fundamental for constructing a sane global society and the perception of the highest hierarchical level of expression of Consciousness (Life Field, *Brahman*, the Absolute Subjectivity). This bestows non-dual Subjectivity, or non-dual and non-conceptual awareness. It resolves all human challenges facing mankind, and usher in a new era of non-violence, peace and bliss. It is the greatest vocation in the world. This is the consummation of human life. This only is the social, psychological, and even the political revolution.

We want a revolution, psychological revolution, in which the whole human race has to participate and every man is an equal partner. This revolution will regenerate or rejuvenate human mind so that it perceives the new dimension of an integral and all-inclusive Consciousness, that can guarantee human survival along with all fauna and flora.

The cultural transformation or the evolutionary mutation of this magnitude and depth cannot be prevented by short-term socio-economic-political activities. This provides mankind the strongest faith in the Order, the Wisdom, the Intelligence, and future of mankind along with all fauna and flora.

Man is his own greatest enemy. Knowing fully well, the path of *Light*, yet he wants to live in ignorance and darkness. If man escapes his responsibility at this critical juncture, he does so at his own peril.

References

1. George Wald, *Synthesis of Science and Religion*, Bombay: The Bhaktivedanta Institute, June 1988, pp. 20, 18.
2. Hideki Yukawa, *Creativity and Intuition*, Tokyo: Kodansha International Ltd., 1973, p. 203.
3. Eugene Wigner, *Two Kinds of Reality*, Monist, 1964, pp. 248-64.
4. Roger Sperry, in *Synthesis of Science and Religion*, Bombay: The Bhaktivedanta Institute, June 1988, p. 199.
5. Joan Walton, *Scientists Tune into Consciousness*, New Science, Winter 1996.
6. Arthur Eddington, in Swami Ranganathananda, *Science and Religion*, Calcutta: Advaita Ashrama, 1979, p. 125.
7. Pere Teilhard de Chardin, *The Phenomenon of Man*, London: Collins, 1959, p. 55.
8. Lincoln Barnett, in Ranganathananda, *op. cit.*, p. 27.
9. Charles Sherrington, *Man on His Nature*, Pelican Edition, p. 38; in Swami Ranganathananda, *op. cit.*, p. 32.
10. Richard Feynman, in Fred Alan Wolf, *Taking a Quantum Leap*, Harper and Row, 1981, pp. 227-28.
11. Henry Stapp, *Are Superluminal Connections Necessary*?; in Gary Zukav,

The Dancing Wu Li Masters, Fontana/Collins, 1982, pp. 87-88.

12. Evans Walker, in Gary Zukav, *ibid.*, p. 88.
13. Richard Feynman, in Fred Alan Wolf, *op. cit.*, p. 197.
14. Sampooran Singh, *Dynamic Interplay Between Science and Religion*, Jodhpur: A Jainsons Publications, 1978, pp. 357-86.
15. Sampooran Singh, AVM Manjit Singh (Retd), Dr (Mrs) Kanwaljit Kaur and Paramjit Singh, *God in Science*, Chandigarh: Vishvas Publication Pvt. Ltd., July 1999, p. 2.
16. Le Compte du Nouy, in Rohit Mehta, *The Eternal Light*, Adyar: The Theosophical Publishing House, 1961, p. 5.
17. Charles-Eugene Guye, in *ibid.*, p. 6.
18. Sampooran Singh, *Science, Human Values, Value Education and Peace (SS: 12)*, New Delhi: New Man and New World Movement (India), September 1993.
19. Swami Ranganathananda, *ibid.*, p. 26.
20. Spencer Brown, *Laws of Form*, New York: Julian Press, 1972, pp. 104-05.
21. J.W.N. Sullivan, *The Limitations of Science*, New York: Mentor Books, 1949, p. 140.
22. E.N. da C. Andrade, *An Approach to Modern Physics*, New York: Doubleday Anchor Books, 1957, p. 255.
23. Erwin Schrodinger, *What is Life*? and *Mind and Matter*, London: Cambridge University Press, 1969, p. 137.
24. Werner Heisenberg, *The Physicist's Conception of Nature*, Harcourt, Brace, 1958, p. 24.
25. Erwin Schrodinger, *My View of the World*, London: Cambridge University Press, 1964, p. 62.
26. Niels Bohr, in Fred Alan Wolf, *op. cit.*, p. 85.
27. A. Commins and R. N. Linscott, ed., *Man and the Universe, The Philosophers of Science*, New York: Washington Square Press, 1969, p. 453.
28. Alan W. Watts, *The Ways of Zen*, New York: Vintage Books, 1957, p. 10.
29. Deepak Chopra, *How to Know God*, London: Rider 2000, pp. 270-71.
30. Sampooran Singh, *Global Education and Human Survival*, New Delhi: Sanbun Publishers, March 2002, p. 77.
31. Ervin Laszlo, *Emerging Concepts of the Cosmos*, *The Scientific and Medical Network*, Newsletter No. 49, August 1992, p. 28; in Sampooran Singh, *The Pilgrimage to the Third Millennium*, Chandigarh: Faith Publishers, 1995, p. 31.
32. Sampooran Singh, *God in Science*, *op. cit.*, p. 124.
33. David Bohm, in Gary Zukav, *op. cit.*, p. 326.
34. Max Planck, *The Philosophy of Physics*, New York: Norton, 1936, p. 83.
35. G.F. Chew, in Gary Zukav, *op. cit.*, p. 331.
36. David Bohm, in *PsychoScience*, Spring 1994, Vol. 1/No. 1, p. 27.
37. Sampooran Singh, Dr. (Mrs) Kanwaljit Kaur and Paramjit Singh, *Global Values Education*, Chandigarh: Faith Publishers, January 2001, pp. 63-64.
38. Vaclav Havel, President of the Czech Republic, *Speech at Philadelphia, July 4, 1994.*
39. Dag Hammarskjold, Secretary General of the UN, in Robert Muller, *My Testament to the UN*, USA: World Happiness and Cooperation, 1993.

40. Andre Malraux, *ibid.*, *My Testament to the UN.*
41. George Wald, Nobel Prize in Medicine, 1967, in Kishore Gandhi, *The Evolution of Consciousness*, New Delhi: National Publishing House, 1983, p. 14.
42. Jacques Monod, *Chance and Necessity*, New York: Vintage Books, 1972, p. 171.
43. Joan Walton, Scientists Tune into Consciousness, New Science, Winter, 1996.
44. Vimala Thakar, *Through Silence to Meditation*, Ahmedabad: Vimal Prakashan Trust, 1998, pp. 22-27; and *Silence*, Holand Handbook Bookfund, Vimal Thakar, 1989, p. 6.
45. *Ibid.*, *Mutation of Mind*, Ahmedabad: New Order Book Co., 1972, pp. 122, 171.

15
Exploring Consciousness : Two

Abstract

Modern science has provided experimental evidence that the world is fundamentally a discontinuous and quantum jump world. An atom can exist only due to quantum jump. The biological autonomous processes – seeing or hearing, smelling or breathing, waking or deep-sleep state are movements in the realm of non-duality (timelessness) and these are functioning in choiceless quantum energy levels of supreme Intelligence. At the psychosocial evolution, there is discontinuity between two thoughts, between intellect and intuition (insight). Life is a self-organising systems – self-maintenance and self-transcendence – and this can only happen in a limitless, unbounded and unindividuated energy processes, events and happenings – and this we call Life Field, or Unmanifest, or Cosmic Consciousness.

We live simultaneously in two dimensions – the objective and the subjective. We have to restructure science so that it takes in its ambient both the objective and the subjective dimensions; and this deals with the wholeness of life. We have to restructure education so that it enfolds both the sensual and the trans-sensual dimensions. The restructured science-education observes the conditioned thought-structure, and the latter goes to abeyance. This leads to quantum jump of knowledge (information) to attentivity and awareness. This flows to understanding, learning and right action. This gives a perception of the fact, which blossoms to non-action of the mind, or the silence.

We need urgently symbiosis of science and spirituality. This heals the diseased mind-brain-body system, which is a dimensional transformation or mutation. It is the Cosmic Consciousness or Life Field that brings in mutation in the psyche. This guarantees human survival and excellence.

Human consciousness is a multi-dimensional activity and it functions in two dominant modes: (a) the *known*, the time-space-causation matrices, the conditioned energy as experience and knowledge (information), the manifest, and (b) the *Unknown*, the Unconditioned Energy, the Unmanifest. The biological autonomous processes – seeing or hearing, smelling or breathing –

are movements in the realm of non-duality; and these are functioning in an interrelated, intertwined, choiceless, complex and yet homogeneous way as expressions of supreme Intelligence which surrounds and interpenetrates us in Nature. The *known* processes gather experience and knowledge, store these in the brain, and act as ego, the "me". This is mechanistic, repetitive and dualistic process. The *Unknown*, the choiceless energy of Unconditioned Consciousness or Awareness, which is an energy born of the emptiness of Silence, which is born by the non-action of conditioned consciousness, and this functions without any volition, or ego, or I-consciousness. The movement of organic Intelligence in the complexity of cosmic Life is a movement of non-duality and conducted in non-duality. It is an energy that permeates the whole being, vibrates in the whole being, that becomes the source of perception and response and has no relation to the willing, the volition, the choice. So Awareness is choiceless.

The *known* is at the lowest frequency of consciousness, or lowest quantum energy potential of Consciousness. The *Unknown* is at the highest frequency, or highest quantum energy potential. Each higher stage is synergistic to its junior components; it includes them but is more than them. When the higher stage bombards the lower stage, it dissipates the lower stage which goes to abeyance. The higher illuminates the lower, and not *vice versa* – this lays foundation of a creative principle. The mutation or radical transformation of the human psyche *happens* when it is bombarded by higher consciousness, so mutation is governed by Consciousness, or by the Unmanifest.

We explore the mysterious Consciousness. It also discusses knowledge, observation, attentivity, awareness, understanding-learning-right action-meditation, and silence. It also discusses brain, memory and mutation and symbiosis of science and spirituality. It is the dimension of Cosmic Consciousness, or Life Field, or the Unmanifest that has the potential of human survival and excellence.

What is Mysterious Consciousness?

Hideki Yukawa wrote, "It is obvious that unless something rises into the consciousness it cannot become an object of rational thought, of rational consideration. At the same time, everyone will surely agree on the importance of an awareness that living in a world open towards both the exterior and the interior is a special characteristic of man's lot".[1] In the oft-quoted words of E.P. Wigner, "The measurement is not completed until its result enters our consciousness.... (this) last step is, at the present state of our knowledge, shrouded in the mystery and no explanation has been given for it so far in terms of quantum mechanics, or in terms of any other theory".[2] The "measurement .... enters our consciousness" in some way implies that consciousness is Intelligence or Self-Awareness. Fritjof Capra defined, "And by consciousness I mean the state of mind characterized by self-awareness, where one is aware that he is aware. We know that we know.... But science, at present, does not know how to describe the nature of consciousness.... And this is very much in agreement with spiritual views".[3]

We are all human beings, we are the rest of mankind, we are the rest of humanity. Your brain is not yours, it has evolved through thousand and thousand years. I am neither individual, nor collective, but humanity. The memory or thought-structure is shared by mankind. Thinking is a playback of thought coupled with feeling and response of the memory – cultural, racial, religious, communal – which is called willing. Willing flows to reaction. Thinking is common to all of us, so thinking is the movement of all mankind, it is not individual thinking. You are concerned with global thinking. You are concerned with the world, with all humanity. We live in the past and the mind can project itself into future. We are conditioned. We have constructed an aggressive and violent society. There is global disorder, the threat of wars, poverty, fear, and so on – we are responsible for that. We live in pain, misery and conflict. Conflict wears out the brain, makes the mind old. We have gathered psychological imbalances and ecological challenges. Life is movement in relationship – man with nature, man and man, man and other organisms, and man and environments. The thought structure cannot have right relationship with others.

The answer to this paradox is embedded in psychological search, seeking out inwardly and finding out for oneself a tremendous area that is beyond all thought. It is to transcend thought or ego, be aware of the *fact* in the non-dual realm, the "understanding of life directly", learning and to explore the all-inclusive dimension of Consciousness or Life Field. It bestows understanding, intuition (insight), spontaneity, right action and meditation. That action is born of intelligence – intelligence, love and compassion, all go together. This is one harmonious movement of the dimension of Consciousness or Life Field. This is really a sacred thing and the beauty is in itself and not beyond it.

Knowledge

Human consciousness is a multidimensional activity. In the common waking state, it expresses itself at the lowest hierarchical expression of consciousness. At this level, the mind cuts itself into the subject and the object, and the subject gathers experience and knowledge (information) and stores it in the brain as frequency and amplitude information – called memory. Knowledge is always an indirect activity, so knowledge is invariably fragmentary, divisive and dualistic.

The conditioned I-consciousness has voluntarily accepted the role of a pseudo-observer, so it reacts to a challenge. Observing and reacting are incompatible. Observing, evaluating and assessing are incompatible.

The thought-energy acts as per its motives, desires, acquisitiveness, psychological needs. Thought constructs habit patterns, so it reacts as personal preferences, or prejudices. Thought accepts ideologies. The thought-energy is controlled, shaped, refined, sophisticated by civilisation, by culture.

Our daily living is nothing but a chain of reactions to the compulsions that surround us – socially, culturally, economically and politically. I accept the authority of the past, of the compulsions and pressures. So I live in the past with pain and sorrow.

The measurements, the symbols representing the measurements, are the substances (matter) of the past. There is no such thing as psychological evolution. There is nothing to learn in the psychological field. Psychologically, inwardly, there is only the *fact*, which is in the non-dual realm. To be a light to oneself requires great freedom, not a conditioned brain.

Thinking is the movement of all mankind, it is not individual thinking. Experience and knowledge stored in the brain, as memory, is thinking.

Observation

The first stage prior to recognition and naming, identifying, judging or the association of ideas and emotions, is a state of non-verbal or thought-free perception.[4] To observe non-subjectively, to observe without any evaluatory comparison or judgement. The occurrence of seeing or hearing, smelling or breathing is a movement in the realm of non-duality. There is no subject-object division and no movement born of the tension of duality. There is no time gap between observation and action; it is a kind of simultaneity which seems to be the principle of Cosmic Life.

Sullivan wrote, "We cannot observe the course of nature without disturbing it".[5] Andrade stated, "Observation means interference with what we are observing .... observation disturbs reality".[6] George Wald stated, "The very crux of twentieth-century physics is the recognition that the observer cannot be excluded from his observation; he is an intrinsic participant in them".[7] The observer and the observed, at this level, are one and the same.

When I observe my own psyche, I see that I am a part of the psychological load and I participate in its activity at every moment, so I am the disorder, the violence. In this observation, the subject and the object get integrated, and the ego, the I-consciousness goes to abeyance. This brings about a humility. If you observe the racial thought, it flowers, blossoms and withers away, and intelligence expresses itself. Humility is the prerequisite of learning; humility is the soil in which intelligence flowers. Intelligence is the unconditioned energy of the whole being.

Human consciousness is a multidimensional activity; it implies that it can express itself at different quantum energy potentials of consciousness. It can express itself to inquire into the meaning of Life, the nature of the Reality, the Truth sustaining facts. Each quantum energy potential of Consciousness can knock down a structure at a *lower* quantum energy potential, so it bestows unpolluted, uncontaminated and non-evaluatory perception.

Human interference introduces discontinuities, it is fundamentally a discontinuous and quantum jump world. Niels Bohr stated that discontinuity is a fundamental truth.[8] The discontinuity happens when the energy from the Unmanifest flows to the manifest raising its quantum energy potential, or the immutable law of nature – Unmanifest → manifest – expresses itself.

The restructured science sees both the objective and the subjective dimensions of life, or the wholeness of life. The Subject of Cognizance is non-variable attentivity, or reaction-free attentivity, or objectless attention, it is also called true-subject, or a mind with a scientific temper, or the state of observation. It is tuned to perceive the *fact*, the *true*, the *present*, the *what is*!, the *non-duality*. It bestows a perception of "understanding of life directly", or in other words, it bestows intuition (insight), spontaneity and right action. The pure perception bestows non-violence, harmony, human values, truth, love and compassion, so it is in communion with the Consciousness, the Life Field.

Attentivity

Observation without an observer and an observed is attentivity. The whole being gets tuned into a state of attentivity, where there is neither subject, nor object. It is a subject-free and object-free attentivity. It is not an activity on the plane of duality. It is just sensitivity of the whole being that becomes attentivity.

Awareness

When attentivity is established over a long chronological time, it makes another quantum jump and it is called awareness. It is awareness of the Wholeness, the Oneness of Life.

Awareness energises the supreme Intelligence. The movement of organic Intelligence in the complexity of cosmic Life, is a movement born of non-duality and conducted in non-duality. There is no association with volition. It is non-voluntary. Awareness brings mutation or transformation of the whole being. Mutation is a radical change in the quality of the mutant, in the substance of the mutant. It is not a peripheral change. It is a radical change in the whole of the mutant. Mutation leads to non-verbal communion between what is outside me and what is within me.

Understanding, Learning, Action and Meditation

Understanding is not in the symbolic-dualistic frame of reference, but it is embedded in the non-dual frame. It is timeless perception. Understanding comes only when the mind is very quiet. Perception has nothing to do with knowledge (information). The perceptions are born of non-duality and responses are born of spontaneity. Spontaneity bestows right action. It is insight alone that can decondition the brain.

Understanding, learning, action and meditation are one harmonious movement. These belong to the realm of non-duality. If this is clearly understood, there is no wastage of energy. To learn there must be total freedom. This one movement is also called meditation. Meditation, or the transcendence of the conditioned mind, is not the result of physical or mental activity. It is in the relaxation, it is in the non-movement, it is in the non-action of the conditioned, that the non-conditioned begins to operate.

Meditation: The roots of every action, every thought and every emotion, are in the subconscious. And that subconscious regulates, directs and controls the nature of the conscious action. The scientific mode of mind, all-inclusive attention, observes the objective challenge, the nature of subjective reactions and the causes of the subjective reactions simultaneously, and such an awareness is always ahead of the reactions. When such reactions face the awareness, they lose their momentum and I am not carried by the reactions. The reaction exhausts its momentum and arrives at the simple, innocent silence. It results in freedom from the momentum of reactions.

A response which is free from the momentum of reaction is a natural response. So natural response is a response which results out of understanding of the wholeness of challenge, reactions and the roots of reactions. The natural response is in the realm of non-duality. Natural response emerges when reaction comes face to face to the light of attention, so we expose the reaction to our attention, to our sensitivity, to our alertness.

The simultaneous awareness of the total movement is meditation. It is a new approach to life, in which we observe the objective challenge, the subjective reactions and the causes of the subjective reactions simultaneously without identifying oneself to any of them.[9]

Silence

It is the total renunciation of the movement of conditioned mind, so the mind goes to non-action. The discontinuity of the conditioned ego takes one to a different dimension altogether, where knowing or experience may not happen at all; but perception, understanding, learning and action – a new dynamics of behaviour – may get activised. The non-action of the I-consciousness – or in other words, *silence* of the conditioned mind – is a different dimension, where the highest frequency of Consciousness bathes the brain cells, so there is no identification, no fear, no violence, no psychological imbalances, no ecological challenges, and so on. Silence is totally unconditioned, the emptiness is unconditioned. It is a nudity of Consciousness.

The motionlessness of speech, the motionlessness of the thought structure, the motionlessness of the subject (the ego, the self), then that relaxation is the emptiness of consciousness. Thought-freeness of consciousness shall release Awareness. If Awareness is stabilised over a long chronological time, it blossoms to perception of Spirituality, the Wholeness of Life.

It is an inner voyage of self-discovery from self-observation through attentivity to Silence. It is a journey from human consciousness (successive consciousness) to Simultaneous Consciousness, and living in the dimension of Silence. It is an enquiry on behalf of the human race for an alternative way of living, an alternative culture, an alternative dynamics of human relationships.

The living and moving Silence settled down in the neurochemical system is called state of meditation. Meditation implies to see and live in the wholeness of life. Meditation is an alternative way of living, living in the Emptiness.

Brain and Memory

Brain: My brain is the brain of humanity. Every cell in the human brain holds memory of the million years of man. So memory or thought-time is of the humanity.

Memory: It has introduced a subtle illusion, which was first spotted by St. Augustine, and recently confirmed by Schrodinger and Watts. Bertrand Russell summarised the viewpoint of St. Augustine in words, "past and future can only be thought of as present: 'past' must be identified with memory, and 'future' with expectation, memory and expectation being both present facts".[10] Watts wrote, "From memories you infer that there have been past events. But you are not aware of any past events. You know the past only in the present and as part of the present".[11] Schrodinger wrote, "Mind is always *now*. There is really no before and after for Mind. There is only a *now* that includes memories and expectations".[12] You are not really aware of the actual past at all. You are looking at a present trace of the past, or trace of the future projection; and surely this comes out of the timelessness. Nature always works in an absolute present. The past trace or future projections happens in a few millionth of a second.

Sri Aurobindo issued a grim warning to mankind in words:

> The evolution of mind working upon life has developed an organization and use of matter which can no longer be supported by human capacity without an inner change. The problem is fundamental and in putting its evolutionary Nature is confronting man with a critical choice which must one day be solved in the true sense if the race is to arrive or even to survive.[13]
>
> A catastrophe may intervene and interrupt and destroy what is being done, but even then the final result is sure .... a new spirit of oneness will take hold of the human race.[14]

Karl Pribram believes memories are encoded not in neurons, or small groupings of neurons, but in patterns of nerve impulses that crisscross the entire brain. It has been estimated that the human brain has the capacity to memorize something of the order of 10 billion bits of information during the average human lifetime. It has been demonstrated that one cubic centimeter of film can hold as many as 10 billion units of information. Our brains mathematically construct "*hard*" reality by relying on input from a frequency domain (Life Field) that transcends time and space. The brain is a hologram, interpreting a holographic universe.[15]

Somewhat identical warnings have been issued by Jacques Monod, Nobel prize winning biologist, wrote in 1972, "What we see opening before us today is an abyss of darkness".[16] George Wald, Nobel prize in Medicine, 1967, wrote, "Our society has taken a wrong turn .... By now, the problem is not transcendence, but survival".[17]

A Critique on Human Survival

(a) The challenges facing mankind are not in the external field, but are embedded in the human psyche, and have assumed global dimensions. The basic challenge is past thought structures, or memory, or cultural belief structure. The memory is meta-stable. It spontaneously disintegrates and the by-product is called thought. So memory or thought-time is of humanity.

(b) Krishnamurti and David Bohm discussed the quality of brain and mind and their interactions in words:

> I have a million years of experience, and it has given me a certain capacity. And I realize at the end of it all there is no relationship between me and truth.... Is it a shock to discover that your brain, and your mind, your knowledge, are valueless? All your examinations, all your struggles, all the things that you have gathered through years and years, centuries, are absolutely worthless?.... What you have done or not done is absolutely of no value.... The ground says, whatever you have done 'on earth' has no meaning.... It has given me a tremendous shock to discover the truth, that all the churches, prayers, books, have absolutely no meaning – except how we can build a better society, and so on.... So there is only one thing, and that is to discover that *all* that I have done is useless – ashes![18]

It is a shock to discover that whatever man has gathered (the known, thought, memory) in centuries is worthless, ashes. It is evident that humanity took a wrong turn. This has to be rectified by an inner journey of self-discovery from self-observation through self-knowledge to Silence. Man has to be a light unto himself. Truth is a process of self-discovery. No one can give the truth to another. Only the truth can liberate the mind from its own ideation.

(c) Krishnamurti wrote, "Every night the brain has been breaking and entering into something immense .... It is a state in which the mind, the brain was no longer in operation – only that source was in operation".[19] It implies that during deep sleep, the Unmanifest, the Immense, the Source takes over the mind and the brain. There is discontinuity between waking state and deep sleep state. When one awakes after deep sleep, the brain, the mind reverts back to successive consciousness -- the old memory, the time.

(d) Thought has no place in the investigation of the inner because thought is a part, limited, fragmentary, divisive and dualistic. When you see that the subject is the object, there is direct perception of the fact. Krishnamurti prophetically wrote, "India has lived all these centuries on ideas. She has to move from a life based on ideas, which are non-facts, to seeing the fact only". He added, "Something had been destroyed in India, but a presence waited in the land .... The benediction of the timeless is waiting, the ground is ready. Can we create something that is of this benediction".[20] My brain is the brain of humanity. The brain says, in words of Krishnamurti, "I am humanity, and what humanity has done, I have done? And I see that it does not produce benediction and therefore all that has been done, has no meaning".[21]

Answer to Human Paradox: (a) We have to learn to observe or listen the conditioned psyche where the subject is restructured science-education. This bestows on us understanding and learning. Understanding of life directly is a perception of the *fact*, the *true*, the *present*, the *what is*!, the non-dual perception. When you listen deeply, or you see instantly or directly, the seed has been sown. And that fact, if it is left alone, is operating; you do not have to do a thing about it. The fact itself is the operator, not you operating on the fact. The fact grows and matures. If you leave it alone, it has no time. When the fact is left completely it blossoms instantly. Let the fact operate, let it live, as the seed that has been planted in the womb, is living. Let the fact live.[22]

(b) The observation or listening leads to mutation of the whole psyche. It is psychological revolution of the whole memory content, it is also called radical transformation or mutation of the whole. It implies to transcend or decrease the matter content of consciousness. Mutation is growing into a new dimension of Consciousness, or in other words, a new quality of consciousness with its pristine glory.

(c) We need psychological revolution, or radical transformation, or mutation of the whole memory content. It implies to transcend or decrease the matter content of consciousness, or in other words, a new quality of consciousness with its pristine glory.

(d) The Unknown energy is a cosmos, the known energy is a chaos. We have to transform the known to the Unknown through higher expression of the Unknown. It is the Consciousness that can transform the chaos to cosmos. To live a life rooted in attention and awareness adds a new dimension to the quality of daily perception and living at every moment; it is living in communion with the immutable law of nature.

(e) Spirituality is the science of life and living. It has the potential to rectify the inner structural crisis, or the crisis of perception. We need urgently the symbiosis of science and spirituality. This is a pointer to grow into the vertical dimension of Consciousness.

Symbiosis of Science and Spirituality

Science is the search for truth in the objective world. A selfless, unfaltering devotion to truth and the use of knowledge for man's upliftment are the ideals of a scientist. Truth is what all earnest scientists seek. Science deals with the objective world (outer), it ignores the subjective world (inner), so its knowledge is invariably fragmentary. It adds to conditioning of the psyche.

Modern science has shown that the outer and the inner are not two separate movements, but a unitary movement, one whole. The objective science must *restructure* itself so that it takes in its ambient both the objective and the subjective dimensions of life. The restructured science sees the wholeness of life, so the subject of cognizance is non-variable attentivity, or reaction-free attentivity, or objectless attention. It is called a Mind with a scientific temper; it is tuned to perceive the *fact*, the *true*, the *present*, the *what is!*, the *non-dual*. This reveals the psychodynamics and limitations of the conditioned psyche; the conditionings go to abeyance; and bestows a perception of intuition (insight), spontaneity and true social action. The pure perception bestows non-violence, harmony, human values, love and compassion; it is in communion with the Universal Consciousness, the Life Field, the spiritual realm.

Spirituality is a science that deals with the whole of life – the inorganic world, the inanimate and animate organisms and the individual man.[23,24] Man is a multi-dimensional organism and it explores the field of self-awareness, the field of his consciousness, his ego, his being the subject and not the object; and above all the expressions of the unconditioned energy especially perception and energising intelligence.

Some aspects of spirituality are as under:

(a) It begins with the awareness of the whole. It proceeds from the awareness of the wholeness, to analyse the particular as organically related to the whole; so with the awareness of the whole, we observe the particular. It is a personal discovery of the truth.

(b) Rational knowledge is derived from the sensory experience we have with objects and events in our everyday environment. The subject is the recalled memory and it gives mutilated perception – partial experience and fragmentary knowledge. In case of spirituality, the subject is awareness of the whole, and it bestows pure perception. Perception is in the trans-sensual (or trans-psychological dimension) realm. Perception transcends knowledge. In spiritual science, the emphasis is upon the act of perception, the quality of the instruments employed for perception, and the quality of consciousness behind the perception.

(c) Perception is subjective, so no experimental evidence can be advanced to verify it. But if a seer/saint has he same (or higher) quantum energy potential of Consciousness, he verifies that perception.

(d) Both science and spirituality have non-authoritarian approach – an approach not conditioned by the authority of the past. Spirituality is exploring the conditioned energy and the unconditioned energy on behalf of the whole human race. It is to perceive the truth and the reality behind the truth. The potential of the human race is infinite as life is infinite.

(e) The quantum energy potential of the unconditioned energy is independent of potential of rationality. Rationality depends on the minimum quantum energy potential and the unconditioned energy has the highest potential of Consciousness. The unconditioned energy is fundamental and energy of intellect/reason is a derivative.

(f) The restructured science transcends I-consciousness and thought structure along with its volitions, so there seems to be a basic demand for freedom, which naturally blossoms to truth, love and compassion. Freedom of the inner being implies fearlessness, non-violence and peace. The institutionalised religion, or organised sects are a hindrance to the progress of spirituality.

(g) Life is a mystery. It is Eternal, Infinite and Immutable. Its dimensions are homogeneous, undifferentiated and dimensionless. Life is unindividuated, unparticularised. Every event/existence/form emerges out of the *emptiness* of space; all forms grow, bloom and decay; and re-absorbed in the emptiness. This is the ground of existence which we call nothingness or formlessness. The emptiness or the silence is a tension and pressure-free state of your being. Being in silence is awareness of the wholeness of life. The silence is the emptiness of your consciousness when all the paraphernalia that human race has collected is brushed aside. Spirituality is physics of consciousness. The awareness of the wholeness is the perfume of spirituality, which is the essence of spirituality. The absolute ground of existence is called *divine.*

(h) Nature invariably expresses itself as pure perception or bare cognition, so pure perception is a mirror-image of reality. It recalls past memory (conation) and grafts it on the pure perception, and this grafting is called impurity or imbalance.

(i) Spirituality confers the sense of reverence for life, for living; and science gives you the precision, the accuracy with which you can be related to the particular and use them for your needs, to provide for your needs.

(j) The human being seems to be born with the faculty of self-awareness and understanding of life directly. That understanding which is not-knowing has a perceptive sensitivity. Understanding is not a cerebral movement. Understanding comes when cerebral activity goes to abeyance voluntarily. In that state, the energies concealed in silence get activised; or in other words, the intelligence awakens.

(k) There is nothing to be acquired; everything is to be understood as it is and to enable ourselves to live in the light of understanding which we carry in ourselves.

Concluding Comment

Life is a multi-dimensional phenomenon and human beings who are expressions of that phenomenon are themselves multi-dimensional organisms. Life is self-generated, self-sustained, self-perpetuating and self-regulated phenomenon. Life is Infinite, Eternal and Immutable. Life is an expression of Supreme Intelligence.

Life, or Universal Consciousness expresses itself in two dominant modes: (a) the *known*, the time-space-causation matrices, the manifest, the conditioned energy; and (b) the *Unknown*, the Timelessness, the Unmanifest, the Unconditioned Energy. As a symbol, we may compare the known to a drop of water and the *Unknown* to a fathomless and boundless ocean. The known is limited, finite and transient, and the Unknown is immutable, infinite and eternal.

We have brought the human race to an explosive situation and our life is poised on razor's edge. From deep within the human psyche (known) there has developed a terrible power (conditioned psyche) that threatens the extinction of all life-supporting systems in a foreseeable period of time. The conditioned psyche is leading the human race to self-annihilation or self-destruction.

Life oscillates between different quantum energy potentials of Consciousness. When higher energy state encounters lower energy state, the latter goes to abeyance. Human being seems to be born with two dominant faculties – (a) faculty of self-knowledge, and (b) faculty of self-awareness and understanding of life directly. Self-knowledge is in the field of duality, time, the manifest, and understanding is in the realm of non-duality, timelessness, the Unmanifest. It is the Unmanifest that governs the wholeness of life.

Life is the Divinity, the Wholeness. To understand life directly, we are going to undertake a journey from fragmentation to homogeneity, from partialness to Wholeness, from the movement of knowing to the action of understanding, from the slumber of knowledge and experience to the state of being Awake or Awareness. This involves a major conceptual shift to complementarity in place of competition, convergence in place of conflict, holism in place of hedonism. We must heal the split within the human psyche; we must gather the diverse fragments of human consciousness and weld them into a glowing whole; we must effect the transition to a holistic world-view to replace the present fractured and fragmented consciousness of the human race. It is the birth-right of every human being to grow into that dimensional transformation, an entirely new way of living. This will guarantee human survival and excellence.

References

1. Hideki Yukawa, *Creativity and Intuition*, Tokyo: Kodansha International Ltd., 1973, p. 203.
2. D.S. Kothari, *Some Thoughts on Science and Religion*, The University of Delhi, 31st Oct. & 3rd Nov. 1977, p. 26.
3. Fritjof Capra, in *Synthesis of Science and Religion*, Mumbai: The Bhaktivedanta Institute, June 1988, p. 275.
4. R.P. Kaushik, *Towards a New Consciousness*, Delhi: Darshan Yoga Sansthan, January 1974, p. 63.

5. J.W.N. Sullivan, *The Limitations of Science*, New York: Mentor Books, 1949, p. 140.
6. E.N. da C. Andrade, *An Approach to Modern Physics*, New York: Doubleday Anchor Books, 1957, p. 255.
7. George Wald, *Synthesis of Science and Religion*, *op. cit.*, p. 19.
8. Niels Bohr, in Fred Alan Wolf, *Taking the Quantum Leap*, Harper & Row, 1981, pp. 21, 85.
9. Vimala Thakar, *Mutation of Mind*, Ahmedabad: New Order Book Co., 1972, pp. 122, 150-51, 164, 171.
10. St. Augustine, in Bertrand Russell, *A History of Western Philosophy*, New York: Simon and Schuster, 1945.
11. Alan W. Watts, *The Wisdom of Insecurity*, New York: Vintage, 1968, p. 82.
12. Erwin Schrodinger, *What is Life*? and *Mind and Matter*, London: Cambridge University Press, 1967, p. 145.
13. Sri Aurobindo, in *The Odyssey of Science, Culture and Consciousness*, New Delhi: Abhinav Publications, 1990, p. 41.
14. *Ibid.*, *A Message on August 15, 1947*, in The Tribune, August 15, 1997.
15. Karl Pribram, in Author Unknown, *The Holographic Universe* (received on January 1, 2004).
16. Jacques Monod, *Chance and Necessity*, New York: Vintage Books, 1972, p. 171; in *Synthesis of Science and Religion*, Mumbai: The Bhaktivedanta Institute, June 1988, p. 127.
17. George Wald, in Kishore Gandhi, *The Evolution of Consciousness*, New Delhi: National Publishing House, 1983, p. 14.
18. J. Krishnamurti & David Bohm, *The Ending of Time*, San Francisco: Harper and Row, 1985, pp. 105, 107, 110.
19. J. Krishnamurti, in Pupul Jayakar, *Krishnamurti: A Biography*, San Francisco: Harper and Row, 1986, pp. 391-92.
20. Pupul Jayakar, *Krishnamurti: A Biography*, *op. cit.*, pp. 379-80.
21. *Ibid.*, p. 389.
22. J. Krishnamurti, in *Within the Mind: On J. Krishnamurti*, KFI, 1982, pp. 148-149.
23. Vimala Thakar, *Science and Creativity*, Ahmedabad: Vimal Prakashan Trust, 1988, pp. 1-39.
24. Dr (Mrs) Kanwaljit Kaur, Paramjit Singh and Sampooran Singh, *Symbiosis of Science and Spirituality*, Vedanta: Ramakrishna Vedanta Centre, U.K., September-October 2001, No. 301, pp. 224-31; & November-December 2001, No. 302, pp. 261-70.

16
Symbiosis of Science and Spirituality

Abstract

Spirituality is defined as a science of life, which deals with the awareness of the wholeness around you and within you. It is to understand the whole of existence, not just a part of it. It explores the fact, the true, the present, the what is! In spiritual science, the emphasis is on the perception, the quality of instruments employed for perception and the quality of Consciousness behind the perception.

Albert Einstein stated that "Objective observing and understanding" leads to "peace and serenity". It leads to "the understanding of life directly, instead of in the abstract, linear terms of representational thinking". "The understanding of life directly", or "direct perception", or "awareness of the fact" are synonyms.

When the mind-brain-body system [non-dual and non-conceptual frame of reference of Mind Continuum] is bathed by the Cosmic Consciousness, the Life Field, it resolves all psychological imbalances, ecological crisis and other human challenges; and bestows order, harmony, nonviolence, peace and bliss. The human survival [and excellence] is embedded in the highest quantum energy potential of Cosmic Consciousness. It is to understand and live according to the immutable laws of nature.

Humanity's ethical and spiritual development has not kept pace with the accelerative progress of material technology, and this imbalance threatens to destabilize and destroy human civilization. Robert Muller, in the Foreword to the "SourceBook for the Parliament of Religions, Chicago, 1993", wrote:

> Nevertheless, one dimension has been missing from this extraordinary journey, a dimension lamented by Secretaries-General Dag Hammarskjold, U Thant and Javier Perez de Cuellar: the spiritual dimension, the highest, deepest, most common, universal and binding dimension of all. What science, politics, economics and sociology were trying to achieve, the religions knew long ago by virtue of transcendence, elevated consciousness, and union with the universe and time. This dimension

is still missing, yet is urgently needed in world affairs.... Hence the crucial importance of the Parliament of the World's Religions. The common heritages and their spirituality can be the greatest single contribution to a better world. Though an atheist, Andre Malraux has said that "the third millennium will be spiritual or there will be no third millennium". Dag Hammarskjold, a rational economist and world observer, said "I see no hope for permanent world peace. We have tried and failed miserably. Unless the world has a spiritual rebirth, civilization is doomed".[1] Albert Einstein, the greatest physicist of the age, in an essay "Out of My Later Years" perceived an accelerative conditioning of the human psyche during 1900-1950, and he summed up his prophetic insight in words:

And yet I know that, all in all, man changes but little, even though prevailing notions make him appear in a very different light at different times, and even though current trends like the present bring him unimaginable sorrow. Nothing of all that will remain but a few spiritual pitiful pages in the history books, briefly picturing to the youth of future generations the follies of its ancestors.[2]

To quote Cyril Hinshelwood from his address (1962) to the Royal Society (London), "To deny the reality of the inner world is a flat negation of all that is immediate in existence; to minimise its significance is to depreciate the very purpose of living, and to explain it away as a product of natural selection is a plain fallacy".[3] George Wald, Nobel Prize in medicine, 1967, concluded, "By now the problem is not transcendence, but survival. If we are to survive there must be fundamental changes in direction, not only organizational, but in our ways of thinking. We have to think different thoughts and to want different things".[4] Jacques Monod, Nobel-Prize winning biologist, wrote in 1971, "What we see opening before us today is an abyss of darkness".[5] Vaclav Havel, President of the Czech Republic, in speech at Philadelphia, July 4, 1994, said:

I think there are good reasons for suggesting that the modern age has ended. Today, many things indicate that we are going through a transitional period, when it seems that something is on the way out and something else is painfully being born. It is as if something were crumbling, decaying and exhausting itself, while something else, still indistinct, were arising from the rubble.[6]

The Expression of Supreme Consciousness

The supreme Consciousness has a dual role in the universe. In the world of the quantum, it is both the awareness and the creation of experience. It is the *being* and the *knowing* of experience. With the stroke of the twentieth-century quantum eraser, the dividing line between ontology (theory of being) and epistemology (theory of knowing) is rubbed out.[7] The fusion of ontology, epistemology and psychology is beginning to emerge.

Since atoms with consciousness are looking at atoms with consciousness, we have the same experience of wonder and mystery whenever we look at atoms with consciousness. We are looking at ourselves. It is this process which we call the universe. Richard Feynman wrote:

> How long it takes for the atoms of the brain to be replaced by other atoms, to note that the thing which I call my individuality is only a pattern or dance. The atoms come into my brain, dance a dance, then go out; always new atoms but always doing the same dance, remembering what the dance was yesterday.[8]

It is apparent that there are atoms with consciousness, matter with curiosity. The mind constructs a time-space matrices with quick succession, and we live in this time-space construct – a synthetic world which is illusory; and where the Reality is divorced. Obviously there is nervous tension and chemical pressure which expresses itself as chaos, strife, pain, misery, and so on.

Fred Alan Wolf wrote, "We are beginning a new age of awareness, the age of quantum consciousness, the age of the conscious atom".[9]

Investigation of an Atom: When an atom makes a transition, or a 'jump', from one stationary state to another the actual process cannot be visualised or even imagined. There are no intermediate stages. We cannot breakdown the process into its components: The entire process of transition from one state to another is a non-visualisable, unanalysable, unity. *It is indivisible.* Its parts cannot even be imagined. It is a 'quantum jump', as we call it.[10] In dealing with atomic phenomena we have to recognise that the act of observation is accompanied by an inevitable disturbance which alters the state of the observed system in an unpredictable manner. What is observed is different from what it was before the act of observation. This is inherent in the nature of things. It cannot be eliminated or allowed for".[11]

In an attempt to understand the mystery of the world of atoms, quantum physics is forced to the view that the act of observation projects, as it were, the atomic object from an "unmanifest" mathematical-space into the "manifest" *physical-space* of our experience. Between two successive acts of observation the atomic object follows its course in the unmanifest mathematical space, called Hilbert space (after the mathematician D. Hilbert).[12]

Investigation into Thought Process: As we 'think' – in any sequence of thoughts – there is always a sudden, discontinuous jump from one thought to the next. We are not aware at all of any intermediate stages. If we were so aware, no thinking would be possible. There would be an almost infinite sequence of 'sub-thoughts' between two thoughts, like an infinity of points between the two ends of a line. *Quantum jumps for thought appear to be as much a necessity for the conscious self as these are for the atom.*

The Challenge of Thought Structure Coupled with I-consciousness: Thought is the past. It can project itself in the future. It is a movement in time. It brings out division, fragmentation, chaos and strife; it is always dualistic.

Thought has created the wars and the instruments of war. It has created all human psychological imbalances, ecological crisis and other challenges. Its every activity leads to increase of entropy, so it leads to death, self-annihilation.

Whenever there is a challenge, whenever life is threatened, the sympathetic nervous system – concerned with defense mechanisms – comes to the aid of the organism. In the animal, the response to a challenge comes through fear. This sympathetic activity leads to an increase in heart rate and breathing, and more blood is pumped to the muscles so that the animal can either run or fight. The regulation of the sympathetic nervous system and thus of the fight-or-flight mechanism is brought about on the instinctual level. It operates whenever there is a challenge; once the challenge is met, the nervous systems returns to normal. But for the human being, because of the development of the intellect in an unbalanced way, even after the challenge is over the mind can go into tension and anxiety from memory of the event alone. So in man, fear is usually related to thinking, not to an actual threat to survival – and therefore the sympathetic nervous system has become unsympathetic to human survival.

The challenge which thought creates can only be met by observing thought, trying to understand what thought means, what it implies from moment to moment. To understand thought, to see it from moment to moment is total meditation, in which no area of the brain is inhibited or suppressed – even thought is not suppressed. There is only a movement of the understanding of thought and thinking. When thought is understood, thought stops, because it is only performing a function. Thought is a function of the brain which points to the need for security. It only arises as a response to a challenge – it shows you that something remains unresolved, incomplete. Even when it creates fear by thinking of the past or future, it compels you to see what fear is, so that fear can be ended. So when thought creates these challenges of fear, envy, jealousy, hate and violence, these challenges have to be seen. Only when they are seen can they be resolved, and when thought has performed its function, it comes to an end. A mind which can understand thought is free of thought.[13]

Electroencephalograph (EEG) recordings during meditation and psychotechnologies give concrete evidence of physical changes : higher levels of integration in the brain itself, more efficient processing, different 'harmonics' of the brain electrical rhythms, shifts in perceptual ability.[14] When one is thinking, there is beta rhythm which implies that one is more attentive to the external world than to inner field. When thinking is replaced by quiet contemplation, there is alpha rhythm which suggests a relaxed state.

When thinking is going on, the brain cells discharge their potential asynchronously – not together but haphazardly, in conflict. Thought itself is a form of conflict and therefore it lowers the amplitude of the brain waves. So thinking is not the highest form of activity. The production of alpha waves is an indication that there is synchronous discharge of energy from the brain cells – they are all discharging this electrical potential at more or less the same time.

The emission of thought is not a continuous stream of energy but in *discontinuous bits of* energy and the energy associated with each thought is called thoughton. As thought is emitted in a discontinuous stream, so thought does a discontinuous scanning of the event (the perceptual-conceptual image) at its surface in one-dimensional system and in one plane, and further the X-coordinate is also changing non-randomly. An artist's view of the limited field which thought scans is shown in Fig. 1, where the firm line represents live-time and dotted line as blank-time. Let AB represent the live time of first thought and BC the blank time between the first and the second thought. Let CD be the live-time of the second thought and DE the blank-time between the second and the third thought. The same applies to the third, the fourth thoughts, and so on.[15]

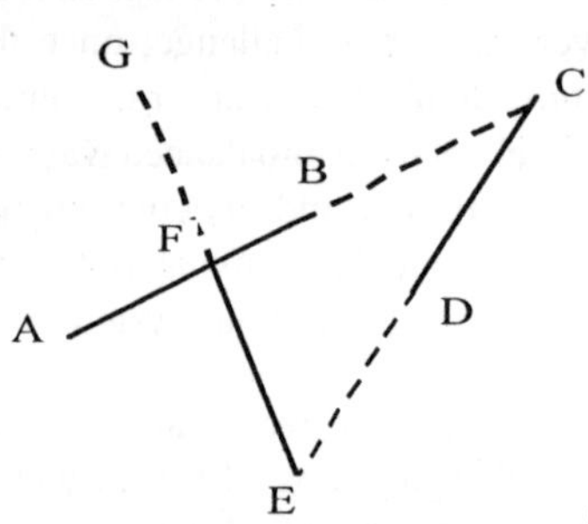

Fig. 1 An artist's view of the scanning of a perceptual image (event) by thought. The perceptual space-time section is within the perceptual image, firm lines are the live-time and dotted lines the blank-time. ABC, CDE and EFG are the different X-coordinates scanned by thought.

Thought is capable to do discontinuous scanning of the perceptual-conceptual image at its surface in one-dimensional system and in one plane; so our effort is one-dimensional and limited. The natural world, however, is not linear and sequential. It is one of infinite varieties and complexities, a multidimensional world, where things do not happen in sequences, but all together. As a one-dimensional system cannot perceive the multidimensional world, so thought cannot determine the results; it is nature that determines the results. This means that man is the director of effort only, which is always one-dimensional; while result, which is multidimensional system, is controlled by nature or Universal Consciousness. We should put in our maximum effort but should not worry about result.

A one-dimensional thought can never give a total solution of a multi-dimensional problem. Thought cannot see the completeness and fullness of the problem. Thinking or intellectual approach has not really solved and cannot solve any of our human problems.

Rational knowledge and rational activities certainly constitute the major part of scientific research, but it must be complemented by the intuitive insights that makes a scientist creative. Albert Einstein, in almost the last interview

that he gave to the international magazine LIFE said as follows:
The mind can proceed only so far upon what it knows and can prove. There comes a point where the mind takes a leap – call it intuition or what you will – and comes out upon a higher plane of knowledge, but can never prove how it got there. All great discoveries have involved such a leap.
The mind follows logical and scientific reason so far as it can guide, but when reason or intellect realises its limitations, then the conscious-mind makes a quantum jump to a higher plane and intuition tends to come suddenly, and characteristically.

The Limitations of the Intellect: Recently, many eminent scientists – Werner Heisenberg, Edwin Hubble, Max Planck, G.F. Chew, Bertrand Russell, Lincoln Barnett, Max Born, and so on – have concluded that we cannot impart value education at reason or intellect level.[16]

Discontinuity: Our symbolic-dualistic mind is continuous, and our non-dual and non-conceptual frame is discontinuous. Sooner or later our activity introduces discontinuities in whatever we are observing. These discontinuities are fundamental to the new physics of the twentieth century. To Niels Bohr, discontinuity was a fundamental truth.[17] Recently, scientists have witnessed the proverbial quantum jump. I.I. Rabi stated, "The atom is in one state and moves to another state, and you can't picture what is in between, so you call this a quantum jump. Negourney said, "You see the trapped ion blinking on and off, and each blink is a quantum jump. It is a striking illustration that things occur discontinuously in nature. Nature is discontinuous in its core, more like a staircase than a ramp.

The Mutation of the Mind-Brain-Body System

We have constructed a society which is dominated by ephemeral and transient values to the neglect of abiding spiritual principles. We have learnt to live in a construct of the conditioned mind-brain-body system, or in successive consciousness or immanent quantum energy potential of consciousness, where we have divorced the Reality totally. All this suggests a profound imbalance in our culture – in our thoughts and feelings, our values and attitudes and our socio-economic-political structures.

When an entity with a lower quantum energy potential of consciousness encounters one with a higher energy quantum energy potential of Consciousness, the lower gets disintegrated and annihilated. The thought-structure and I-consciousness, which are at a lower quantum energy potential than true-subject [restructured science-education-health], the lower goes to abeyance. Mutation is not in any part, but it is in the whole mutant. It is the Cosmic Consciousness, or Life Field that governs the mutation of the conditioned psyche. It activates the Intelligence, which expresses as order, harmony, nonviolence and peace.

Consciousness expresses itself as awareness and the creation of experience, in the conditioned psyche and the Unconditioned Consciousness.

Fig 2 shows an artist's view of the states of mind as a function of the expression of consciousness. The different states are discussed as under:

(a) Psychological Mode of Mind: Mind cuts itself into the observer (subject) and the observed (object). Both are at the same quantum energy potential of successive consciousness, so there is interchange of energy. The observer is the past, who has known what violence is. Inwardly, we are very primitive, almost barbarous, howsoever cultured we may seem outwardly. It is the past, it is knowledge, it is experience, it is all the stored up memories. The observer is not different from that which he is observing. Psychologically, inwardly, there is only the fact. The fact is, one is violent and jealous, and so on. The psychological mode of mind is violent and aggressive. Wars have never solved human problem and they never will. We are not aware of the *fact*, the *true*, the *what is*!; but we are aware of the *non-fact*, the *false*, the *what should be*? We cannot end violence by any intellectual approach, because the violence and the intellect are at the same quantum energy potential.

(b) Scientific Mode of Mind: Science aims at nothing but making true and adequate statements about its object. The scientist only imposes two things, namely truth and sincerity, imposes upon himself and upon other scientists. Erwin Schrodinger calls the scientific observation as "the principle of objectivation".[19] Without being aware of it and without being rigorously systematic about it, we exclude the Subject of Cognizance from the domain of nature that we endeavour to understand.

Schrodinger prophetically wrote:

> Mind has erected the objective outside world of the natural philosopher out of its own stuff. Mind could not cope with this gigantic task otherwise than by the simplifying device of excluding itself – withdrawing from its conceptual creation. Hence the latter does not contain its creator.[20]

Charles Sherrington wrote:

> The material world has only been constructed at the price of taking the self, that is, mind, out of it, removing it; mind is not part of it; obviously, therefore, it can neither act on it nor be acted on by any of its parts.[21]

Schrodinger also quoted C.J. Jung and B. Spinoza who had the same opinion.

The mind splits itself as the subject and the object; the subject is at the immanent quantum energy potential of Consciousness, while the object is at the successive consciousness. There is no transfer of energy between the subject and the object, so there is awareness of the objective values fact. It implies that the mind gives an answer to the problem put to it as 'modified use of the present material' to enhance the sensitivity. The scientific mode of mind does not deal with the Unknown.

Science gives us awareness of the *fact*, the objective values; it leads us to modify the known and then construct an artifact with a higher sensitivity. The object is in the external world, so it gives objective values. Science can design better aeroplanes, better television set, better air conditioner, but it

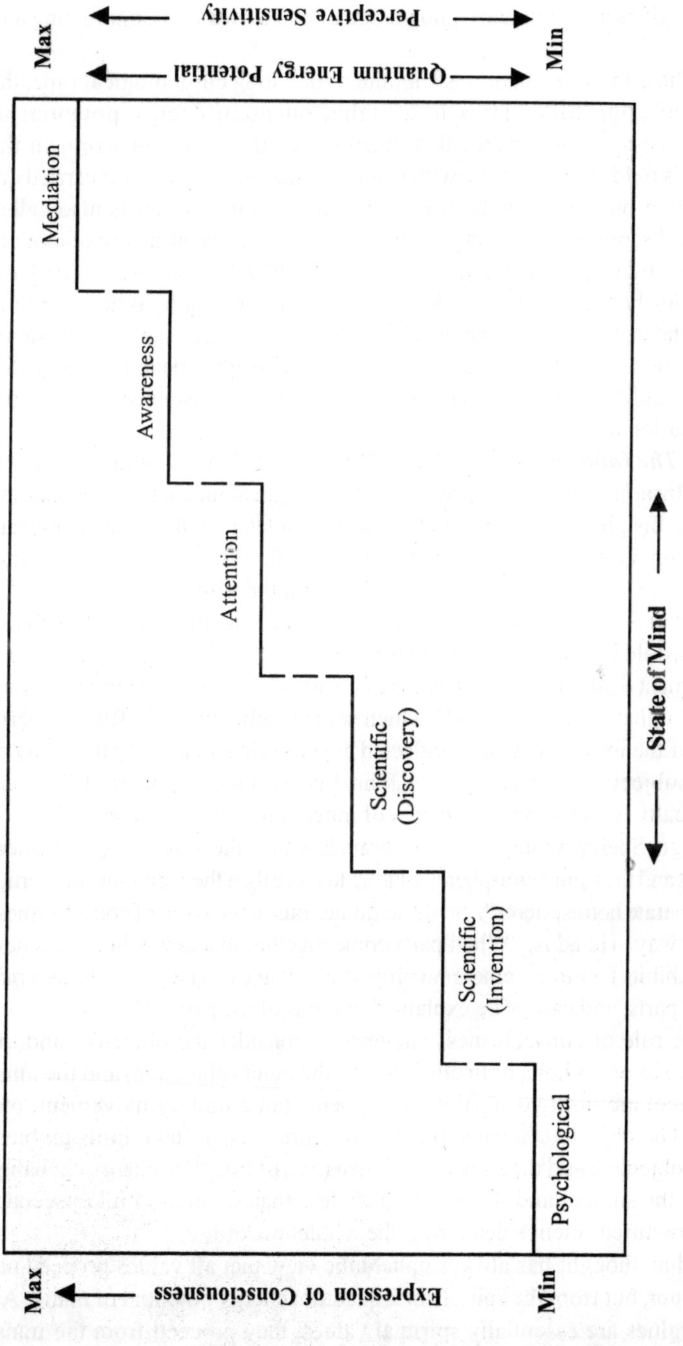

Fig. 2. The States of mind as a function of the expression of quantum energy potential of consciousness. The dotted lines show the quantum jumps between states of mind.

cannot resolve any psychological challenge. This is called scientific invention. This also gives a rudimentary knowledge of one's own thinking, feeling and action.

If the art of observation is continued for a long chronological time, the subject of cognizance rises to a higher quantum energy potential of consciousness. It is physics that presupposes the observed world in the observer's field. Our dialogue with nature will be successful if it is carried out from within nature. This bestows subjective values, which is also called scientific discovery. Discovery implies to perceive a new structure of thought not known till that moment. It is subjective truth. All great discoveries have arisen from the realm of the timelessness, so scientific enquiry is not a cerebral activity and gathering information, but a quest of the Timelessness, the Divinity, the Life Field. A selfless, unfaltering devotion to truth and use of knowledge for man's uplift – higher hierarchical expression of Consciousness – are the only vocation of a scientist.

(c) The Intuitive Mode of Mind: Discoveries come as intuitive (Insight) flashes. Insight is a timeless perception. It is insight alone that can decondition the brain. Insight is complete and from that completeness there can be logical, sane, rational thought. Only then there can be flowering of the brain. Only then you can have a complete relationship with the Mind.

Erwin Schrodinger wrote, "Scientific attitude would have to be rebuilt, science would be made anew. Care is needed".[22] Karl Pribram wrote, "Brain science must deal with the awareness of awareness. It could no longer afford to shut out that part of the world which we call subjective".[23] Roger Sperry "spoke of the importance of a neglected area of science, namely the study of human subjective experience".[24] John Eccles also emphasized "to deal scientifically with the neglected area of inner conscious awareness".[25]

Roger Sperry wrote, "When the brain is whole, the unified consciousness of the left and the right hemispheres adds up to more than the individual properties of the separate hemispheres". So these studies raised the issue of consciousness in a new way. He adds, "When parts come together in a new whole, this new whole exhibits features – emergent properties – that can't be predicted as a rule from the parts, and cannot be explained in terms of the parts".[26]

The role of consciousness suggests to consider the objective and the subjective as one whole, or in other words, the outer (objective) and the inner (subjective) are not two separate movements but a unitary movement, one whole. The objective science must restructure itself to take in its ambient both the objective and the subjective dimensions of life. The energy exchange between the subject and the object takes less than a hundred microsecond. The restructured science deals with the wholeness of life.

Indian thought has always upheld the view that all values proceed not from matter, but from the spirit (inner quantum energy potential of man). All ethical values are essentially spiritual values, they proceed from the trans-sensual dimension of human personality. Recently, many eminent scientists – Werner Heisenberg, Edwin Hubble, Max Planck, G.F. Chew, Bertrand Russell, Lincoln Barnett, Max Born and so on – have concluded that we cannot impart value education at reason or intellect level.[27]

All diseases are maladjustment of the human mind-brain-body system, or conditioned human psyche, or crisis of perception. Health is a complete physical, mental, psychological and social well-being; it is a dynamic state which is equipoised and balanced. A living in the *present*, the *now* heals; and living in the *past* and *duality* suppresses man's perceptive sensitivity and enhances disease. The curative functions of medicines have to be supplemented with understanding and living the "symbolic-dualistic frame of reference of mind spectrum (successive consciousness)" and the "non-dual and non-conceptual frame of reference of Mind Continuum (Simultaneous Consciousness)".

The new dimension of science is "science-education-health". It deals with the wholeness of life. The subject is at the highest hierarchical quantum energy potential. The thought-structure coupled with volition is not in the external world but it is within the inner world. Ilya Prigogine wrote, "It is physics that presupposes an observer within the observed world. Our dialogue with nature will be successful only if it is carried on from within nature".[28] When thought-structure coupled with volition encounters the consciousness of the restructured science, it goes to abeyance. This flows to intuition or insight. It is the insight that has the capacity to resolve all individual and collective (humanity) challenges which includes all physical calamities, or catastrophes, or tsunamis.

When we are actually observing, without the observer, without thought, without any desire, without memory, there is complete attention. When there is complete attention, one is in a state of love. The state of love bestows intuition or insight and that resolves all human challenges.

(d) State of Attentivity: J. Krishnamurti stated, "At the moment of attention all the conditionings disappear, all the image-building comes to an end; it is only when you are not sensitive that the whole thing begins".[29]

Observation without an observer and an observed is attentivity. The whole being gets tuned into a state of attentivity, where there is neither subject, nor object. It is a subject-free and object-free attentivity. It is not an activity on the plane of duality. It is just sensitivity of the whole being that becomes attentivity.

The restructured science sees both the objective and the subjective dimensions of life, or the wholeness of life. The Subject of Cognizance is non-variable attentivity, or reaction-free attentivity, or objectless attention, it is also called true-subject, or a mind with a scientific temper. It is tuned to perceive the *fact*, the *true*, the *present*, the *what is*!, the *non-duality*. It bestows a perception of "understanding of life directly", or in other words, it bestows intuition (insight), spontaneity and right action. The pure perception bestows non-violence, harmony, human values, truth, love and compassion, so it is in communion with the Consciousness, the Life Field.

The attentivity is also defined as seeing the fact, the observed without the pseudo-observer. The subject of cognizance is pure consciousness, so it gives a perception of ethical values and meanings. It is born in the state of observation and get finer and acute in the state of attention.

(e) State of Awareness: In the ocean of Silence, the Mind makes a quantum jump to "Active Reality (Vibration)", which has dynamism and illimitable potency to manifest as many; and "Passive Reality (Non-Vibration)". Both are different states of the Supreme Reality, the obverse and the converse affirmations; both are Pure Existence and Conscious Force, or Self-Existence Bliss and Supreme Intelligence, and are One. It is the Active Reality which manifests as the perfect freedom and omnipotence of an eternal divine activity in innumerable galaxies as inanimate world and plant and animal kingdoms.

When you begin to see that which is false, then there is the beginning of awareness, of intelligence. Awareness is to see the false as the false and the true as the true; this brings out transformation or mutation in the psyche. It is the truth that liberates. So awareness, intelligence, the truth, the transformation are shades of the same. Truth is not cumulative. It is from moment to moment.

It is perception without the word, which is without thought. It is with the brain (the heart) and with all the senses – it is with the pure mind, it is called the total perception, or the perfect perception, or the holistic perception. Holistic perception dissolves the fact. Perception and understanding are not of time. It is with the height and depth of the immense. There is a movement of great ecstasy which is not to be confounded with pleasure. It is this ecstasy which gives to the eyes, to the brain, and to the heart, the quality of innocency. It is born in the state of attention and gets finer and acute in the state of awareness. It is born when the non-dual state of mind is presented as an object to pure consciousness, the subject, and it reveals itself. It is the awareness of the awareness itself, so it can be realised through direct perception of the unified field in the self-referral state of consciousness. Holistic perception implies that the mind is totally free, to be totally unattached to everything that man has put together. It implies to be aware of the wholeness, of the Fundamental Vibration.

When attentivity is established over a long chronological time, it makes another quantum jump and it is called awareness. It is awareness of the Wholeness, the Oneness of Life.

Awareness energises the supreme Intelligence. The movement of organic Intelligence in the complexity of cosmic Life, is a movement born of non-duality and conducted in non-duality. There is no association with volition. It is non-voluntary. Awareness brings mutation or transformation of the whole being. Mutation is a radical change in the quality of the mutant, in the substance of the mutant. It is not a peripheral change. It is a radical change in the whole of the mutant. Mutation leads to non-verbal communion between what is outside me and what is within me.

(f) State of Meditation: The Active Reality bestows the perception of one Self of all, the Brahman. In an ocean of Silence, the vibrations of the cosmic energy also end, and this bestows the perception of the one Self of all, the Brahman.

The characteristics of perception are dependent on the 'quantum energy potential' of the mind. The 'quantum energy potential' increases as one travels from the waking state, to the state of observation, to the state of attention, to the state of awareness, and to the state of meditation.

Knowledge is of time. Intuition (insight) is timeless and free from the bondage of time. Mankind urgently needs a quantum jump from time to timelessness, from reason to intelligence (wisdom). There must be freedom from knowledge to understand that which is eternal and timeless. The sacred Energy is eternal, infinite and immortal. It is the greatest vocation of the world.

Concluding Comment

Life is a multi-dimensional phenomenon and human beings who are expressions of that phenomenon are themselves multi-dimensional organisms. Life is self-generated, self-sustained, self-perpetuating and self-regulated phenomenon. Life is Infinite, Eternal and Immutable. Life is an expression of Supreme Intelligence.

Man had been busy in gathering fragmentary experience and partial knowledge and storing these as memory for about a million years or more as frequency and amplitude information. This has brought the human race to an explosive situation and our life is poised on a razor's edge. For the past 10,000 to 30,000 years, man has been trying to resolve the cultural crisis and evolutionary crisis, but has utterly failed. We address the fundamental question as, "Has humanity taken a wrong turn?", and "What needs to be done within the consciousness of man to preserve our very existence and bring about a new culture?" Many saints-sages and many scientists have explored this crisis and have provided an answer to this paradox in words: "We need Symbiosis of Science and Spirituality" *here* and *now*. We need a new paradigm, a new vision of Reality – a fundamental change in our thoughts, perceptions and values.

We have to introduce the psychodynamics of mind-brain-body system in scientific faculty and transpersonal psychology or deep ecology in art faculty as a part of educational curricula at University/College/School levels and in technical institutions. We have to share this perennial wisdom with every human being on the globe. Man has to learn himself to perceive higher perceptive sensitivity, or higher quantum energy potential of Consciousness, of Life Field. This energises the Intelligence which resolves all human challenges.

The present crisis is unprecedented because it is in the field of ideation, conceptualization, data and symbols, time-space matrices. The answer to this paradox is a timeless action, an action which is not based on idea, on system. The essence of the action plan is – (a) to educate the physical organism, psychological structure and explore the Unmanifest, the Life Field; (b) epistemological revolution through scientific temper; (c) the restructuring of science, the restructuring of education, the restructuring of health, and the awareness of human values; (d) to inculcate the art of observation (seeing) and listening (hearing); and (e) live in the light of that understanding and learning.

It is the Cosmic Consciousness or Life Field that governs the mutation of the conditioned psyche. It activises the Intelligence, which expresses as order, harmony, non-violence, peace and survival. This is the only vocation of human life. This only is the social; psychological, and even the political revolution.

References

1. Robert Muller, in *Foreword to the SourceBook for the Parliament of World's Religions*, Chicago, 1993.
2. Albert Einstein, *I Believe: Nineteen Personal Philosophies*, London: Unwin Books, 1962, pp. 30-31.
3. D.S. Kothari, *Some Thoughts on Science and Religion*, The University of Delhi, 31st October and 3rd November 1977, p. 3.
4. George Wald, in *The Evolution of Consciousness*, New Delhi: National Publishing House, 1983, p. 14.
5. Jacques Monod, in *Synthesis of Science and Religion*, Mumbai: The Bhaktivedanta Institute, June 1988, pp. 126-27.
6. Vaclav Havel, President of the Czech Republic, *Speech at Philadelphia*, July 4, 1994.
7. Fred Alan Wolf, *Taking the Quantum Leap*, San Francisco: Harper and Row, 1981, p. 229.
8. Richard Feynman, in Fred Alan Wolf, *ibid*., p. 228.
9. Fred Alan Wolf, *ibid*., p. 248.
10. D.S. Kothari, *Atom and Self*, New Delhi: National Publishing House, 1983, p. 212.
11. *Ibid*., *Some Thoughts on Science and Religion*, *op. cit*., p. 25.
12. D. Hilbert, *ibid*., p. 25.
13. R. P. Kaushik, *Light of Exploration*, Journey Publications, March 1977, pp. 57-63, 34-47.
14. Marilyn Ferguson, *The Aquarian Conspiracy*, Los Angeles: J. P. Tarcher, 1980, pp. 67, 79.
15. Sampooran Singh, *The Holistic Education*, New Delhi: Siddharth Publications, 1987, pp. 79-81.
16. Sampooran Singh, Dr (Mrs) Kanwaljit Kaur, and Paramjit Singh, *Global Education and Human Survival*, Sanbun Publishers, March 2002, pp. 46, 35-36.
17. Niels Bohr, in Fred Alan Wolf, *op. cit*., pp. 85, 21.

18. I.I. Rabi, in *The Invincible*, November-December 1987, Vol. IV, No. 6, pp. 6-10.
19. Erwin Schrodinger, *What is Life*? and *Mind and Matter*, Cambridge University Press, 1967, pp. 126-37.
20. *Ibid*., pp. 131-32.
21. Charles Sherrington, in Erwin Schrodinger, *ibid*., p. 128.
22. Erwin Schrodinger, *op. cit*., p. 132.
23. Karl Pribram, in Ken Wilber, *The Holographic Paradigm and Other Paradoxes*, Boulder: Shambhala, 1982, p. 17.
24. Roger Sperry, in *Synthesis of Science and Religion*, Mumbai: The Bhaktivedanta Institute, June 1988, p. 199.
25. John Eccles, in *Synthesis of Science and Religion*, *ibid*., p. 199.
26. Roger Sperry, *Bulletin of the Theosophy Science Study Group*, India, February and April 1988, p. 23 and June and August 1988, p. 27.
27. Sampooran Singh, Dr (Mrs) Kanwaljit Kaur and Paramjit Singh, *Global Education and Human Survival: A Scientific Appraisal*, New Delhi: Sanbun Publishers, March 2002, pp. 46, 35-36.
28. Ilya Prigogine and Isabelle Stengers, *Order Out of Chaos*, New York: Bantam Books, 1984, p. 218.
29. J. Krishnamurti, *Krishnamurti in India*, 1970-71 (India: KFI), 1971, p. 69.

Part-III
Science, Education and Ethics

There is a questioning today of the basic postulates of the education structure and its various system in India and in the rest of the world. At all levels there is a growing realisation that the existing models of education have failed and that there is a total lack of relevance between the human being and the complex contemporary society. Many eminent scientists – Werner Heisenberg, Edwin Hubble, Max Planck, G.F. Chew, Bertrand Russell, Lincoln Barnett, Max Born, and so on – have concluded that we cannot impart value education at reason or intellect level. The response to the turbulence caused by failures of our existing models of education has led to the following recent events:

(a) The University of Delhi, Delhi, has decided on May 16, 2005 to celebrate the Birth Centenary of late Prof D.S. Kothari in 2006. The aim is to achieve "a synthesis of science, ethics and education" which is likely to be one of the frontier areas of 21st century science. The plan is to have National and International Symposia.

(b) The Prime Minister Dr Manmohan Singh has constituted a "National Knowledge Commission" on June 2, 2005. "The Commission will advise the P.M. on how India can produce excellence in the education system to meet the knowledge challenges of the 21st century…."

(c) The UNESCO has declared the year 2005 as the year of Physics. President Dr Abdul Kalam, in an address on March 31, 2005 at IIT, Delhi, suggested that we should "become innovators and creators of science and technology …."

I consider the above events as a turning point.

The world situation/challenges demands a reappraisal of our lives, our science, our values, our religions, and our philosophies. It requires restructuring of our dimension of science-education-ethics-health. The only way of resolving our present crisis is mutual "understanding, learning, action" through observing (seeing) and listening (hearing).

The purpose *in* human life is a venture for our survival through the spiritualization of matter. Matter gets expressed at the lowest quantum energy potential of consciousness (successive consciousness) and spirituality at the highest quantum energy potential of Cosmic Consciousness. At the spiritual (timelessness) the matter goes to abeyance and there is direct perceptual expression of the Consciousness. This energises Intelligence, which resolves all human challenges.

The major unforeseen events or happenings in the world are not play of the human intellect; these are parts of a great symphony, eternal music, order, harmony, coherence and enacted by nature through man to set right his vision because the answer is embedded in intuition/insight. The invisible force of nature moves to guide mankind and fulfil the high purpose of his existence. I humbly submit that the present monograph is not written by me at the intellect level, but *through me* at the intuitive realm. The object of nature is to apprise man of his follies and set right the art of living. Nature guides man to "direct perception" and intuition. It is for man to "understand and act", and not "thinking and reacting" depending on his state of mind-brain-body system. We intuitively see that at this critical time, "if we move one step towards God (Cosmic Consciousness), God advances a hundred steps to receive us". This is a critical time for man to make a quantum jump from psychosocial evolution to spiritual evolution.

17

Science, Education, Ethics and Health

Abstract

The University of Delhi, Delhi is celebrating the Birth Centenary of late Prof D.S. Kothari – a great visionary and a renowned physicist – in 2006, who had produced remarkable work for achieving "a synthesis of science, ethics and education" which is likely to be the one of the frontier areas of 21st century science. This paper explores the synthesis of science, education, ethics and health.

The Prime Minister Dr Manmohan Singh has constituted "a National Knowledge Commission" on June 2, 2005. We intuitively see that the synthesis of "science, education, ethics and health" is fundamental to provide excellence in the education system and promoting creativity in science and technology in different laboratories/institutions.

We intuit that the mandate of the Commission and of Prof D.S. Kothari Centre at the University of Delhi are complementary to each other – in other words, these events are not constructs of the intellect, but nature's benediction to guide man for his survival and excellence. It is for man to "understand and act" or "thinking and reacting" to the nature's message.

Fritjof Capra, author of "The Tao of Physics" wrote, "The rational part of research [partial knowledge, information] would, in fact, be useless if it were not complemented by the intuition that gives scientists new insights and makes them creative. The objective and the subjective dimensions of life are one dynamic flow of life. The values must be accepted as a gift of intuitive insight, because the basis of our perennial values of life can directly affect our brain states and brain physiology. Intuition is fundamental and reason is a derivative.

Each individual is a part of humankind, and mankind, in turn, is part of the global ecosystem. Consciousness is a singular, of which the plural is unknown. There is Oneness of life.

We intuitively see that the primary responsibility of the Commission and KCSEE is to educate oneself for creativeness of science and Symbiosis of Science and Spirituality.

1. Introduction

Many scientists of international repute – Sir C.V. Raman, Chandrasekhar Subramaniam, Homi Jehangir Bhabha, Vikram Sarabbhai, S.N. Bose, J.C. Bose, Meghnad Saha, and Srinivasa Ramanujam have appeared on the Indian scene during the last half of the twentieth century. D.S. Kothari, an eminent physicist and educationist – who is considered an architect of defence science in India – undertook an exploration of conditioned mind-brain-body system, the Unconditioned Energy, value education, mutation of psyche and energising Intelligence/Wisdom.

Kothari was concerned with mind-brain-body challenges. He once said, "Life is a mystery and not reducible to science". He was deeply concerned with the moral dimensions of man, with philosophy, and with the relation between science and religion. He explored the wholeness of life. Dr. Dossey wrote, "Dr. Kothari spoke the language of modern physics and the ancient Upanishads with equal profundity and grace".

This article discusses symbiosis of science and spirituality, where the dimension of science is [science-education-ethics-health], and this deals with the wholeness of life. This will go a long way in generation of innovation in science and technology, or generate creative leaders of excellence. This will lead to emergence of a new human being who is free from all conflict, and conflict is violence, will and assertiveness. It will create a historical revolution and we must be in the driver's seat to contribute towards the societal change. This is Kothari's unique contribution to the service of mankind. It regenerates a revolution or "nature's evolutionary impulse" for human survival and excellence.

This article is fundamentally based on 'intuitive wisdom' of Albert Einstein [1879-1955], the greatest physicist of the age, and of D.S. Kothari [1906-1993], an eminent physicist, and supplemented by other scientists. The aim is the regeneration of innovation in science for human survival and excellence.

2. The Unmanifest-Manifested

(a) The ancient seers of India, about 10,000-30,000 years ago, perceived a state of true illumination. There is One Reality, the Unmanifest-Manifested. The nature of the Unmanifest is 'Deep-Silence', 'Inactivity (Non-Vibration)', 'Passivity', 'Immutability', 'Eternal'; and also 'Activity (Vibration)', dynamism and illimitable potency to manifest as many. Activity (Active Reality) expresses which is self-hidden in the Silence (Passive Reality); both are different states of the Supreme Reality, the obverse and converse affirmations; both are Pure Existence and conscious Force, or Self-Existent Bliss and Supreme Intelligence, and are One.

(b) The Active Reality is a powerful, highly directional, monochromatic (same frequency) and coherent (the wave is in one phase) beam of energy, or a state of highest frequency (higher frequency means higher energy), or high quantum energy potential. The fundamental vibration has much shorter wavelength as that of X-rays, gamma rays and cosmic rays. The Active

Reality or Life Field expresses itself as innumerable existences/events and governs the whole cosmos.

(c) Man is an amphibian who lives simultaneously in two worlds – (i) the outer (the objective dimension), time-space-causation matrices, the transient and ephemeral, it is a construct by the mind-brain-body system, a man-made social cosmos, the successive consciousness; and (ii) the inner (subjective dimension), the timelessness, a construct of the immanent quantum energy potential of consciousness, the Simultaneous Consciousness, the Eternal, the Infinite, the Immortal. The objective dimension is a world of data and symbols, and it offers no guidance to understand the more fundamental problem of the relationship of man with Timelessness and Eternity.

(d) Erwin Schrodinger concluded, "Consciousness is never experienced in the plural, only in the singular .... Consciousness is a singular of which plural is unknown".[1] This suggests oneness in the cosmos. The dimensions of the Simultaneous Consciousness are homogeneous, undifferentiated and non-dimensional. The dimensions of successive consciousness are heterogeneous, differentiated and multi-dimensional. Consciousness alone can bestow fullness or wholeness to every aspect of life, to fulfilment at every level of life.

(e) Schrodinger concluded, "The world extended in space and time is but our representation".[2] *The construct of the mind is the only reality where the transcendent potential has gone to abeyance and its only linkage is the immanent potential. It represents relative field of consciousness, or successive consciousness.*[3]

(f) Fred Alan Wolf wrote, "Consciousness performs a dual role in the universe. In the world of quantum, it is both the awareness and the creation of experience. It is the *being* and the *knowing* of experience. With the stroke of the twentieth-century quantum eraser, the dividing line between ontology (theory of being) and the epistemology (theory of knowing) is rubbed out".[4] So consciousness expresses itself at different hierarchical levels of expressions of quantum energy potential. The fusion of ontology, epistemology and psychology is beginning to emerge. Being is at the highest quantum energy potential. Knowing is at the lowest quantum energy potential (successive consciousness).

3. The Mind-Brain-Body Interactions

(a) The Psychological Mode of Mind: For the past million years or more, man has been gathering fragmentary experience and partial knowledge (information), and storing these as memory in the brain. This led the nature's evolutionary impulse to express as the subjective dimension. Spencer Brown, in a most spectacular passage, explained:

> The world we know is constructed in order to see itself .... Evidently it must first cut itself up into at least one state which sees, and at least one other state which is seen. In this severed and mutilated condition, whatever it sees is *only partially* itself.[5]

As both the object and the subject are at the same quantum energy potential, so there is energy interaction – called subject-object interaction. The mind-brain-body system gathers the object, the fact; recalls the past corresponding memory; there is coherent superpositions of the recalled memory on the fact (energy exchange); this results in mutilated fact and mutilated memory; the modified memory is recorded; and one is aware of the mutilated fact. This is written as :

$$\text{Fact} + \begin{matrix}\text{Recalled}\\ \text{Memory}\end{matrix} \rightarrow \begin{matrix}\text{Multilated}\\ \text{Fact}\end{matrix} + \begin{matrix}\text{Modified}\\ \text{Memory}\end{matrix}$$

The interaction is a biochemical interaction or a dance. The perception of the mutilated fact gives us the response, the dualistic action – the will of action is always dualistic. The mind is the past and it is always aware of mutilated perception. It revels in duality. It constructs chaos, conflict and fragmentation.

This world is marvelously beautiful. It is our world, our earth, to live upon. We are violent, aggressive, greedy, envious, competitive, acquisitive. We are frightened beings, we have no relationship, we live in the past, our every activity is dualistic. The conditioned psyche constructs all psychological imbalances, ecological crisis and other human challenges. We live in non-fact, which constructs conflict, fragmentation. We do not know what it means to live in that ecstatic, blissful sense.

(b) Where is Science Going?: Science is the search for truth. A selfless unfaltering devotion to truth and the use of knowledge for man's upliftment are the ideals of a scientist. Truth is what all earnest scientists seek. Science aims at nothing but making true and adequate statements about its object. The scientist only imposes two things, namely truth and sincerity, imposes them upon himself and upon other scientists.

The psychological imbalances, ecological and pollution challenges, a rising wave of violence and crime, terrorism, religious fundamentalism, ethnic conflicts, corruption, increasing unemployment, poverty, hunger, malnutrition, health crisis, are forcing man inevitably to face the realities of the human situation. All appeals to achieve full disarmament and permanent peace; all declarations by Nobel laureates; all appeals by Pugwash Conferences (since 1957); all appeals by the International Physicians for the Prevention of Nuclear War (IPPNW); by Secretaries-General (Dag Hammarskjold, U. Thant, Javier Perez de Cuellar) of the United Nations; all proclamations of a manifesto by eminent scientists, philosophers, mystics have fallen on the deaf ears of mankind. In investigating the nature of knowledge or of truth or of reality, logical and scientific reason confines itself to the field of the 'known'; it ignores the 'knower', the subject, or the Self; this explains the limitations of its knowledge, the partial character of the truths it finds, and the relative character of the reality it reveals. Erwin Schrodinger stated, "Scientific attitudes would have to be rebuilt, science must be made anew. Care is needed".[6]

At the level of human species, the immanent potential or successive consciousness is called mind. The mind is the conditioned energy moving through the body, monitored by brain and operating through the senses. It is engaged in gathering fragmentary experience and partial knowledge; and storing these as memory or cultural belief structure (frequency and amplitude information). It is used for thinking, feeling and reaction.

(c) The Miracle of Scientific Mode of Mind (Invention): Brain is a part of the objective world. It can be investigated objectively. Mind is subjective. When for instance, sound waves impinge on our ears, the pressure changes produce electric currents in the nerve fibres which from the ear reach the brain. How do these electric currents in the brain transform in the mind into sensation of sound – into music. The same applies to other sensations. Science provides no answer to this riddle. Because of its very objectivity science can give no answer to this riddle of all riddles.

The retina of the eye sends an electrical impulse to the sight sensorium in the brain where the electrical impulse is converted to a perception of the object which we wanted to see. How do these electrical signals transform into perception of the object? George Wald, a Nobel laureate, in medicine, 1967, spent most of his scientific life studying mechanism of vision. He wrote:

> One can put together everything that we have learned, and add to it everything that workers in this field hope to learn; and none of it comes anywhere near, or even aims in the direction of what it means to see.[7]

George Wald added, "Seeing – the event in consciousness – seems to lie in another universe, unapproachable by science. Science offers no way of testing for the mere presence or absence of consciousness".[7]

Charles Sherrington wrote:

> The mental is not examinable as a form of energy …. Thoughts, feelings and so on are not amenable to energy (matter) concept. They lie outside it. Therefore they lie outside natural science …. Our mental experience is not open to observation through any sense organ …. Mind, for anything perception cannot compass, goes through in our spatial world more ghostly than a ghost. Invisible, intangible, it is a 'thing' not even of outline. It is not a 'thing'. It remains without sensual confirmation, and remains without for ever.[8]

The mechanics of mind which constructs continuity out of discontinuity is not known. Eddington says:

> The physiologist can trace the nerve mechanism up to the brain (inner sensorium); but ultimately there is a hiatus which no one professes to fill up. Symbolically we can follow the influence of the physical world up to the door of the mind; they ring the bell and then depart.[9]

Wilder Penfield, the renowned neurosurgeon, concluded that the fact that the mind can survive brain trauma and function under anesthesia points very strongly to the separate existence of the mind. He concluded that, "it is the mind which experiences and it is the brain which records the experience .... our minds and our brains are not at all the same".[10] He again wrote, "To suppose that the consciousness or the mind has location is a failure to understand neurophysiology".[11] The mind appears to be everywhere. It is observing on the scale of atoms and molecules, neurons, cells, tissues, muscles, bones, organs – in other words, it is observing on all scales of physical existence. The time required for a signal and its perception is about five thousandths of a second.[12] Wolf wrote, "We are beginning a new age of awareness, the age of quantum consciousness, the age of the conscious atom. By looking within ourselves, we may be able to solve the problems facing us on the final frontier – the frontier of the human spirit".[13]

(d) Creativity in Science: Intellect gives us objective truth so it ignores the subjective world. Intellect is invariably dualistic, so it has no capacity to gather *true* knowledge, and it is aware of *distorted* or *mutilated* knowledge. It leads to invention in the technological field. It is an improvement in the field of the known to the known. It suggests improvement in the field of higher sensitivity of matter, and thus science can construct better aeroplane, cars, frigidaires, air-conditioners, more efficient use of war machines, and designs and constructs artifacts with a greater sensitivity. Science deals with man-made social cosmos or synthetic construct of the mind-brain-body system.

Science deals with gathering of fragmentary experience and partial knowledge (information), so it has no potential to resolve human challenges. Maurice H. Wilkins, Nobel Prize in Medicine, 1962, wrote:

> Science makes progress largely by a combination of two aspects: open-mindedness and the following of abstract concepts. Abstraction involves excluding and narrowing down, which is very opposite of open-mindedness. Science always puts the most emphasis on its abstraction and narrowing down, on analytical procedures and on the role of intellect as distinct from intuition and feeling.[14]

Abstraction is generally intellectual, it sees a part. Abstraction can be regarded as a form of violence. It deals with the objective values and rejects the subjective dimension. Open-mindedness sees the wholeness of life. It deals with both the objective values and the subjective values. Abstraction deals with invention and open-mindedness bestows discoveries.

Objective science leads to objective values, so it suggests materials that can be used by technology to construct artifacts with higher sensitivity. Objective science is concerned with creating new knowledge and opening new frontiers. Technology is the systematic orchestration of all knowledge and experience to lead to something practical and commercially useful. Objective science deals with inventions, and not discovery. It plays a crucial role of innovation in the growth of technology or industry.

If "objective observing and understanding" is continued for a long chronological time, the subject of cognizance makes a quantum jump to higher quantum energy potential of consciousness and this bestows awareness of the wholeness of life, or intuitive wisdom, or subjective dimension of man. This leads to a discovery of a truth in nature hitherto unknown to man. The quantum jump takes place through discontinuities and not through a linear path – the discontinuities leads to the non-dual and non-conceptual path while the symbolic-dualistic frame is the linear path. The discovery is subjective, inspirational, intuitive, insight and often irrational; it is not a very structured activity. Genius prefers homogeneity of an individual and not the heterogeneity of a group. Such individuals are called creative scientists. We as a nation will have to innovate to survive.

Fritjof Capra wrote:

> Rational knowledge and rational activities certainly constitute the major part of scientific research, but are not all there is to it. The rational part of research would, in fact, be useless if it were not complemented by the intuition that gives scientists new insights and makes them creative. These insights tend to come suddenly and, characteristically, not when sitting at a desk working out the equations, but when relaxing in the bath, during a walk in the woods, on the beach, etc. During these periods of relaxation after concentrated intellectual activity, the intuitive mind seems to take over and can produce the sudden clarifying insights which give so much joy and delight to scientific research".[15]

Houston stated, "In high-level creativity the mind can do weeks of work in a short time".[16] He added that, "we are perhaps on the verge of a golden age of brain and mind research". It implies a quantum jump from symbolic-dualistic mind to non-dual frame of reference of Mind Continuum, from reason (intellect) to intuition. The intuition or insight bestows a creative Mind and in this is embedded human survival and excellence.

(e) Intuition: The "understanding of life directly, instead of in the abstract, linear terms of representational thinking" bestows intuition, spontaneity, sensitivity and creativity. These are metaphysical (quantum logic) insight, falling outside the realm of duality, or the intellectual construction. *Understanding, learning and action are a cyclic process, which results in a spiral going upwards in a vertical dimension; it is the growth in vertical dimension. The higher the understanding, the higher the expression of consciousness in the human mind-brain-body system.* The higher sensitivity or creativity leads to a perception of pure consciousness.

(f) Intelligence: Intelligence perceives life in its wholeness, and responds to it out of its own wholeness. The perception through intelligence, through sensitivity is a perception in non-duality. In love, beauty there is instantaneous perception. The perception and the response have no time lag between them.

The communion with the Divine, the Eternal, the Infinite is a happening; it is not the result of my doing. The uncovering of the secret is not a human achievement. It is something that happens, when the human being puts all his conditionings into relaxation, and remains before the mystery of Life in utter humility, in utter nakedness of his psyche. The consummation of human growth is the manifestation of the Divine, the Self-Unfoldment of man. Life is consummation of human growth. Then the mystery unfurls itself, uncovers itself. It uncovers itself into us.

All great discoveries have arisen from the realm of the timelessness, so scientific enquiry is not a cerebral activity and gathering information, but a quest of the timelessness. It also suggests that observation (seeing) is direct, listening (hearing) is direct, understanding and learning of life is direct, all action arising from understanding and learning is direct; but thinking is indirect. All direct relationship is in communion with the immutable laws of nature; and indirect relationship is antagonistic to the laws of nature.

4. Historical : The Emergence of the Conditioned Psyche

(a) Succeeding generations have found themselves confronted by the inner conditioning of mind which expressed itself as outer violence. Plato more than 2,000 years ego, realised this predicament and wrote, "What is happening to our young people? They disrespect their elders. They ignore the laws. They riot in the streets inflamed by wild notions. Their morals are decaying. What is to become of them".[17]

(b) Albert Einstein, in prophetic words, indicated the conditioning of the psyche in the first five decades of the twentieth century, in words:

(i) Einstein – **Out of His Later Years** – wrote, "The confidence in the stability, even the very basis of existence, of human society has largely vanished. One senses not only a threat to man's cultural heritage but also that a lower value is placed upon all that one will like to see defended at all costs .... Life in human society meant dangers of all sorts .... even standards of virtue, aspiration and practical truth were taken for granted as an inviolable heritage, common to all civilised humanity .... The sanctity of life vanished .... The lie was raised to the dignity of a political instrument. It was a manifestation of weakness of character and lack of responsibility on the part of the leaders in the affected countries". This presents evidence that the human psyche during the first five decades of twentieth century got heavily conditioned leading to lowering of its sensitivity.

(ii) Einstein – *Out of His Later Years* – wrote, "*Man changes but little .... though current trends like the present bring him unimaginable sorrow. Nothing of all that will remain but few pitiable pages in history books, briefly picturing to the youth of the future generations the follies of its ancestors*". Perhaps, the benign turn of destiny, who have faith in the mysterious will be saved.

(iii) The accelerative conditioning of the psyche continued from 1950 to date. The recent challenges are terrorism, organised crime, religious fundamentalism, ethnic conflicts, corruption and so on. The conditionings have gone from the conscious to the sub-conscious level. We have reached a critical stage in conditioning the psyche. Einstein painted a very grim picture of the future of mankind based on his perception of the Reality. And man has to face this situation. Many eminent scientists – George Wald, Nobel laureate in Medicine, 1967; Hideki Yukawa, a Japanese NL in Physics; Jacques Monod, NL in Biology, also talked that the world is heading towards self-destruction or self-annihilation. The fundamental cause as to why science and technology has failed mankind to lead to 'peace and serenity' is that he started living in the external world of time-space-causation (objective world) and rejected the Reality (the subjective dimension).

This implies that knowledge (information) proliferates while wisdom atrophies. This is a serious crisis. This crisis has enhanced between 1950 to date. At present, at all levels, there is a growing realisation that the existing models of education in India and in the rest of the world have failed and there is a total lack of relevance between the human being and the complex contemporary society. The increasing human challenges are forcing man inevitably to face the realities of the human situation. Many eminent philosophers-mystics have pointed out that the challenge is not only addressed to the structure of education, but to the nature and quality of man's mind, his way of living, consciousness and life.[18]

(c) J. Krishnamurti met Jawaharlal Nehru and Acharya Kripalani in 1931 and said, "If they ignored inner regeneration in the struggle for independence, India would lose its way". It appeared that the Indian mind was degenerating and this is continuing at an accelerative pace during the last 74 years and it appears that now we have reached an abyss. India must wake up from deep slumber to regenerate or rejuvenate the Indian mind.[19]

(d) Pupul Jayakar, in an essay entitled "*Changing Horizons, 1962-1977*" wrote, "A general destructive violence, a fragmentation at all levels, and a growing lack of sensitivity was permeating the landscape in India. J. Krishnamurti said, "One observes this (violence) …. happening around us socially, what is happening at the periphery of life, but also from what is happening inwardly".[20]

(e) C. Rajagopalachari, while in jail for civil disobedience made entry in his diary on January 24, 1922, in prophetic words, "Elections and corruption, injustice and the power and tyranny of wealth, and inefficiency of administration, will make a hell of life as soon as freedom is given to us? He also suggested that "Hope lies only in universal education by which right conduct, fear of God and love will succeed in this that Swaraj will mean happiness".[21]

(f) V. R. Krishana Iyer also remarks, "Inside each country there is societal entropy. Crimes are escalating".[22] The mankind is engulfed in the crisis of the human psyche. The crisis is tending to produce social dinosaurs, unfit to live in an evolving cosmos. Man has taken a path that leads to sterility of mind and heart.

It appears that the human psyche is being conditioned for more than 2,000 years; it has taken accelerative pace since the beginning of twentieth century. Every activity of the conditioned psyche leads to increase of entropy, which is death, or self-destruction, or self-annihilation. As far as we are aware, no steps have been taken by mankind collectively to end violence.

5. Education

For more than 2,000 years, man has been storing fragmentary experience and partial knowledge, as frequency and amplitude information, in the brain. These are called the conditionings which society has imparted to it. It is very narrow area where that intellect has been cultivated. It has the awareness of the distorted mutilated past. This we call as our present education which is merely accumulation of knowledge, information. It is nothing but an illusion, a mirage. It appears that humanity has taken a wrong turn which has created the chaos in the world?; and what needs to be done within the consciousness of man to preserve our very existence and bring about a new culture? His sensitivity and clarity through "objective observing and understanding" can bring a new insight in the order.

At all levels there is a growing realisation that the existing models of education in India and the rest of the world have failed because these deal with adding to memory regarding small part of life and these lead to adding of knowledge (information) and culturing the psyche. There is a total lack of relevance between the human being and the complex contemporary society.

Education must be concerned with the cultivation of the total human being, the nurturing of the mind-brain-body system. It should have two aims. The primary aim is to give the child the capacity to be and to remain awake to the dynamics of his or her own mind. The secondary aim is to teach the child the various skills and disciplines. We have to introduce the science of the mind-brain-body system in the science curriculum and transpersonal psychology in the arts. We have to teach the child to learn to read the book of life through the art of listening, the art of observing, the art of learning, and the art of living in a way that he or she is meaningfully connected to the context of the social and biological nexus and of the overall subtext of Reality. The educational structure must help generate human values through the enhancement of such awareness. We must learn to view the world holistically. This leads to the awakening of Intelligence which arises out of observation, self-knowledge, and compassion or Love. Intelligence, Truth, or Love, which are expressions of the same feelingful or intuitively felt Reality, alone has the capacity to resolve all human psychological problems.

The right education must be concerned with the cultivation of the whole being, the nurturing of the mind-brain-body system. It must deal with understanding. Albert Einstein, the greatest physicist of all time, wrote:

> Who would have thought around 1900 that in fifty years time we would know so much more and understanding so much less. It is in fact,

nothing short of a miracle that the modern methods of instruction have not yet entirely strangled the holy curiosity of inquiry; for this delicate plant, aside from stimulation, stands mainly in need of freedom: without this it goes to wrack and ruin without fail.[23]

In the words of Ramana Maharishi, "Illiteracy is ignorance and education is learned ignorance. Both are ignorant of the true Aim".[24] Ken Wilber wrote, "The 'true Aim' is the non-dual mode of knowing, while being literate or illiterate concerns the dualistic and symbolic mode of knowing. Ignorance, in other words, is ignorance of the non-dual and non-conceptual mode of knowing, which would instantly reveal the universe to be Mind-only".[25]

Education must encourage the right kind of self-observation, which flows to scientific temper. It is to perceive the wholeness of life at every moment, which is not to give significance to the '*part*', to the '*me*' and the '*mine*', but to help the mind-brain-body system to go above the psychological mode and to understand the *real*, the *fact*.[26]

Education deals with the sensual dimension and trans-sensual dimension of life. The sensual dimension is often called the objective field and the trans-sensual dimension as subjective realm. The restructured science and restructured education are a single dynamic flow, one unitary movement, and we call it "restructured science-education".

Fundamental Principles of Education: The right kind of education is as under:

(a) To understand life is to understand ourselves and that is both the beginning and end of education.

(b) Education is not merely acquiring knowledge, gathering and correlating facts; it is to see the significance of life as a whole.

(c) Intelligence is the capacity to perceive the essential, the *what is*!; and to awaken this capacity, in oneself and in others is education.

(d) Understanding comes only through self-knowledge, which is awareness of one's total psychological process. This education in the true sense, is the understanding of oneself. The dimension of understanding is embedded in the non-dual and non-conceptual frame of reference of Mind Continuum.

(e) Education in the true sense is helping the individual to be mature and free, to flower greatly in love and goodness.

(f) The highest function of education is to bring about an integrated individual who is capable of dealing with life as a whole.

(g) In the perception of the wholeness there is no conflict and suffering.

(h) Truth comes into being when there is complete cessation of thought; and thought ceases when the self is absent, when the mind has ceased to construct time-space-causation matrices; that is it is no longer caught in its own pursuits. When the mind is utterly still without constructing any psychological imbalances, ecological crisis. When the self goes to abeyance, then there is creation or rejuvenation of the Mind.

(i) Einstein wrote, "Education is *objective observing and understanding* which leads to *peace and serenity*". He wrote, "We want the creative and impressionable individuality, the personality" and this is the basis of creativity in science. Education is to nurture the basis of creativity in science.

(j) The child must understand and learn to live according to the immutable laws of nature. It is to bring home to the child that all activities in terms of rupture between man and nature must go to abeyance. The art of living is a dialogue with nature when carried on from within nature. It is by allowing Nature to work that we get the benefit of the gifts she has bestowed on us.

(k) Discovery is the beginning of creativeness; and without creativeness, do whatever we may, there can be no peace or happiness of man.

(l) Sri Aurobindo wrote, "The true teaching is that nothing can be taught". The secret of teaching is "the understanding of life directly".[27] The self-unfoldment and the higher quantum energy potential of Consciousness are the means and the process.

(m) The mutation happens in a brain through insight, not through knowledge, not through accumulation. That insight is the ability to observe a truth directly, a direct perception of the truth, and education must foster that.

(n) The foundation of renaissance of education is to regenerate the unexplored and hidden dimension of Universal Consciousness, so Silence is the gateway and the only way. In Silence, the mind relaxes into non-action. When the mind-brain-body system is equipoised the intuition and spontaneity blossoms. Intuition bestows value-oriented education or holistic education.

Action Plan: (a) The Action Plan has to be multifaceted and multidimensional. The fundamentals are "to educate the educator or training of teacher", and writing of books both for the teachers and students; and naturally these are at different levels, and in different languages. The teaching must be confined to *how to think* and not *what to think*. *How to think* implies awareness of the mind-brain-body system, or awareness of the wholeness of life. 'How to think' involves "observing". This is a stupendous task.

(b) We have to introduce psychodynamics of mind-brain-body system in science faculty and transpersonal psychology in art faculty as a part of educational curriculum at University/College level. We have to share through open dialogues the psychodynamics of conditioned mind and the dimension of Consciousness. The aim of these dialogues is "exploration of a spiritually inspired understanding of life" and "holistic living".

(c) To conduct courses at all levels – research scholars at doctoral level or post-graduate level especially in the science faculties of different universities and at lower levels in schools and colleges. To conduct seminars at the parents level.

(d) The principle governing restructuring of education is fundamentally based to assist nature for an evolution of life. The evolution of mankind can only happen when there is understanding and spontaneity; dynamic interplay of intuition and action; and awareness of the transcendent dimension of life. The real goal for each one of us is awaken and activise in ourselves perfect spiritual consciousness. A true seeker is the one who with his ideal in view, goes on purifying his feeling, thoughts and actions. We have to learn to empty our mind of all worldly thoughts and wait silently for the vision of the universal consciousness.

(e) The above programme is at national level and also at international level. We suggest that workshops be fundamentally based on symbiosis of modern science and spirituality.[28]

6. Global Ethics

Roger Sperry, who won the Nobel prize in 1981, stated that our values, beliefs and goals can directly affect our brain states and brain physiology.[29] So man-made social cosmos glorifies all sensual values – the values of the products of mind and senses, and the values of things made by hand and the machine. The more significance we give to the sensual value of things, the greater the confusion, turmoil and strife. All values in the sensual world can only be mere intellection, so every value is false to itself. These lead to conflict and confusion, fragmentation and misery, and the mind is enclosed, in idealistic or materialistic values.

Virtues, social values and moral values are constructs of the intellect. These belong to relative frame of reference of the mind spectrum, these refer to the psychological mode of mind spectrum. These imply time-space-causation matrices constructed by the symbolic-dualistic thought.[30]

Recently many eminent scientists – Werner Heisenberg, Edwin Hubble, Max Planck, G.F. Chew, Bertrand Russell, Lincoln Barnett, Max Born, and so on have concluded that we cannot impart value education at reason or intellect level.[31] We have constructed a society, where the intellect rejects ethical and spiritual values.

(a) Ethical Values: A mind that has set aside seeking in future and has denied its past is an ethical mind. It refers to non-dual and non-conceptual frame of reference in the Mind Continuum and it arises from the depth dimension of man. An ethical mind perceives the fact, the 'what is'; it is also called the "perception of the inner perceptual construct", "direct perception of actual fact", an "inward perception (i.e. through the mind), an "insight or intuition"; it gives "spontaneous response", which bestows the capacity to meet the present challenge.

(b) Spiritual Values: All knowledge is structured in consciousness. This recognises that consciousness alone can be totally self-referral, so the pure mind or non-dual state of mind is presented as an object to pure consciousness, then it reveals itself; or in other words, the non-dual mind perceives the spiritual value. The ultimate ethical value is called spiritual value. Spiritual value is the awareness of awareness itself. These refer to the transcendent quantum energy potential.

Max Delbruck said in 1980, "Further course of history will not be decided by further discoveries in science, but by these questions of human values".[32] The ethical/spiritual values emerge from the depth dimension of man, so these emerge from the Cosmic Consciousness.

(c) The Declaration of Global Ethics (Human Values): Albert Einstein on 20 Nov. 1950 wrote:

> The most important human endeavour is the striving for morality in our action. Our inner balance and even our very existence depend on it. Only morality in our actions can give beauty and dignity to life .... To make this a living force and bring it to clear consciousness is perhaps the foremost task of education.

He again wrote, "Our task must be to free ourselves from this prison by widening our circle of compassion to embrace all living creatures and the whole nature in its beauty".

Max Born at the third Pugwash conference stated, "clever, rational ways of thinking are not enough. The danger of mass slaughter .... can only be overcome by moral conviction, by a determination to replace national prides and prejudices with human love".[33] Bertrand Russell in England and Sorokin in USA have warned that modern civilisation cannot survive unless it is inspired by ethical and spiritual values. Among these values the most important which they stressed, was love. According to Einstein, the foundation of values must be independent of repetitive and mechanistic dynamics of the conditioned mind.[34]

Global ethics is defined as the perception in a non-dual and non-conceptual frame of reference of the Mind Continuum. It is the foundation of all ethical and spiritual values. The universal declaration leading to the emergence of global ethics constitutes a call to mankind that human mind must make a quantum jump from dualistic frame to non-dual frame of reference of Mind Continuum. And one who is wholly consecrated to Truth, who wants to live the Truth, serve the Truth, would know at each moment the ethical values; it would be a kind of intuition or revelation or insight which would make you know the truth, the global ethics at every moment. To perceive the ethical values or global ethics, man has to make a quantum jump from psychosocial evolution to spiritual evolution. The future of mankind is hinged on welding of Eastern Spirituality and Western Science, and out of that will emerge a new dimension of all-inclusive integral Consciousness that leads to ending of all human crisis, pain and misery, and bestows freedom, peace and bliss on human life. This is the only vocation of the human race.[35]

7. Ending of Violence

(a) We have built a society which is violent. Man has emerged as a violent species. Technologically, man has advanced incredibly, yet in his nature (mind-brain-body system) he remains as he was for thousands of years ago – all our education, science, organised religion, socio-economic-political reforms have proved to be sterile in transformation of human psyche. If the present conditioning continues, and we continue with our present way of living, we will be as violent even after one century as we are today.

(b) All psychological imbalances (violence, greed, fear, anger and so on), all ecological crisis, the challenges of poverty, hunger and so on, all religious fanaticism (terrorism, organised crime, corruption, etc.) are constructs of the waking state of mind, or relative frame of reference in the mind spectrum.

(c) Violence is an indication of psychological immaturity. When the scientific mode of mind sees the cause of psychological immaturity, all violence goes to abeyance and nonviolence self-expresses itself. We conclude that violence is a construct by symbolic-dualistic frame of reference, and nonviolence is a direct percept in the non-dual and non-conceptual frame of reference of the Mind Continuum; where the non-dual frame is synergistic or at higher quantum energy potential to symbolic-dualistic frame. This lays foundation of 'humanist technology', or 'science of life', or 'subjective research'.[36]

(d) The nonviolent state cannot exist where there is violence because nonviolence (as thought) and violence are at the same quantum energy potential of consciousness. Nonviolence is an ideal, a concept. Nonviolence and peace require the mutation of the individual psyche. These self-express themselves in a non-dual and non-conceptual frame of reference of Mind Continuum. A quantum jump from violence to nonviolence happens when you understand, learn and live the whole of existence, and not just a part of it. The perception of the whole ends all distortion, contradiction, conflict, and it establishes true relationship between the observer and the observed. We have to learn to observe our own thoughts without condemning, without justifying, without choosing.

(e) You cannot impose the order on the top of disorder. You cannot impose nonviolence on a consciousness which is violent because that violence has deep-rooted causes in the psyche. Unless these causes are eliminated there is no ending of violence. It is only when violence ends that one can be nonviolent.[37]

When you encounter a mind inwardly with ethical values, there is the ending of violence. The ending of violence in our consciousness is ethical values.

(f) Global Disarmament: Violence has many hideous faces – animosities and antagonisms, conflicts and conflagrations. As long as there is violence in the psyche, there is no hope for global disarmament. We have tried this since 1988, but the spectre of weapons of mass destruction continued to loom large in different parts of the world.

(i) In 1932, after a World Disarmament Conference in Geneva, Einstein, in an unusual bout of anger, exclaimed, "they (politicians and statesmen) have cheated us, they have fooled us. Hundreds of millions of people in Europe and in America, billions of people have been cheated, traded and ticked out of their lives, health and well being". These remarks are equally valid today. In principle man is deceiving himself.

(ii) The famous statement, issued in 1955, jointly by Bertrand Russell, Einstein, Yukawa and others is, if any thing, even more relevant today. The declaration of the Pugwash Movement on the occasion of the 25th Pugwash Anniversary Conference in Warsaw (August 1982), signed by amongst others, nearly 100 Nobel Laureates, concludes:

> There lies before us, if we choose, continual progress in happiness, knowledge, and wisdom. Shall we, instead choose death, because we cannot forget our quarrels? We appeal, as human beings to human beings: Remember your humanity and forget the rest.

(iii) Had I known that the Germans would not succeed in producing an atomic bomb, I would not have lifted a finger.

(iv) I know not with what weapons World War III will be fought, but World War IV will be fought with sticks and stones.[38]

As long as there is violence, fear and other psychological imbalances, the "Dialogue for Survival of Civilisation" continues and there can never be sustainable peace.

8. The Resolution of the Human Paradox

A knowledge-based society adds to materialism, adds to comfort, a sense of security and pursuit of pleasure. The future of India, and of mankind is not embedded in a knowledgeable society. Knowledge (information) leads to economic prosperity, technology, production. This enhances the conditionings which ultimately leads to death, self-annihilation. The knowledgeable society is not related to creativity in science and creative living which leads to nurturing technology and also ending violence, fear, greed and so on. It is for man to choose to "understand and direct action", with a view to follow the path of *Buddha*; or "thinking and reacting", and direct his onward march to an abyss.

Many eminent scientists-mystics have talked about convergence of Nano, Bio and Information technologies and opined that "this will create a historical revolution" and proclaimed that these studies will transform India to evolution of creative leaders in another one to two decades. Albert Einstein, in a pithy statement said, "We shall know a little more than we do now. But the real nature of things – that we shall never know, never". Einstein pleaded that we must explore our "inner security", or "Symbiosis of Science and Spirituality".

To sum up, the "economic prosperity, technology, productivity" cannot produce creative leaders even in a century, but it is exploration of inner world that can produce creative science and creative scientific leaders.

(a) Restructuring of Science: Karl Pribram wrote, "Brain science must deal with the awareness of awareness. It could no longer afford to shut out that part of the world which we call subjective".[39] Roger Sperry "spoke of the importance of a neglected area of science, namely the study of human subjective experience".[40] John Eccles also emphasized "to deal scientifically with the neglected area of inner conscious awareness".[41]

Roger Sperry wrote, "When the brain is whole, the unified consciousness of the left and the right hemispheres adds up to more than the individual properties of the separate hemispheres". So these studies raised the issue of consciousness in a new way. He adds, "When parts come together in a new whole, this new whole exhibits features – emergent properties – that can't be predicted as a rule from the parts, and cannot be explained in terms of the parts".[42]

The role of consciousness suggests to consider the objective and the subjective as one whole, or in other words, the outer (objective) and the inner (subjective) are not two separate movements but a unitary movement, one whole. The objective science must restructure itself to take in its ambient both the objective and the subjective dimensions of life. The energy exchange between the subject and the object takes less than a hundred microsecond. The restructured science deals with the wholeness of life.

(b) Ethics: It implies to take in its ambient the sensual and the trans-sensual dimensions of life.

(c) Health: All diseases are maladjustment of the human mind-brain-body system, or conditioned human psyche, or crisis of perception. Health is a complete physical, mental, psychological and social well-being; it is a dynamic state which is equipoised and balanced. A living in the *present*, the *now* heals; and living in the *past* and *duality* suppresses man's perceptive sensitivity and enhances disease. The curative functions of medicines have to be supplemented with understanding and living the "symbolic-dualistic frame of reference of mind spectrum (successive consciousness)" and the "non-dual and non-conceptual frame of reference of Mind Continuum (Simultaneous Consciousness)".

There is an urgency to restructure medical education which takes in its ambient the psychodynamics of the mind-brain-body system, or in other words, the wholeness of life. We have to jettison the illusion of the *past*, and to live in the *present*, the *now*. We have to restructure the biomedical model and adopt the holistic ecological perception. The restructured medical education takes in its ambient the wholeness of life.

(d) Ending of I-consciousness: We transfer what is happening at the biological level and transfer it to the psychological level; and this we call ego, or I-consciousness, so the 'I' is not an entity, but a concept. I-consciousness has no factual reality. It has a conceptual reality. It has an ideational reality. We graft the subjective on the perceptual level and transform it to conceptual level.

The I-consciousness is at the successive consciousness. The subject of cognizance or true-subject is at the immanent quantum energy potential. When I-consciousness (thought coupled with various volitions) encounters the true-subject, the I-consciousness is knocked off because it is at the lowest quantum energy potential. This suggests that (i) the movement of thought-time is independent of I-consciousness; (ii) it is the movement of the humanity, and (iii) both the I-consciousness and thought-structure have a conceptual reality.[43,44]

(e) Ending of Thought-Time: Every partial experience and fragmentary knowledge is stored in the human psyche, and it is called the cultural belief structure. It is meta-stable and disintegrates to thought structure. When thought-time comes in the realm of true-subject, the photons meet an 'impregnable wall', or "event-horizon", or "frequency-barrier", and are dissipated, which implies that thought-time goes to abeyance and what is left behind is the *fact*, the *true*, the *present*, the *what is*![45] When thought-time subsides, it is a discontinuity in the mental stream, and in philosophical language, it is called "no-mind" state, which comes gracefully and voluntarily. There is no fear, no acquisitiveness, no desire for ownership and possession, no desire to dominate over others. There are no psychological imbalances, no ecological crisis, no human challenges. A new human race, with a new culture and a new dynamics, a new foundation for human relationship will be there. Suddenly there is an inner stillness. This state is highly alert and aware, but no thinking.

(f) Ending of the Fact: If we continue observing or listening to the *fact*, without justification, or condemnation, or motivation, we observe that there is only the *fact* and nothing else. The fact itself is the operator and not you operating on the fact. If you leave the fact alone, it has no time. The fact is free from fear. The fact grows, which implies that there is release of tremendous energy and the fact moves to higher quantum energy potential. This is expressed as Pure Consciousness → Fact.

(g) Discovery: *All great discoveries have arisen from the realm of the timelessness, so scientific enquiry is not a cerebral activity and gathering information, but a quest of the timelessness. It also suggests that observation (seeing) is direct, listening (hearing) is direct, understanding and learning of life is direct, all action arising from understanding and learning is direct; but thinking is indirect. All direct relationship is in communion with the immutable laws of nature; and indirect relationship is antagonistic to the laws of nature.*

(h) Mutation: We have to restructure the dimension of science – which implies that it takes the subjective dimension in its field. The object is within the field of nature. When such a dialogue happens, the object is knocked off and we are aware of the wholeness of life, the oneness of life. The mutation is in the brain through insight, a direct perception of the truth.

9. Scientific Knowledge versus Spiritual Wisdom

(a) Mike King wrote, "Scientific knowledge is quite different from the 'final knowledge' (Spiritual Wisdom) because once the scientific community knows it, it does not have to be rediscovered, whereas the final knowledge of enlightenment has to be rediscovered every single time for each aspirant".[46] He added, "Having delivered us to the shores of the territory marked 'spirituality', the momentum of the scientific paradigm seems to be carrying the intrepid explorers *inland*, that is to colonise the unfamiliar territory".[46]

(b) Eddington wrote, "Scientific knowledge is only an empty shell – a form of symbols. It is knowledge of structural form, and not knowledge of content".[47] Spiritual wisdom cannot be stored, it is from moment to moment, it is a happening.

(c) Discovery is a flash, an insight. Kothari wrote, "A rational account of a discovery, the proof, follows and not precedes the discovery". Discovery is a happening in the timeless moment. All great discoveries have arisen from the realm of the timelessness, so scientific enquiry is not a cerebral activity and gathering information, but a quest of the Timelessness, the Divinity, the Life Field.[49] All scientific discoveries are spiritual wisdom. The conditioned psyche is only capable for a minor invention.

(d) A scientific theory is always provisional. It is a guide to further development, to its own replacement. Scientific knowledge can never be 'final', it is cumulative. It is communicable through reason.[48] The spiritual wisdom is the 'final' truth, it is not communicable by language.

(e) In the epistemology of science, there is 'successive consciousness' (consciousness at the lowest quantum energy potential). The epistemology of Spiritual Wisdom is beyond brain; the Universal Consciousness (at the highest quantum energy potential) will never be found by any search, but it expresses Itself to Itself, so it self-manifests which is a 'happening'.

(f) When turbulence – memories, thoughts, desires, emotions, mental images – go to abeyance, which implies ending of matter content in the mind-brain-body system; this state is called "mental vacuum", or "the vacuum states of consciousness". In this state, the "Unknown" expresses itself at higher frequencies or higher energies. Alan Wallace has drawn a parallel between outer vacuum of physical science and the "mental vacuum", the inner world.[50] Buddhist define the nature of consciousness as empty (of all matter content) and luminous (higher quantum energy potential of consciousness).

(g) It is a shock to discover that your brain, your mind (successive consciousness), your knowledge are valueless because these have no relationship with the immensity.

(h) Knowledge gives physical comfort, a sense of security, economic prosperity, but it is not related to the wholeness of life.

10. Science, Character Building and Humanism

Science is a part of human relations and the modern scientific work involves him and other scientists with society to an even greater extent than other people. More than political measures, treaties and professional goodwill are necessary; a new way of thinking about human values (ethical/spiritual) is needed. And the foundation of human values is right education.

Albert Einstein wrote, "All our lauded technological progress – our very civilisation – is like the ax in the hand of a pathological criminal".[51] C.P. Snow wrote, "We are walking with complacency into a situation more ominous than any in recent history .... worse than any the human race has ever known".[52] The dangers facing us are real. Yet all of them are man-made.

The most severe indictment of our educational system came from Lord Wavell. In his farewell letter to King George VI, he wrote:

> Education is the thing we have done worst in India, I believe, because we have provided education for the mind only and not the character. As a result, the average educated Indian has little character and no discipline. The Indians will have to learn both if they are every to become a nation.[53]

The editor of the Tribune concludes, "We have been calling ourselves a nation for 37 years now. But if still we do not look like one, it is mainly (if not entirely) because we have little character and less discipline". Radhakrishnan emphasised repeatedly civilisations are not built with brick and mortar, steel and machinery. They are built with men and women, their quality and character.

The character building, or devotion to humanism is concerned with the ethical/spiritual values – but not with psychological mode of mind, or scientific mode (invention) or religious dogma, or institutionalized religion. It is according to the immutable laws of nature or direct perception (and not mutilated perception).

Nature's methods of working are very mysterious. It tries to steer man's boat to safety through stormy waters, but always to security and without any error. Albert Einstein stated, "Our task must be to free ourselves from this prison by widening our circle of compassion to embrace the living creatures and the whole nature in its beauty". To widen "our circle of compassion" is called devotion to humanism. An intuition comes when mind sees discontinuity between any two consecutive continuities. A silent mind sees the false as the false and the true as the true. The true expresses itself as character.

The secret of character building, or devotion to humanism, or permanent peace is embedded in the transformation of the very foundation of perception and response. This transformation is possible through holistic education and science with values.

Let every human being take a small step forward in understanding the dynamics of his own mind and live in the light of that understanding; and cumulatively it will become a giant leap of mankind. We shall surely survive if we understand and prove that we are worthy of survival.[54]

11. Freedom and Peace

(a) Freedom: We define freedom as transcendence of thought with its entanglement with volitions, motives and desires. In other words, freedom is "cessation of I-making tendencies", or abeyance of "the machinery of image production". It implies that freedom of thought and speech and action emerge from a non-dual and non-conceptual frame of reference of Mind Continuum.

Freedom is pure observation without direction, without fear of punishment and reward. Freedom is without motive. In scientific observation one begins to discover the lack of freedom. Freedom is found in the choice of awareness, in the attentiveness of our daily existence and activity. Freedom comes, when even momentarily, thought or pseudo-subject goes to abeyance and non-variable attentiveness takes over as the true-subject. Pure observation bestows insight, which is without any shadow of past or time. The timeless insight brings about a deep radical mutation in the conditioned psyche or mind-brain-body system.

(b) Peace: It is a total way of living. As long as there is inherent violence in the human psyche, there can never be sustainable peace in the world. Peace requires the mutation of individual psyche. It is only holistic education that helps man to grow from knowledge to understanding, from confrontation to cooperation, from rationality to intuition, from intellect to intelligence.

We are on the crossroads of destiny; we must take the road from information gathering and experience to understanding and learning, from violence to non-violence, to motivation of sharing rather than dominating, complementarity and not competition, harmony and not discord; and in the final analysis education alone that can bring this about.

There can be no lasting happiness and sustainable peace as long as we accept a social order in which there is endless strife and antagonism between man and man. As long as we cling to security and are conditioned by the dogmas, there will be strife and misery both within ourselves and in the world. If we want peace, we must be aware of the ways of mind-brain-body system and this transcends thought structure coupled with I-consciousness. We have to learn observing (seeing) and listening (hearing) and to understand ourselves, and this bestows a non-dual and non-conceptual frame of reference of Mind Continuum. If this is sustained over a long chronological time, this flowers and blossoms to joy and peace.

Peace cannot be achieved by patchwork reform, nor by a mere rearrangement of old ideas and superstitions. Modern civilisation is based on violence and is therefore courting death. There can be peace only when we understand what lies "beyond a world divided", and thereby stop this wave of destruction which has been unleashed by our own aggressiveness and fears. A living in peace is the only panacea to save ourselves and our children from collective suicide.

12. Love and Compassion

In every human being, there is a hidden capacity to transcend thought-structure coupled with I-consciousness, or in other words, the capacity to breakthrough bondage, so all humankind is in the lineage of impersonal compassion. Compassion is care and concern for all life and living creatures.

Love is not different from truth. Love is that state in which the thought process, as time, has completely ceased. Where love is there is transformation or mutation. Without love revolution has no meaning, for then revolution is merely destructive, decay, a greater and greater ever mounting misery. Mankind needs to live in love and compassion.

13. Science and Religion

The basic core of religion in India is not believing in a creed of dogmas, but experimenting and experiencing the truth about God, the Life Field. The training of the senses and of the mind into effective instruments in the search for truth and character-excellence is education; and religion is only continued education.

Albert Einstein stated, "Science without religion is lame, and religion without science is blind". There is no conflict between science and pure religion. Both are complementary disciplines and non-authoritative approaches. What we need today is a dynamic interplay of pure religion and modern science, or in other words, symbiosis of science and spirituality.

Roger Sperry, Nobel prize in 1981, raised one of the most thoughtful questions, in words:

> What form religion and the teachings of Christ, Muhammed, Buddha, Confucius, and other founders might have taken, if Copernicus, Darwin, Einstein, and all the rest had come before their time instead of after?[55]

Every one of the founders of great religions would have taught us to feel responsible for the long-range effects of all our actions, the grandeur of the interaction of our nervous system with that of universe, and accept Cosmic Consciousness or Life Field. Perhaps they would have taught us to learn "objective observing and listening, understanding and questionings" with a view to "initiate an open dialogue with an open mind". The basic problems are: (a) to integrate into the solid, broader framework of religion and human "intuitive wisdom"?; (b) how can the merger of science and "human perennial values" become a self-evident part of the majority's mind-set?; (c) health technology; and (d) to observe "direct perception" in our every activity. The answers to the above problems must be based on the hard facts of modern science emerging from "objective observing and understanding".

14. Observation (Seeing) and Listening (Hearing)

Observation and listening happen in a non-dual and non-conceptual frame of reference of Mind Continuum, it is without accepting or denying, without conformity or resistance. In observation and listening, the object goes to abeyance, so non-duality ends, and the subject of cognizance sees itself as "objective truth", or the fact.

(a) Objective Scientific Mode: The objective scientific mind is very factual. It leads to invention. Such a mind sees from fact to fact. The spirit of science has nothing to do with individual conditions, such as nationalism, with race, with prejudice. But the scientific mind and its inventions are used and exploited by the nationalistic mind that is India, by the mind that is America. Scientific inventions are utilized and exploited by sovereign states and continents. It indicates that the achievements of objective science are limited – it adds to human comfort, leisure and pleasure; but it also contributes to competition, motivation and disorder, which flows to selfish, ambitious, greedy and possessive. It constructs a world which is insane, unhealthy and corrupt.

(b) Direct Perception: We have to restructure scientific mode, which implies a dimension of [science-education-ethics-health], which has the capacity to observe the wholeness of life. The restructured science takes in its ambient: (i) both the objective and the subjective dimensions, (ii) the sensual and the trans-sensual dimensions of life, (iii) perennial values of life embedded in the depth potential of life, and (iv) the wholeness of health. It is a perception of the whole process to a profound depth in which you observe the whole as insight, and this has nothing to do with the accumulation of knowledge about yourself. Some philosophers-mystics call it the true religious mind – which does not belong to any cult, to any group, to any religion, to any institutionalised religion. The restructured scientific mind and the true religious mind are synonymous. In scientific parlance, we call the restructured science as the highest quantum energy potential of Cosmic Consciousness. It is a creative mind. Discovery is its mission, its perception. It is ever busy in the art of observing and perceives the insight. The insight is direct perception. It is the cultivation of freedom in search of truth.

(c) Nature's Direct Perception: Nature invariably expresses itself as insight or direct perception. It bestows love and compassion. Love offers freedom. Love is not different from truth. Love is that state in which the thought process, as time, has completely ceased. Where love is there is transformation. We define nature as restructured scientific mode of mind, which expresses as the wholeness of life. This is the consummation of human life.

(d) To sum up – (i) every activity of the psychological mode of mind is dualistic, so it is a waste of energy; (ii) in the scientific mode of mind, there is no energy transfer between the subject of cognizance and the object, so there is awareness of the Fact, the *true*, the *present*, the *what is*?; and it bestows modified use of the known to the known; (iii) the restructured science regenerates the intelligence, which resolves our all challenges, leads us to order, rhythm, nonviolence and peace; and (iv) it guarantees our survival and excellence.

15. Concluding Comment

(a) Mind has the power to create illusion: and without understanding its ways, to seek inspiration is to invite self-deception. Albert Einstein wrote that man needs "objective observing and understanding" where both "observing and understanding" are in the non-dual and non-conceptual frame of reference of Mind Continuum and as such are in the non-authoritarian dimension. The subject of cognizance has the dimension of (science-education-ethics and health), which is the wholeness of life. And this leads to "peace and serenity". It is the Cosmic Consciousness or Life Field that governs the mutation of the conditioned psyche. It activises the Intelligence, which resolves all human challenges.

(b) Mankind urgently needs a quantum jump from reason to intuition. Discovery is an inspired flash of insight, a momentary intuitive knowledge. Discoveries come as intuitive flashes. Intuition is fundamental and reason is a derivative. It is only in that silent state that the old neural paths are short circuited and new neural paths are energised and it renews itself so that it is made fresh, young, decisive, not confused. When thought goes to abeyance, the energy of thought gets transformed into the energy of subject of cognizance, and this is responsible for creativeness of the Mind Continuum. Any pursuit of materialism cannot bestow creativeness of the Mind. The creative state of mind is the noblest.and the most mysterious faculty of man – the ability to transcend itself and perceive Itself in the act of direct perception. It is the expression of the Unknown, the Life Field, the Mysterious, the Timelessness, in the brain. It is invariably the *truth*, the *present*, the *fact*, the *what is*!

(c) Many eminent scientists believe that our journey through the turbulence of unforeseen events will succeed best if guided by: (i) education of the mind; (ii) objective observing and understanding; (iii) sustained observing (seeing), listening (hearing) the mind for a long chronological time, as that bestows understanding of Consciousness, the wholeness of Life; (iv) living in the light of that understanding; and (v) direct perception of the Oneness of Life. All this is a single dynamic flow of Cosmic Consciousness and leads to mutation of individual psyche which bestows love and compassion at every moment. The individual then acts as a boat to ferry others across the turbulent ocean infested with crocodiles and sharks and guides them to the other shore.

(d) Erika Erdmann and David Stover wrote, "The hour is late, the problems confronting us immense, and inhumane, catastrophic solutions become ever more likely the longer we delay. Our only hope lies in replacing a laissez-faire mentality with a willingness to shoulder responsibility for the future .... [We have to move] from a blind rush toward extinction to an enlightened advance toward yet unrevealed further wonders of evolution.[55]

(e) It is the Cosmic Consciousness or Life Field that governs the mutation of the conditioned psyche. It activates the Intelligence, which expresses as order, rhythm, nonviolence, peace and survival.

References

1. Erwin Schrodinger, *What is Life*? and *Mind and Matter*, Cambridge University Press, 1967, pp. 94, 95.
2. *Ibid*., p. 146.
3. Sampooran Singh, *The Pilgrimage to the Third Millennium*, Faith Publishers, January 1995, p. 19.
4. Fred Alan Wolf, *Taking the Quantum Leap*, San Francisco: Harper & Row, 1981, p. 229.
5. Spencer Brown, *Laws of Form*, New York, Julian Press, 1972, pp. 104-05.
6. Erwin Schrodinger, *op. cit*., p. 132.
7. George Wald, in *The Evolution of Consciousness*, New Delhi: National Publishing House, 1983, pp. 6, 7.
8. Charles Sherrington, *Man on His Nature*, Cambridge University Press, 1951, pp. 228-257; in D.S. Kothari, *Some Thoughts on Science and Religion*, The University of Delhi, October 31 & November 3, 1977, pp. 45-46.
9. Arthur Eddington, *The Nature of the Physical World*, 1928; in Sampooran Singh, *The Holistic Education*, New Delhi: Siddharth Publications, 1987, p. 41.
10. Wilder Penfield, in Deepak Chopra, *How to Know God*, Rider, 2000, pp. 217, 214.
11. *Ibid*., *The Mystery of the Mind*, Princeton: Princeton University Press, 1975, p. 109; in Fred Alan Wolf, *Taking the Quantum Leap*, Harper and Row, 1981, p. 244.
12. Fred Alan Wolf, *ibid*., p. 236.
13. *Ibid*., p. 248.
14. Maurice H. Wilkins, *Synthesis of Science and Religion*, Mumbai: Bhaktivedanta Institute, June 1988, p. 30.
15. Fritjof Capra, *The Tao of Physics* (Second Edition), New York: Bantam Books, 1988, p. 18.
16. Houston, in *Science*, 9 March 1973, Vol. 179, No 4077, pp. 982-83.
17. Plato, in Sampooran Singh, *The Survival of Mankind and Quantum Jump from Violence to Peace*, New Delhi: Books India International, October 1999, p. 223.
18. Albert Einstein, *Out of my Later Years*, *I Believe: Nineteen Personal Philosophies*, London: Unwin Books, 1962, pp. 29-31.
19. Pupul Jaykar, *J. Krishnamurti: A Biography*, San Francisco: Harper and Row, 1986, pp. 397-98.
20. *Ibid*., p. 314.
21. C. Rajgopalachari, in Sampooran Singh, Ed., *The Survival of Mankind and Quantum Jump from Violence to Peace*, *op. cit*., p. 224.

22. Krishna Iyer (Justice), *ibid.*, pp. 225-26.
23. Albert Einstein, Kennth Brecher, Nature, 15 March 1979, Vol. 278, pp. 216-17
24. Ramana Maharishi, in Ken Wilber, *The Spectrum of Consciousness*, USA: The Theosophical Publishing House, 1982 (Third Quest Book), p. 110.
25. Ken Wilber, *ibid.*, p. 110.
26. Sampooran Singh, Dr (Mrs) Kanwaljit Kaur, and Paramjit Singh, *Global Education and Human Survival: A Scientific Appraisal*, New Delhi: Sanbun Publishers, March 2002, p. 26.
27. Sampooran Singh and Dr (Mrs) Kanwaljit Kaur, *Science, Education, Health and Consciousness*, New Delhi: Kalpaz Publications, in press.
28. Sampooran Singh, *A Review on Symbiosis of Science and Spirituality*, Dagshai: The Atam Science Trust, October 7, 2002, p. 33.
29. Roger Sperry, in Erika Erdmann and David Stover, *Beyond a World Divided*, London: Shambhala, 1991, Cover Jacket.
30. Sampooran Singh, *Science, Human Values, Value Education and Peace*, New Delhi: NM&NWM (India), September 1993, pp. 84, 85; and Sampooran Singh, Dr (Mrs) Kanwaljit Kaur and Paramjit Singh, *Global Value Education*, Chandigarh: Faith Publishers, January 2001, pp. viii + 116.
31. Sampooran Singh, Dr (Mrs) Kanwaljit Kaur and Paramjit Singh, *Global Education and Human Survival*, New Delhi: Sanbun Publishers, March 2002, pp. 46, 35-36; and Sampooran Singh, in *The Temple of Understanding*, Delhi: Shipra Publications, 2003, pp. 86-107.
32. Max Delbruck, in *Beyond a World Divided*, *op. cit.*, p. 87.
33. Max Born, in Hideki Yukawa, *Creativity and Intuition*, Tokyo: Kodansha International Ltd., 1973, pp. 189-90.
34. Sampooran Singh, in *Science, Human Values, Value Education and Peace*, NM&NWM (India), September 1993, p. 84.
35. Sampooran Singh, *ibid.*, pp. 84-85.
36. Sampooran Singh, Ed., *The Survival of Mankind and Quantum Jump from Violence to Peace*, New Delhi: Books India International, 1999, pp. 221-338.
37. P. Krishna, *Education, Science and Spirituality*, The University of California, April 3, 1993, p. 15.
38. Albert Einstein, in *Bridging Science and Spirituality*, Mount Abu: Brahma Kumaris World Spirituality University, 1988, pp. 67-68.
39. Karl Pribram, in Ken Wilber, *The Holographic Paradigm and Other Paradoxes*, Boulder: Shambhala, 1982, p. 17.
40. Roger Sperry, in *Synthesis of Science and Religion*, Mumbai: The Bhaktivedanta Institute, June 1988, p. 199.
41. John Eccles, in *Synthesis of Science and Religion*, *ibid.*, p. 199.
42. Roger Sperry, *Bulletin of the Theosophy Science Study Group*, India, February & April 1988, p. 23 and June & August 1988, p. 27.
43. Vimala Thakar, *Through Silence to Meditation*, Ahmedabad: Vimal Prakashan Trust, 1998, pp. 22-27.
44. *Ibid.*, *Silence*, Holland Bookfund, 1989, p. 6; and *ibid.*, *The Invincible*, July-August 1987, Vol. IV, No. 4, p. 14.
45. Deepak Chopra, *How to Know God*, *op. cit.*, pp. 270-71.

46. Mike King, *The Scientific and Medical Network Review*, No. 78, April 2002, pp. 2-7.
47. Arthur Eddington, in Swami Ranganathananda, *Science and Religion*, Calcutta: Advaita Ashrama, July 1979, p. 125.
48. D.S. Kothari, *The Perception of Truth in Science and Philosophy*, in *The Odyssey of Science, Culture and Consciousness*, New Delhi: Abhinav Publications, 1990, pp. 49, 53.
49. Sampooran Singh, AVM Manjit Singh (Retd), Kanwaljit Kaur, Paramjit Singh, *God in Science, Symbiosis of Modern Science and the World Religions*, Chandigarh: Vishvas Publications Pvt Ltd, July 1999, Abstract, p. xv.
50. Alan Wallace, *The Scientific and Medical Network Review*, No 77, December 2001, pp. 21-25.
51. Erika Erdmann and David Stover, *op. cit.*, p. 6.
52. *Ibid.*, p. 3.
53. Sampooran Singh, *The Survival of Mankind and Quantum Jump from Violence to Peace*, *op. cit.*, p. 224.
54. *Ibid.*, p. 338.
55. Roger Sperry, in Erika Erdmann and David Stover, *op. cit.*, pp. 177-80.

18
Symbiosis of Science, Education, Ethics and Health

Abstract

This research article entitled, "Symbiosis of Science, Education, Ethics and Health" presents an exploration with a view to initiate "an open dialogue with an open mind". This transformation is addressed to all global scientists because there is a need for new relations between man and nature and between man and man, which in turn can create a new civilization and a new society. This is the key to ending all violence, terrorism, conflict from human psyche, and this points to human survival and excellence. Perhaps this may be of some assistance to the "National Knowledge Commission" constituted by the Prime Minister of India on June 2, 2005 and launched the 'Commission' on August 2, 2005.

Man had an evolutionary origin. He emerged on the planet Earth at about 2.5 million years ago. Man is an amphibian who lives simultaneously in two worlds – (a) the successive consciousness, the *outer* world of time-space-causation matrices, the objective materialistic world called objective dimension, and (b) the Simultaneous Consciousness, the *inner* world of timelessness, the subjective world. But the *outer* world offers no guidance in regard to the much more fundamental problems of the *inner* world. The outer is limited, finite, transitory and ephemeral, while the inner is Eternal, Infinite and Immortal.

Roger Sperry, Nobel Prize in 1981, discussed that our values, beliefs and goals can directly affect our brain states and brain physiology. Each individual is part of humankind, and humankind, in turn, is a part of the global system. The insight gives Facts and Values which cannot be separated.[1] If civilization is to survive and continue on a forward course, we have to educate our brain and its perception to perceive the spiritual values. Science did not create our present dilemma, it is the lack of perception of spiritual values or right education that has misdirected the application of science in technological and social problems. In philosophical language, we have to regenerate the Cosmic Intelligence to resolve our challenges. This article explores the regeneration of Intelligence, which is fundamental for our survival.

1. Has Humanity Taken a Wrong Turn?

At all levels, there is a growing realization that the existing models of education in India and in the rest of the world have failed and that there is a total lack of relevance between the human being and the complex contemporary society. The cause of the failure of education models are as under:

(a) The recorded history as paintings in the caves established that man was violent about 30,000 to 50,000 years ago. Plato, more than 2,000 years ago; Swami Vivekananda, about a century ago; Albert Einstein, in 1900 to 1950; C. Rajagopalachari in 1922; Lord Wavell and Mahatma M.K. Gandhi also perceived violence in the human psyche.[2] It implies that experience and knowledge, or cultural belief structure, or memory, is recorded as frequency and amplitude information in the brain. The memory is metastable, so it disintegrates and the by-product is called thought-time. Thought has brought about tremendous conflict between human beings, not only between individuals but also collectively. Every activity of thought-time leads to increase of entropy which is leading to self-annihilation or self-destruction.

(b) Man started living in the external time-space matrices, called objective dimension, and neglected the Reality, the Life Field, called subjective dimension. Nature thus provides that knowledge of one-half of the world will ensure ignorance of the other half.

(c) In modern physics, a very difficult situation has now developed. Physicists have come to see that all their theories of natural phenomena, including the 'laws' they describe, are creations of the human mind, properties of our conceptual map of reality, rather than of reality itself; or in other words, parts of the map, not of the territory. The fragmentation of the mind leads to duality, the object and the subject interactions. This leads to mutilated perception. All psychological imbalances, ecological crisis, and other human challenges are constructs of the conditioned psyche, and not by nature. All chaos and strife, all pain and suffering, all nervous tension and chemical pressures, are constructs of the mind.

(d) We live in a globally interconnected world, in which biological, psychological, social and environmental phenomena are all interdependent. To describe this world we need the ecological perspective which the Cartesian world view does not offer.

(e) The present educational model lays emphasis on materialistic facet, and ignores the spiritual view. To illustrate this, let us take an example of biological world where a seed is put in a suitable soil and we water it adequately. The seed sprouts in about 3 weeks. The young sapling has time and space matrices and the growth is governed by Consciousness. We see the effect and remain ignorant of the cause. The educational model is not tuned to understand the time-space-causation of the sapling; and in understanding this the intelligence is embedded.

(f) We have not taken any steps to end violence in the human psyche. This has now penetrated deep in the subconscious, and is expressing itself as terrorism, ethnic conflicts, corruption, drug trafficking, and so on. We may be in a position to suppress this violence temporarily, but cannot eradicate it from the human psyche.

(g) We are faced with fundamental questions as, "Has humanity taken a wrong turn which has created this chaos in the world?" and "What needs to be done within the consciousness of man to preserve our very existence and bring about a new culture?"[3]

2. Right Education

(a) Albert Einstein (1879-1955), the greatest physicist of all ages, wrote, "objective observing and understanding" leads to "peace and serenity". He again wrote, "Our task must be to free ourselves from the prison by widening the circle of compassion to embrace all living creatures and the whole nature in its beauty". He added that we have to energise our "creative and impressionable individuality, the personality" to understand the whole drama of existence. This has been aptly described by Ilya Prigogine, Nobel Laureate, 1977, in words, "There is a need for new relations between man and nature and between man and man".

(b) Many eminent scientists have stated that the outer (objective) and the inner (subjective) are not two separate movements. These are one, a unitary movement, one whole. It implies that we have to restructure the science so that it takes in its ambient both the objective and the subjective dimensions of life, the sensual and the trans-sensual world, and the wholeness of health. We may write this as: <science-education-ethics-health>. The subject of cognizance deals with the wholeness of life, the dialogue of the subjective and the objective is carried on from within nature; as the object approaches the subject, it is knocked off and only the subject is there, which observes Itself.

(c) Understanding comes only through self-knowledge, which is the awareness of one's whole psychological process. Thus education, in the true sense, is the understanding of oneself.

(d) In 1950, Roger Sperry wrote:

> At the core of all metaphysical problems stands the mind-brain relationship, real understanding of which has vast influence on all the ultimate aims and values of mankind.[4]

Roger Sperry raised one of the most thoughtful questions:

> What form religion and the teachings of Christ, Muhammad, Buddha, Confucius, and other founders might have taken, if Copernicus, Darwin, Einstein and all the rest had come before their time instead of after.[5]

(e) Roger Sperry and Jonas Salk opine that "insight (bestows) facts and values cannot be separated".[6] We opine that in classical or objective science, the *fact* is in the field of time and perennial values are independent of the fact and we have no awareness of the values. The objective knowledge has no capacity to give us values. However when the subject of cognizance is insight, we are aware of the *Fact* and perennial *Values* which cannot be separated because both are in the realm of the timeless. The Values emerge from the depth potential of man, or the highest quantum energy potential of Consciousness.

3. Reformation of Education

(a) Pandit Jawaharlal Nehru in 1959 stated, "Can we combine the progress of science and technology with the progress of the mind and spirit (spirituality) also? We cannot be untrue to science because that represents the basic fact of life today". The International Commission on Education set up by UNESCO in 1972 stated, "Link education to life, associate it with concrete goals, establish a close relationship between society and economy, invent or discover an education system that fits its surroundings, surely this is where the solution must be sought....".[7] The above quotations are laudatory, but, in principle, we have utterly failed even to touch the core of holistic education or value education, and we are moving on the periphery on the intellect level.

(b) The first priority is to educate the educator – that is, to have him understand himself. This is also termed as "teachers training course". The UGC may call for volunteers at post-graduate level from different universities and Teachers Training Colleges for advanced training. The teaching must be confined to *how to think* and not *what to think*. How to think implies awareness of the mind-brain-body system, and awareness of the wholeness of life. How to think implies "observing (seeing), listening (hearing) and questioning", it leads to "understanding, learning and living". It is synergistic to "what to think".

(c) We have to publish monographs for teachers and students, and naturally these are at different levels and in different languages. This is a stupendous task.

(d) We have to introduce psychodynamics of mind-brain-body system in science faculty and transpersonal psychology (or deep ecology) in art faculty as a part of educational curriculum at University/College level.

(e) We must observe (enquire, watch) with great care the dynamics of mind, listen to what is *within* and *without*; out of this arises sensitivity and in sensitivity there arises insight. Attentivity (insight) gives you the perception of the *fact*. That insight alone will wipe away sorrow.

(f) Insight is instantaneous. Insight happens in the non-dual and non-conceptual frame of reference of Mind continuum. Insight breaks down the patterns of thought structure. Perception of totality is an instant perception. The moment you have perception into the *fact* you are free of the *fact*.

(g) If you follow each thought to its completion with an attentive mind, you will see that at the end of it there is silence. From that silence (Unmanifest), there is renewal. The new approach is to bring *time* to an end. So man has to work for quantum jump from symbolic-dualistic frame of reference to non-dual and non-conceptual frame of reference of Mind continuum.

(h) The enquiry within is infinite. You must be alone, stripped, then you take the journey into the unknown. It is a journey of the alone (conditioned psyche) to the Alone (Universal Consciousness). Modern science has brought man on the control panel of transformation of human psyche. In a recent (1996) scientific conference at Moscow, the Russian scientists proposed, "A Life Field, beyond our physical concepts of time, space and the known electromagnetic and gravitational fields". In another conference at Cambridge, it was proposed, "A new scientific approach would include subjective experience as valid information and address all forms of human experience – including intuition, spirituality and mystical encounters".

(i) We have to share through "open dialogues with an open mind" the psychodynamics of conditioned mind and the Unconditioned Energy. The aim of these dialogues is "exploration of a spiritually inspired understanding of life" and "holistic living".

(j) The above programme is at national level and also at international level.

(k) Roy L. Walford, a medical doctor, wrote:

> What we need is a jump or mutational event in the social structure, a discontinuity, an abrupt historical change in man's whole orientation towards himself and his problems.[8]

This is called insight, which is a state of the highest quantum energy potential of consciousness, it is stillness in which there is no conflict, no contradiction. It is direct perception of the Life Field (Cosmic Consciousness).

(l) There will be need to create some infrastructure for conducting courses. Without Government active support, any action plan will prove to be a failure. And if Government escapes its responsibility, it will jeopardize the future of India and mankind.

4. The Resolution of Human Paradox

(a) Ending of I-consciousness: We transfer what is happening at the biological level and transfer it to the psychological level; and this we call ego, or I-consciousness, so the 'I' is not an entity, but a concept. I-consciousness has no factual reality. It has a conceptual reality. It has an ideational reality. We graft the subjective on the perceptual level and transform it to conceptual level.

The I-consciousness is at the successive consciousness. The subject of cognizance or true-subject is at the immanent quantum energy potential. When I-consciousness (thought coupled with various volitions) encounters the true-subject, the I-consciousness is knocked off because it is at the lowest quantum energy potential. This suggests that (i) the movement of thought-time is independent of I-consciousness; (ii) it is the movement of the humanity, and (iii) both the I-consciousness and thought-structure have a conceptual reality.[9,10]

(b) Ending of Thought-Time: Every partial experience and fragmentary knowledge is stored in the human psyche, and it is called the cultural belief structure. It is meta-stable and disintegrates to thought structure. When thought-time comes in the realm of true-subject, the photons meet an 'impregnable wall', or "event-horizon", or "frequency-barrier", and are dissipated, which implies that thought-time goes to abeyance and what is left behind is the fact, the true, the present, the what is![11] When thought-time subsides, it is a discontinuity in the mental stream, and in philosophical language, it is called "no-mind" state, which comes gracefully and voluntarily. There is no fear, no acquisitiveness, no desire for ownership and possession, no desire to dominate over others. There are no psychological imbalances, no ecological crisis, no human challenges. A new human race, with a new culture and a new dynamics, a new foundation for human relationship will be there. Suddenly there is an inner stillness. This state is highly alert and aware, but no thinking.

(c) Ending of the Fact: If we continue observing or listening to the fact, without justification, or condemnation, or motivation, we observe that there is only the fact and nothing else. The fact itself is the operator and not you operating on the fact. If you leave the fact alone, it has no time. The fact is free from fear. The fact grows, which implies that there is release of tremendous energy and the fact moves to higher quantum energy potential. This is expressed as Pure Consciousness → Fact.

6. An Overview of "National Knowledge Commission"

Our education is concerned merely with the culture of the outer. Violence, terrorism, corruption, conflict, sorrow cannot be resolved, cannot be understood as long as we live in the outer time-space-causation matrices; so we go through revolt after revolt. This can only end when we are engaged in "objective observing (seeing), listening (hearing) and enquiring"; and this blossoms to "understanding, learning and living". The fundamental challenge is reformation of [***science-education-ethics-health***], and we need total understanding of man and not emphasis on one fragment of life. This establishes harmonious relation between man and nature, and between man and man.

There is an urgency for a quantum jump from reason to intuition (insight), from thinking to perception, from knowledge to intelligence, from time to timelessness. So we need creative scientists instead of repetitive technicians, or mechanical golems. Freedom, love and compassion are the acts of perception. Love offers freedom. A creative scientist is one who lives in the non-dual and non-conceptual frame of reference of Mind Continuum. He is gifted with "direct perception, discovery, spontaneous action, and intuitive awareness of the oneness of life", and his every act is "perfect and fulfilling". Fritjof Capra wrote, "As the turning point approaches, the realization that evolutionary changes of this magnitude cannot be prevented by short-term political activities provides our strongest hope for the future".[12]

The Prime Minister on August 2, 2005 stated, "India cannot afford to lag behind the rest of the world". We suggest the following:

(a) Brain is a part of the objective world. It can be investigated objectively. Mind is subjective and independent of brain. We need mutation of the mind on priority, which blossoms to intuition (insight).

(b) Science does not create our present dilemma or constructs our challenges. It is the lack of perception of human values that misdirects the application of science in technological and social problems.

(c) It is the higher hierarchical level of Consciousness that can enhance the creativity of scientists.

(d) Education is not merely to give you experience and knowledge (information), but also to give you capacity to look at the world objectively, not emotionally, not with a prejudice, not with a bias, not comparing to past thought, but to look at it. Observation (seeing) and listening (hearing) happens in a non-dual and non-conceptual frame of reference of Mind Continuum. In the words of Albert Einstein, we need "objective observing and understanding" that will blossom to "peace and serenity".

(e) We have to restructure science – [*science-education-ethics-health*] - which deals with the wholeness of life. When subject of cognizance is the wholeness of life, then the object is within the subjective realm. When the object comes in the realm of the subjective Consciousness, it is knocked down and what is left is pure Consciousness. It energises Intelligence which resolves all human challenges and one revels in freedom. It bestows love, compassion and security.

7. Concluding Comment

The hour is late, the problems confronting us immense, and inhumane. Catastrophic solutions become ever more likely the longer we delay. We move from narrow egocentric and anthropocentric perspective to a large vision, including not only all mankind, present and future, but the entire interrelated web of life.[13]

The movement of thought-time is independent of I-consciousness, and both have a conceptual reality.

The objective science is governed by the principle of Objectivation, it moves from fact (objective truth) to fact, and is unaware of the perennial values, so fact and values are separate and independent identities. When the subject of cognizance is insight (intuitive wisdom), it is aware of the Fact (subjective truth) and perennial Values, which cannot be separated because both are in the timeless realm. All values pertain to humanity's ultimate aim to survive as a species.

The outer (objective) and the inner (subjective) are not two separate movements, it is a unitary movement, one whole. The science must restructure itself to take in its ambient the outer and the inner, the sensual and the trans-sensual world, and the wholeness of life. The restructured science is: [*science-education-ethics-health*].

The violence in the human psyche can only end through "observing (seeing), listening (hearing), questioning" which results in "understanding, learning and living". Observing implies to see both objective and subjective dimensions of life, not with prejudice, not with evaluation or verbalization, not with any inference, not with any bias, but just observe as it is. It is the highest quantum energy potential of Cosmic Consciousness that can bestow order, rhythm, nonviolence and peace.

Mutation is possible in the brain through insight, not through knowledge, not through accumulation. That insight is the ability to see a truth directly, a direct perception of the truth and education must foster that. P. Krishna said, "We are going towards self-annihilation, and not too much time is left now".[14] George Wald, Nobel Prize in Medicine, 1967, said, "By now the problem is not transcendence, but survival".[15] Jonas Salk said, "Not only survival, but further evolution!".[16]

References

1. Erika Erdmann and David Stover, *Beyond a World Divided*, Shambhala: 1991, pp. 82-83, 74.
2. Sampooran Singh, *The Survival of Mankind and Quantum Jump from Violence to Peace*, New Delhi: Books India International, 1999, pp. 223-29.
3. P. Krishna, *Education, Science, Spirituality*, The University of California, April 3, 1993, Cover Page.
4. Roger Sperry, *Neurology and the Mind-Brain Problems*, American Scientists, 1952, Vol. 40, p. 311; *ibid.*, in Erika Erdmann and David Stover, *op. cit.*, p. 42.
5. *Ibid.*, pp. 177-78, 180.
6. Erika Erdmann and David Stover, *op. cit.*, p. 83.
7. Jawaharlal Nehru, in *Dagshai: The Atam Science Trust*, October 7, 2002, p. 26; and *International Commission on Education*, UNESCO, 1972.
8. Roy L. Walford, in Erika Erdmann *et. al.*, *op. cit.*, p. 99.
9. Vimala Thakar, *Through Silence to Meditation*, Ahmedabad: Vimal Prakashan Trust, 1998, pp. 22-27.
10. *Ibid.*, *Silence*, Holland Bookfund, 1989, p. 6; and *ibid.*, *The Invincible*, July-August 1987, Vol. IV, No. 4, p. 14.
11. Deepak Chopra, *How to Know God*, *op. cit.*, pp. 270-71.
12. Fritjof Capra, *The Turning Point*, London: Wildwood House, 1982, p. 466
13. Erika Erdmann *et. al.*, *op. cit.*, pp. 179-80.
14. P. Krishna, *op. cit.*, p. 12.
15. George Wald, in *Evolution of Consciousness*, New Delhi: National Publishing House, 1983, p. 14.
16. Jonas Salk, in Erika Erdmann, *et. al.*, *op. cit.*, p. 98.

19

Science and Ethics

Abstract

Recently many eminent scientists – Werner Heisenberg, Edwin Hubble, Max Planck, G.F. Chew, Bertrand Russell, Lincoln Barnett, Max Born, and so on have concluded that we cannot impart value education at reason or intellect level.

Man does not live by bread alone: his life, joy and fulfillment has deep relationships with love and compassion, ethical/spiritual perennial values. A silent fundamental revolution is going on, a dimensional change in our cultural transformation.

Nature expresses itself as understanding and learning, and these flow to intuition, spontaneity and right action. Nature responds to the arts of "observing, listening, understanding and questioning", which bestows a harmony with the inner and the outer, with himself and his environment, and who is free from all ideologies and conditionings that now divides humanity.

In the non-dual and non-conceptual frame of reference of Mind Continuum, there is direct input of the Unmanifest in the brain cells without mediation of the sensory apparatus. It is a state of no-mind, it is a state of pure perception where nature expresses Itself as restructured science and ethics.

We have to educate the mind-brain-body system so that it can undertake an inner voyage to perceive perennial human values – which are love and compassion. We have to understand and learn to live according to immutable laws of nature. It is the Unmanifest, or the Life Field, or the Cosmic Consciousness that can guarantee peace and survival.

The aim of this article is to "initiate an open dialogue with an open mind". This may lead to a new "understanding, learning and action", which may be the foundation of an International Symposium on "Science and Ethics" to be held in 2006 by the University of Delhi (Prof. D.S. Kothari Centre for Science, Ethics and Education).

Today, after billions of years of pre-biological evolution and meta-biological evolution, the homo-homonid species emerged at about 4.5 million years ago and homo-sapiens at about 2.5 million years ago and is now at the evolutionary and cultural pyramid. The whole cosmic existence and beyond is governed by Cosmic Consciousness. It expresses itself in two dominant dimensions – first, the successive consciousness, the time-space-causation matrices, the *known*, the objective world; and second, the Simultaneous Consciousness, the Life Field, the subjective world, the *Unknown*. The characteristic of the objective world is finite, transitory, ephemeral, and limited, while the subjective world is Infinite, Eternal, Timelessness, Immortal. As an analogy, we say that if successive consciousness is represented by a drop of water the Simultaneous Consciousness is the boundless and fathomless ocean.

Science is the search for truth. The aim of science is to explore nature and human experience and knowledge (information) objectively. Objectivity and precision, both as to thought and its verbal formulation, are the two important characteristics of the scientific method. The values in science are objective truth, integrity and honesty. A selfless, unfaltering devotion to truth and use of knowledge for man's upliftment – higher hierarchical expression of consciousness – are the only vocation of a scientist.

The accelerative progress of science and technology in the material field and the thrust of psychosocial evolution has led to growth of globalization, but global consciousness is not keeping pace with the globalization.

J. Krishnamurti stated that man was programmed for the last 5,000 years, since the recorded history. Plato, more than 2,000 years ago, observed violence in a child. Swami Vivekananda about a century ago stated that all psychological imbalances in the human psyche have become automatic and concluded that, "It is really a tremendous problem". Albert Einstein, the greatest physicist, also perceived conditioning of the psyche in the first five decades of the twentieth century. There is an inner urge in the human mind to end all psychological imbalances and to live peacefully.

The fundamental questions facing mankind are: (a) "Has humanity taken a wrong turn which has created the chaos in the world?"; (b) mankind has been engaged to end all chaos and conflict at least for the last 10,000 years, but all his efforts have proved to be sterile; and (c) what needs to be done within the Consciousness of man to preserve our very existence and bring about a new culture?

This article explores the fundamental challenges facing mankind: (a) the root of psychological chaos, conflict and imbalances; (b) what is insight which can bring about mutation in the brain-body system; and society; and (c) what is Cosmic Consciousness. It discusses science and complementary science and ethics. The ancient spiritual truths – that underlie the religious impulse – are again beginning to glow with inner luminosity.

There is a convergence between the new physics and the ancient spiritual insights. A "Dialogue of Civilizations" is going on. India is uniquely placed to make a historic contribution toward "Symbiosis of Science and Spirituality".

1. Mind-Brain-Body Interactions

(a) In modern physics, a very difficult situation has now developed. Physicists have come to see that all their theories of natural phenomena, including the "laws" they describe, are creations of the human mind; properties of our conceptual map of reality, rather than of reality itself. These are creations of the mind; parts of the map, not of the territory. Obviously, the chaos, the strife, the nervous tension, the chemical pressure can only end if we explore and understand the whole mind-brain-body system.

> There is no explanation, no clue whatever, as to *how* the interaction between mind and matter occurs. Paradoxically, nothing is more commonplace than the *reality* of this interaction. My writing this sentence, and your reading it, is itself an expression of the body-mind interaction, and yet nothing is more intractable, nothing a greater mystery.[1]

> There is no problem more mysterious than the mind-body interaction. Modern science has not made it less intractable. On the other hand it has added a new urgency and also a new poignancy. The complementarity approach may open up some new possibilities worth exploring.[1]

(b) Brain is a part of the objective world. It can be investigated objectively. On the other hand, mind is subjective. Science provides no answer to this riddle. Charles Sherrington wrote, "The mental is not examinable as a form of energy.... Thoughts, feelings, and so on are not amenable to energy (matter) concept. They lie outside it. Therefore they lie outside Natural Science .... Mind, for anything perception can compass, goes therefore in our spatial world more ghostly than a ghost. Invisible, intangible, it is a "thing" not even of outline, it is not a 'thing". It remains without sensual confirmation, and remains without it for ever".[2]

(c) Candace Pert, in 1989, observed that, "White blood cells are bits of the brain floating round the body".[3] Ultimately, this implies that cognition is a phenomenon that expands through an intricate chemical network of peptides that integrates our mental, emotional, and biological activities.

(d) Roger Sperry argued that our values, beliefs and goals can directly affect our brain states and brain physiology.[4] At the core of all metaphysical problems stands the mind-brain-body relationship, real understanding of which could have vast influences on all the ultimate aims and values of mankind.

2. Dialogues Between J. Krishnamurti and David Bohm

(a) J. Krishnamurti and David Bohm had thirteen dialogues in 1980 and these were published in a book "*The Ending of Time*" in 1985. Krishnamurti prophetically stated:

> I have a million years of experience, and it has given me a certain capacity. And I realize at the end of it all there is no relationship between me and truth.... Is it a shock to discover that your brain, and your mind, your knowledge, are valueless? All your examinations, all your struggles, all

> the things that you have gathered through years and years, centuries, are absolutely worthless? .... What you have done or not done is absolutely of no value .... The ground says, whatever you have done 'on earth' has no meaning.

It has given me a tremendous shock to discover the truth, that all the churches, prayers, books, have absolutely no meaning – except how we can build a better society, and so on.[5]

So there is only one thing, and that is to discover that *all* that I have done is useless – ashes!

This implies that every activity of the conditioned mind is repetition of the past, so it has no newness, no originality; every perception is distorted or mutilated; and there is awareness of the *non-fact*, the *false*, the *past*, the *what should be*?; and all this has no relationship with the Life Field, the Immensity, the Cosmic Consciousness.

(b) In two dialogues, Krishnamurti stated that man has been programmed for the last five thousand years.[6] David Bohm in the *preface* wrote, "He (Krishnamurti) emphasizes that the mind is essentially free of the distorting bias that is inherent in the conditioning of the brain, and that through insight arising in proper undirected attention without a center, it can change the cells in the brain and remove the conditioning".[7]

3. What is Mind?

(a) Wilder Penfield, the renowned neurosurgeon, concluded that the fact that the mind can survive brain trauma and function under anesthesia points very strongly to the separate existence of the mind. He concluded that, "it is the mind which experiences and it is the brain which records the experience .... our minds and our brains are not at all the same".[8] He again wrote, "To suppose that the consciousness or the mind has location is a failure to understand neurophysiology".[9]

(b) The emission of thought is not a continuous stream of energy, but in discontinuous bits of energy. Thought is capable to do discontinuous scanning of the perceptual image at its surface in one-dimensional system and in one plane; so our effort is one-dimensional and limited. The natural world, however, is non-linear and sequential. It is one of infinite varieties and complexities, a multidimensional world, where things do not happen in sequences, but all together. As a one-dimensional system can never give a total solution of a multidimensional problem. Thought cannot see the completeness and wholeness of the problem.[10]

(c) It implies that – (i) mind has a separate existence from the brain, and it is the mind which experiences and the brain that records; (ii) the conditioning of the mind does not happen, or the mind has no capacity to store the experience and knowledge, (iii) if mind is at a higher quantum energy potential of consciousness then the conditionings go to abeyance and this ends the conditionings; and (iv) objective observing and understanding" leads to higher level of awareness which brings mutation of the psyche.

4. Science and Complementarity

(a) We can demonstrate that light is particle-like with the photo-electric effect; that it is wave-like with the double-slit experiment, and that it is both particle-like and wave-like with Compton scattering. Both of these complementary aspects of light (wave and particle) are necessary to understand the nature of light. Wave-like behaviour and particle-like behaviour are not property of light, but of our interactions with light. The electron and X-rays also show the complementarity (wave and particle) properties.

Niels Bohr has considered the particle-picture and the wave-picture as two complementary descriptions of the same reality, each of them being only partly correct and bring a limited range of application. Each picture is needed to give a full description of the atomic reality, and both are applied within the limitations given by the Uncertainty Principle.

The higher frequency means higher energy, so wave-like properties are at a higher quantum energy potential than particle-like properties.

(b) Kothari discussed the far-reaching philosophical and ethical implications in Bohr's Principle of Complementarity. Max Born stated that theoretical physics today is actual philosophy. Erwin Schrodinger, Ernst Mach and Albert Einstein were deeply influenced by Indian philosophical thought.[11]

(c) R.P. Kaushik wrote, "The first stage prior to recognition and naming is a state of non-verbal or thought-free perception (*nirvikalpa jnana* or pure perception). From this pure perception we proceed to thought, recognition and naming (*savikalpa jnana*)".[12] In the psychological mode of mind, the pure perception recalls the past memory and there is coherent superpositions resulting into distorted perception.

(d) An object is small in the absolute sense if the effect of the '*disturbance*' which inevitably accompanies an act of observation is not negligible. An electron, or an atom, is a small object in the absolute sense, because the unavoidable disturbance (inherent in the act of observation) drastically alters the state under observation. The small object has its own dormant consciousness which keeps it in a dynamic state. Quantum physics states that "self-awareness cannot arise in small systems".[13] We have to reexplore that because in the absence of a dormant consciousness, the system is bound to collapse to an amorphous mass, but this consciousness does not interact with other small systems. But when small systems encounters the subject of cognizance with higher hierarchical quantum energy potential of consciousness, the small objects go to abeyance and the subject of cognizance becomes quiet and silenced.

5. The Loss of Values

(a) The Root Cause of Loss of Values: The accelerative progress in the material field has not added even an iota to human happiness, goodness and creativity. On the contrary, it has brought in more emptiness of heart. Eminent social psychologist Erich Fromm writes, "The great promise of unlimited progress, the promise of the domination of nature, of material abundance, of the greatest happiness for the greatest number, and of unimpeded personal freedom, has failed".[14]

T.S. Eliot, in his poem '*The Rock*', with his intuitive insight, knows better:

Where is the life we have lost in living?
Where is the wisdom we have lost in knowledge?
Where is the knowledge we have lost in information?
The cycles of heaven in twenty centuries
Bring us further from God and nearer to dust.

Man does not live by bread alone: he also lives his intuition and reason. Reason gives him emotions and volition. He has become a psychological wreck.

The relationship between the rational and intuitive aspects of the human mind can be seen as a dynamic interaction of the two archetypal poles which are called the *yang* (self-differentiating potential) and the *yin* (self-integrative potential) in Chinese culture. Science and mysticism are two manifestations of these poles. Man lives in a dynamic interplay of his self-differentiating potential (*Yang*) and self-integrative potential (*Yin*), so his relationships are determined by the complementarity of *Yang* and *Yin*. He is in search of happiness, creativity and spontaneity, freedom, non-violence and peace. Only discovering the meaning of life can satisfy the hunger of the human heart. There is a teleological urge in every human being that will not give him rest till he discovers the meaning of life by realizing the 'divinity within him'. Ages ago, our Rishis declared: "Even if man can roll up space as if it were a piece of hide, there will be no end of misery without realizing the Lord".[15]

The root cause of loss of values is that we have started living in a man-made social cosmos or our sensory world, which is interfering with the complex web of relationships, upon which the life of our planet depends. Modern science has brought man to the crossroads: one is leading to annihilation through conditioning of psyche, and the other to 'wisdom' or Buddha through understanding, intuition and spontaneity.

(b) Values in the Light of Indian Philosophy: Swami Ranganathananda, President of the Ramakrishna Mission, in December 1966, stated, "Indian thought has always upheld the view that the values proceed not from matter but from the spirit of man .... All ethical values are essentially spiritual values; they proceed from the trans-sensual dimension of the human personality. This dimension, which stimulates the ethical sense, is beyond the reach of animals and of such humans who, like animals, are content to live at the sensate level". We reject the virtues and the social values of the outer socio-economic-political and uphold the ethical and spiritual values emerging from the trans-sensual dimension of human personality, from the depth potential of man, the inner nature of man, the wholeness of man.

The future of mankind is embedded in a higher level of awareness, higher level of perception, higher level of sensitivity, higher quantum energy potential; in quantum jump from symbolic-dualistic mind to non-dual and non-conceptual Mind Continuum; in an inner voyage, subjective research, from psychosocial evolution to spiritual evolution; in a fundamental shift from successive consciousness to Simultaneous Consciousness; and in symbiosis of science and spirituality (science of life).

(c) Dimension of Values: Observation is to perceive directly without the word, without the reaction, without the memories entering into perception; and this is called direct perception. If we see the psychodynamics of the mind-brain-body system, the reason, the rationality of the activity of memory (recalled past) that is limited, the mind makes a quantum jump from the waking state to the state of observation. When this is sustained for a long chronological time, it makes another quantum jump and is called attentivity. Attentiveness is at a higher quantum energy potential than thought. In common parlance, the attentivity is synergistic to thought, so there is flow of energy from the attentivity to the object.

Attentivity now sees the object at a higher quantum energy potential, so it sees the wave-aspect of thought, the subjective nature of thought, the implicate order of thought, the true nature of thought; which are the values. The wave-aspect of thought is at a higher quantum energy potential than the particle-aspect of thought. The wave-aspect bestows the human values, the ethical values, the moral values.[16,17]

Man must grow from knowledge to understanding, from intellect to intelligence, from successive consciousness to Simultaneous Consciousness. He must learn to perceive the truth at every moment. One has to learn to live his own understanding – the great Buddha said, "Be a light unto yourself". This revolution will regenerate or rejuvenate human mind so that it perceives the new dimension of an integral and all inclusive Consciousness, that can guarantee human survival along with all fauna and flora.

The mind has an in-built mechanism to make a quantum jump from the 'dimension of measure' to the 'dimension of value' and both are complementary to each other. The dimension of measure (objective information content) expresses itself as particle-like behaviour of thought. The dimension of value (subjective information content) expresses itself as wave-like behaviour, and both are complementary. The dimension of value is at a higher quantum energy potential than the dimension of measure.[18]

The wave-aspect is an expression at a higher quantum energy potential than that of particle-aspect. This deep understanding would radically alter the level of expression of consciousness of the observer, reorientating it toward a perception of the 'unbroken' of which everything is a form. An understanding of the psychodynamics of mind-brain-body system results in a journey from intellectual entrenchment to intellectual openness; it leads to a quantum leap beyond rationality to insight, and simultaneously activates the unconditioned energy, the Fundamental Vibration. It leads to a quantum jump from symbolic-dualistic mind to non-dual mind.[18]

(d) The Urgency to Inculcate Human Values: Objective science must restructure itself to take subjective dimension in its fold. There cannot be a global order without a global ethics. And that global ethics will emerge from a non-dual and non-conceptual frame of reference of Mind Continuum.[19] Let us undertake an individual and collective experiment to save ourselves from annihilation. It is the Cosmic Consciousness which can guarantee human survival and excellence.

General Omar Bradley prophetically said, "Ours is a world of nuclear giants and ethical infants. We know more about war than we do about peace, more about killing than we do about living".[20] The Parliament of Religions, Chicago, 1993 issued a *Declaration of a Global Ethic* and concluded, "No new global order without a global ethic! .... Earth cannot be changed for the better unless the consciousness of individual is changed". Gerald O. Barney, a fusion-energy physicist, addressing the Parliament of Religions said, "*At the core of the crisis is a failure of vision*".[21]

The Blaker Foundation, U.K., aims at a transition from our evolving materialistically based culture to a broader spiritually inspired understanding of the world.[22] George Blaker put it succinctly, "Our science of today cannot stop the spiritual advance, but can only delay it for a while, for such delays will automatically drag in their wake calamitous consequences, including geological disasters".[22]

Fritjof Capra, an eminent nuclear physicist, prophetically wrote, "As the turning point approaches, the realization that evolutionary changes of this magnitude cannot be prevented by short-term political (socio-economic-political) activities provides our strongest hope for the future".[23] And if man escapes his total responsibility, he does so at his own peril, at his own self-annihilation.

One has to learn to establish oneself in the non-dual, non-conceptual awareness, which bestows all ethical spiritual values. When ethical values self-express themselves, all psychological imbalances vanish. It is now at the control panel of radical transformation of diseased (conditioned) human psyche. The ethical or spiritual values emerge from the spiritual dimension of human personality. The human survival and excellence is embedded in the symbiosis of science and spirituality. This will enable mankind to take a giant leap towards human survival and excellence, a glorious future.[24]

The real wealth of a nation is its people. And the purpose of human life is to transcend all psychological imbalances and to imbibe the understanding of life directly. This bestows a perception of the Wholeness or Oneness, which self-expresses as human values, nonviolence, harmony, coherence, and so on. This blossoms to self-restructuring of society and bestows the art of living in freedom, peace and bliss.

6. Science and Ethics

(a) The Principle of Objectivation: In 1947, Erwin Schrodinger put forward the principle of objectivation as one of the basis of the scientific method. Schrodinger wrote:

> Without being aware of it and without being rigorously systematic about it, we exclude the Subject of Cognizance from the domain of nature that we endeavour to understand.[25]
>
> Mind has erected the objective outside world of the natural philosopher out of its own stuff. Mind could not cope with this gigantic task otherwise than by the simplifying device of excluding itself – withdrawing from its conceptual creation. Hence the latter does not contain its creator.[26]

Schrodinger quoted Charles Sherrington in words:

> The material world has only been constructed at the price of taking the self, that is, mind, out of it, removing it; mind is not part of it; obviously, therefore, it can neither act on it nor be acted on by any of its parts.[27]

Schrodinger also quoted C.J. Jung and B. Spinoza to support the hypothesis. It implies that there is no energy transfer from the subject to the object and the mind is aware of the "objective information content", and not the "subjective value". The objective information content is in *time*, which is the dimension of measure. It leads to materialism and the human values go to abeyance.

(b) The Principle of Understanding: If we restructure the dimension of the subject – which is education-science-ethics-health – it sees the wholeness of life, the mysterious boundary between the subject and the object breaks down – and the object goes into abeyance and the subject sees itself. The subject is the non-dual and non-conceptual frame of reference of the Mind Continuum. This bestows an understanding of the wholeness of life, the subjective values of life, the ethical/spiritual values of life.

(c) Perception of Human Values: Objective science must restructure itself so that it takes in its ambient the subjective dimension in its realm. The objective and the subjective are one, a unitary movement, not separate, but whole. It implies that the subject of cognizance observes the human problems from a larger perspective, one that brings into focus not only all of mankind, but all life – and beyond that, all of creation. From this vantage point, we see humanity as a small but extraordinary fragment of reality, endowed with the gift of Mind – probably as a part of continuum point of the Cosmic Consciousness – but we are wasting that gift, and indeed on the verge of destroying it.

The larger perspective takes into consideration the objective and the subjective dimensions of life, the sensual world and the trans-sensual world, and wholistic health. So the dimensions of consciousness are education-science-ethics-health. It deals with the wholeness of life. Ilya Prigogine stated, "It is physics that presupposes an observer situated within the observed world. Our dialogue will be successful only if it is carried on from within nature".[28] So the observer and the observed are in one field.

When there is objective observing (seeing) or listening (hearing), the movement of thought-structure coupled with the I-consciousness go to abeyance. In other words, we are aware of every movement of thought and feeling in our relationship with people, with things and with nature, that the mind is open, pliable, not tethered to self-protective demands and pursuits; and only then the thought-structure coupled with I-consciousness go to abeyance, and the mind becomes quiet and silenced.

(d) If the observation is continued for a long chronological time, nature expresses itself in the brain or the mental structure. The perception does not originate in the brain, but the Consciousness uses the brain as a media for ultimate perception. It implies that nature invariably expresses itself in a non-dual and non-conceptual frame of reference of the Mind Continuum as "*direct perception*" and in a superluminal channel (faster-than-light) without signal. The ultimate perception is free from any chaos, conflict, fragmentation. There is perfect order in the perception. It gives the perception of love and compassion and oneness of life – which are the perennial human values. There is rhythm, non-violence, peace. And this bestows human survival and excellence. This only is the social, psychological, and even the political revolution. This is the consummation of human life.

7. Understanding Truth

Nature invariably expresses itself as direct perception in the brain-body system. It implies that nature (the highest quantum energy potential of Cosmic Consciousness) expresses itself as direct perception in a pure brain.

Krishnamurti wrote, "Only the truth can liberate the mind from its own ideation; to see the truth, the mind must realize the fact that as long as it is agitated it can have no understanding .... Stillness of mind comes only when there is no process of isolation through accumulation but complete understanding of the whole process of relationship".[29]

Transformation is not in the future, can never be in the future. It can only be *now*, from moment to moment. Transformation means, "Seeing the false as the false and the true as the true is transformation, because when you see something very clearly as the truth, that truth liberates. When you see something is false, that false thing drops away".[30]

Truth is not cumulative, it is from moment to moment. Truth is *being* from moment to moment. "Love is not different from truth. Love is that state in which the thought process, as time, has completely ceased".[31] Where love is there is transformation. Truth or love expresses itself when we transcend the normal boundaries of our patterns of thought-structures coupled with volitions.

Kothari penned a booklet entitled "Some Thoughts on Truth".[32] He wrote, "The new theorems, unless trivial, are of course first 'discovered' and then proved. Discovery is a flash of insight. It is not a step by step process, but a sort of quantum jump .... A rational account of a discovery, the proof, follows and not precedes the discovery".[32] He again wrote, "The fundamental discoveries which go deep into the source of things are the results of uncommon intuition and faith, not cold logic".[33] Insight is basically an expression of truth in a non-dual and non-conceptual frame of reference of Mind Continuum.

The scientific truths and ethical truths are not contradictory but complementary. There can be no science without ethics and an ethical society. Equally, in the modern world there can be no ethics without science. It seems that the most significant ethical lesson of science is that "experimentation" in ethics is as necessary for its development, its practice and philosophy, as it is for progress of natural science. And the method of ethical science "experimentation" is nonviolence (*Ahimsa*).[34]

The objective world of science and the subjective world of perception appear as two entirely different worlds. Alexander Koyre wrote that we live in a world where "there is a place for everything, there is no place for man .... Two worlds: this means two truths. Or no truth at all. This is the tragedy of the modern mind which 'solved the riddle of the universe', but only to replace it by another riddle, the riddle of itself".[35] This is deeply frustrating. It is agonizing.

The objectivity of science is secured by excluding from its domain all subjectivity. This has led to pursuit of materialism. Objective science is not enough as it contributes to chemical pressures and nervous tensions. It leads to duality, a conflict. And yet the crisis is of our making. There is an inner compulsion in man to live the subjective perennial values of life – as love and compassion, freedom, nonviolence, fearlessness and peace.

If the subject-object interaction ends, or made *zero*, the object goes to abeyance and there is only the subject, which expresses itself as perennial values of the wholeness of life.

Mutation: Transformation or mutation is in the whole psyche, not in a part. Mutation is possible in the brain-body system through insight, not through knowledge, not through accumulation. That insight is the ability to see a truth directly, a direct perception of the truth, and education must foster that. We must not only emphasize knowledge, but emphasize insight. The human mind must come to this no-mind state, a non-accumulative state, a state of awareness, alertness, observation and learning, and then living in that state. Mutation constructs a sane and rational society, free from any conflict, and leads to peace and survival.

8. The States of Mind

As an exegetical convenience, a device to make the complex story of the expression of Consciousness, in the brain-body system, a little easier to tell, we discuss the expression of Consciousness in three phases:

(a) Psychological Mode: (i) The subject of cognizance or the mind or the successive consciousness cuts itself up into at least one state that sees, and at least into one other state that is seen; (ii) as both are at the same quantum energy potential, so there is exchange of energy resulting into mutilation or distortion of both; (iii) we are aware of the distorted perception, the *non-fact*, the *false*, the *past*, the *what should be*?; and (iv) this perception leads to strife, pain, nervous tension and chemical pressures.

(b) Scientific Mode: The subject is synergistic to the object, so there is no exchange of energy between the subject and the object, because the photons cannot cross "the event horizon". We are aware of the objective value of the object, which is basically a pursuit of materialism. It leads to invention, which is an improvement of the known to the known. And this is guiding man to self-annihilation.

(c) Direct Perception: We have to restructure the dimension of science – which implies to take the subjective and the objective dimensions in its field. The object is within the field of nature. When the object encounters the subject, it gets exploded and we are left with the pure perception of the subject, true nature of the subject. The truth comes into being only when there is direct perception of it in your consciousness. A direct perception leads to the awareness of science and ethics, the wholeness of life, the Life Field. It bestows love and compassion, order, rhythm, harmony, nonviolence and peace.

(d) It is the hierarchical level of expression of Consciousness or quantum energy potential of Consciousness that is responsible for different states of Mind. The undisturbed tranquil and silent Mind is bliss, the disturbed state is the world.

9. Concluding Comment

(a) The early record of history – since 5,000 years ago – has established that man started living in the external world and jettisoned the inner world. This led to the conditioning of the psyche, which expressed as chaos, strife, chemical pressure and nervous tension.

(b) The dialogues between J. Krishnamurti and David Bohm suggest that during the last millions of years, all activities of the conditioned psyche have proved to be useless because it has not given us perception of the Immense, of Life Field, except to build a better society and so on.

(c) The objective world of science and the subjective world of perception appear as two entirely different worlds. In principle, science is a dialogue with nature, but this dialogue gets snapped the moment science takes up the role of objectivity. The exclusion of subjectivity from objective science is the basic cause of our strife and pain. And yet the crisis is of our own making.

(d) Man does not live by bread alone. He is in search of joy, creativity and spontaneity, freedom, order, harmony, nonviolence and peace. Herman Weyl wrote, "The inner awareness of myself is the basis for the understanding of my fellow men".[36] He has to realise the "Divinity within him". His fulfilment is embedded in the dynamic interplay of science and ethics. This is the only vocation of man.

(e) We have to educate our mind-brain-body system so that it makes a quantum jump from reason to intuition. Nature responds to the arts of "observing (seeing), listening (hearing), understanding and questioning". If we observe the conditioned psyche for a long chronological time, nature expresses itself as love and compassion, which are the perennial human values.

(f) To grasp this whole movement of life requires intelligence, not the intelligence of thought, or knowledge, but the intelligence of love and compassion with its sensitivity.

(g) We have to restructure science, whose dimensions are the wholeness of life. The subject of cognizance and the observed world are within nature, so when the observed world encounters the subject, it gets annihilated and the restructured science expresses itself as perennial values.

(h) Nature invariably expresses Its truth as pure perception. Human joy and fulfilment is embedded in living the immutable laws of nature. Every event/happening is governed by insight. Voltaire wrote:

> .... everything is governed by immutable laws.... everything is prearranged .... everything is a necessary effect .... We are both toys in the hands of destiny.[37]

We agree completely with voltaire.

(i) It is the Cosmic Consciousness that can guarantee order, harmony, nonviolence, peace and survival.

(j) Fritjof Capra wrote, "As the turning point approaches, the realization that evolutionary changes of this magnitude (a movement from psychosocial evolution to spiritual evolution) cannot be prevented by short-term socio-economic-political activities provides our strongest hope for the future".[38] This regenerates our deepest faith in the Immensity, the Life Field, the Cosmic Consciousness, and guarantees our excellence and survival.

References

1. D.S. Kothari, *Some Thoughts on Science and Religion*, The University of Delhi, 31st Oct. & 3rd Nov. 1977, pp. 28, 41, 45.
2. Charles Sherrington, *Man on His Nature*, Cambridge University Press, 1951, pp. 228-257; in D.S. Kothari, *ibid.*, pp. 45-46.
3. Candace Pert, in Fritjof Capra, *The Web of Life*, Flamingo, 1996, p. 277.
4. Roger Sperry, in Erika Erdmann and David Stover, *Beyond a World Divided*, Shambhala, 1991, Cover Jacket.
5. J. Krishnamurti and David Bohm, *The Ending of Time*, Harper and Row, 1985, pp. 105, 107, 110.
6. *Ibid.*, *The Future of Humanity*, KFI, 1986, p. 38.
7. David Bohm, *The Future of Humanity*, ibid., 1986, Preface.
8. Wilder Penfield, in Deepak Chopra, *How to Know God*, Rider, 2000, pp. 217, 214.
9. *Ibid.*, *The Mystery of the Mind*, Princeton: Princeton University Library, 1975, p. 109, in Fred Alan Wolf, *Taking the Quantum Leap*, Harper and Row, 1981, p. 244
10. Sampooran Singh, *The Holistic Education*, New Delhi: Siddharth Publications, 1987, pp. 79-81; *ibid.*, *A Scientific Guide to the Integration of Man, Society and Mankind*, *ibid.*, 1987, pp. 155-160
11. D.S. Kothari, *Atom and the Self*, Bombay: Bhartiya Vidya Bhavan, 1983, p. 30; in Kishore Gandhi, *The Evolution of Consciousness*, New Delhi: National Publishing House, 1983, pp. 207-241
12. R.P. Kaushik, *Towards a New Consciousness*, Delhi: Darshan Yoga Sansthan, 1974, pp. 63-71

13. D.S. Kothari, *Atom and the Self, op. cit.*, pp. 35, 45
14. Erich Fromm, in *The Vedanta Kesari*, December 1999, Volume 86, p. 453, in Sampooran Singh, *DILIP*, October/December, 2000, p. 17
15. *Svetasvatara Upanishad, vi. 20*; in *Vedanta Kesari, ibid.*
16. Sampooran Singh, (a) *A Renewed Look at the Wave-Particle Complementarity in the Light of Mind-Brain Interactions*, India: Bulletin of the Theosophy Science Study Group, June & August 1988, pp. 35-37; (b) *Holistic Perspective and Human Values*, February 1989, pp. 9-14; and (c) *Humanist Technology, ibid.*, May 1989, pp. 43-48
17. *ibid., Report of Proceedings: National Seminar on Value-Oriented Education*, Ramakrishna Institute of Moral and Spiritual Education, Mysore, October 1986, pp. 36-42
18. Sampooran Singh, *The Survival of Mankind and Quantum Jump from Violence to Peace*, New Delhi: Books India International, 1999, pp. 82, 88
19. Sampooran Singh, *A Critique on a Global Ethic*, Prabuddha Bharata, ibid., pp. 307-312
20. General Omar Bradley, in *Vedanta Kesari*, December 1999, Volume 86, p. 453
21. Gerald O. Barney, in Leo D. Lefebure, *Prabuddha Bharata*, July 1994, pp. 302-306, 312
22. George Blaker, *Network, The Scientific and Medical Network*, UK, April 1999, pp. 2-7, 17
23. Fritjof Capra, *The Turning Point*, London: Wildwood House, 1982, p. 466
24. Sampooran Singh, in *DILIP*, October/December 2000, p. 17 (Footnote)
25. Erwin Schrodinger, *What is Life*? and *Mind and Matter*, Cambridge University Press, 1967, p. 127
26. *ibid.*, p. 131
27. Charles Sherrington, in Erwin Schrodinger, *ibid.*, p. 128
28. Ilya Prigogine and Isabelle Stengers, in *Order Out of Chaos*, London: Bantam Books, 1984, p. 218
29. J. Krishnamurti, *The First and Last Freedom*, London: Victor Gollancz Ltd., 1977, pp. 278-279
30. *Ibid.*, p. 286
31. *Ibid.*, p. 287
32. D.S. Kothari, *Some Thoughts on Truth*, New Delhi: Indian National Science Academy, 1975, pp. 1-22; and *ibid., The Perception of Truth in Science and Philosophy*, in *The Odyssey of Science, Culture and Consciousness*, New Delhi: Abhinav Publications, 1990, pp. 48-49.
33. *Ibid., Atom and Self, op. cit.*, p. 227.
34. *Ibid., The Perception of Truth in Science and Philosophy, op. cit.*, pp. 55-56.
35. Alexander Koyre, in Ilya Prigogine and Isabelle Stengers, *Order Out of Chaos*, New York: Bantam Books, 1984, pp. 35-36.
36. Herman Weyl, in Ilya Prigogine *et al., op. cit.*, p. 311; in *Philosophy of Mathematics and Natural Science*, Princeton, N. J., Princeton University Press, 1949.
37. Voltaire, in Ilya Prigogine, *op. cit.*, p. 257.
38. Fritjof Capra, *The Turning Point*, London: Wildwood House, 1982, p. 466.

20
An Overview of Human Destiny

Brain is objective. Mind is subjective and independent of the brain. Our values, beliefs, and goals can directly affect our brain states and brain physiology. There is no problem more mysterious than the mind-brain-body (subject-object) interactions, and yet our educational structure ignores it completely.

Man is an amphibian who lives simultaneously in two worlds – time (outer) and timeless (inner) dimensions. The cerebral processes in time and symbols offer no guidance to the much more fundamental problem of the relationship of man to man and man to nature.

This article explores the regeneration of Intelligence for human survival. This bestows insight. It is the Cosmic Consciousness or Life Field that governs the mutation of mind-brain-body system. It blossoms to energising intelligence. This flowers to order, harmony, nonviolence and peace. An overview of human destiny is also presented.

The aims and objects of this paper is "to initiate an open dialogue with an open mind", and such sharing may lead to a modified model of "understanding, learning and living". A beginning has to be made, let us do it now and here.

1. Introduction

Man had an evolutionary origin. For many million years, he had been observing, comparing, and seeking to classify his observations – both as partial experience and fragmentary knowledge (information). He had been recording these as memory or cultural belief structure, as frequency and amplitude information (subtle matter). The memory is metastable and it disintegrates to thought-time.

In modern physics, a very different attitude has now developed. Physicists have come to see that all their theories of natural phenomena, including the "laws" they describe, are creations of the human mind; properties of our conceptual map of reality, rather than of reality itself. The time-space-causation is a part of the mind-brain-body system, not of reality. Man started living in a synthetic world, and jettisoned the reality.

There is no objective world independent of the observer. Atoms may not exist without observer of the atoms. The order of the universe may be the order of our minds.

Spencer Brown, in a most spectacular passage, wrote·

> The world we know is constructed in order to see itself. But in order to do so, evidently, it must first cut itself into at least one state which sees, and at least one another state which is seen. In this severed and mutilated condition, whatever it sees is *only partially itself.*[1]

This introduces subject-object dualism. Most immediate to every one of us is the problem of matter-mind (object-subject) interactions and yet our educational structure ignores it completely. There is no explanation, no clue whatever, as to how the interaction between mind and matter occurs. There is no problem more mysterious than the mind-brain interaction. Modern science has not made it less intractable. On the other hand it has added a new urgency and also a new poignancy.[2]

Roger Sperry, Nobel prize in 1981, wrote in 1950:

> At the core of all metaphysical problems stands the mind-brain relations, real understanding of which could have vast influences on all the ultimate aims and values of mankind.[3]

Sperry argued, "Our values, beliefs and goals can directly affect our brain states and brain physiology".

This article explores the conditioned mind-brain-body system, and the expression of Unconditioned Energy. It deals with education of the mind-brain body system which bestows the capacity to observe the world both objectively and subjectively.' It helps to look at the world with intelligence which resolves all challenges of life and to live in order, rhythm, nonviolence and peace. At this critical moment nature is ready to help mankind, which in philosophical language is expressed as:

If we move one step towards God,
God advances a hundred steps to receive us

Our spiritual wisdom has been our strength. Our duty is "Symbiosis of Science and Spirituality". We have to enhance our awareness through "observing (seeing) and listening (hearing)" both individually and collectively. In this is embedded our survival. This only is the social, psychological, and even the political revolution. This is the only vocation of mankind.

2. The States of Mind

(a) Psychological Mode of Mind: R. P. Kaushik wrote, "The first stage prior to recognition and naming is a state of non-verbal or thought-free perception (*nirvikalpa jnana* or pure perception). From this pure perception we proceed to thought, recognition and naming (*savikalpa jnana*)".[4] The subject proceeds to recognition, identification, comparing, naming and stores the mutilated fact (non-fact). The whole process is so fast that we are not aware of the *fact*, the *true*, the *present*, the *what is*!; but aware of the *non-fact*, the *false*, the *past*, the *what should be*? This constructs the psychological imbalances, the ecological challenges, and so on. In the psychological field, both these processes are one and the same. Hence there is no perception but only distortion.

Both the subject and the object are at the same quantum energy potential of successive consciousness, hence there is exchange of energy. The whole process implies an exchange of photons (quanta of light), these are "gentle quantum events", and these bring in a discontinuity at that moment. At the moment of this exchange, the distinction between the subject and the object is lost. One flowing web of frequencies (subject) observes another flowing web of frequencies (object) and both intermingle losing their identities. But the subject and the object become continuities of change, so the subject is the object. We perceive the mutilated fact and record the modified memory, where, in principle, the mutilated fact and the modified memory are the same. The interaction is a biochemical interaction or dance, it remembers what the dance was yesterday, so it is always doing the same dance. The perception of the mutilated fact gives us the response, the dualistic action; the will of action is always dualistic.

The whole mankind has programmed the mind-brain-body system as a single channel computer. We are engaged in acquisition, storing it in memory, possessing, acquiring and using it. The process is mechanical or repetitive, so there is no newness or creativity. It is this false process which is responsible for all human psychological challenges.

(b) Science and Measurement: About 1600 Kepler and Galileo simultaneously and independently formulated the principle that the laws of nature are to be discovered by measurement. All propositions were to be confined to that which was objectively measurable and verifiable.

(c) The quantum revolution was cataclysmic because it attacked not one or two conclusions of classical physics but its very cornerstone, the foundation upon which the whole edifice was erected and that was subject-object dualism. Sullivan put it, "We cannot observe the course of nature without disturbing it".[5] Andrade wrote, "Observation means interference with what we are observing .... observation disturbs reality".[6] The objective measurement and verification could no longer be the mark of absolute reality.

A scientific mind takes up "the role of a non-concerned observer", in the words of Schrodinger.[7] "It is unrelated to our beliefs, our ambitions, our hopes" in the words of Prigogine.[8] It must not construct a pseudo-observer, it must see in the present and reject the past remembrances or future projections. In the art of observation, attentivity takes up the role of the subject. Attentivity leads the brain from order to higher order. It sees the objective challenge as pure perception.

The scientific mind is very factual. Invention is its mission, its objective perception. It sees through a microscope, through a telescope; everything is seen as it is; from that perception science draws conclusions, builds up theories. Such a mind moves from fact to fact.

3. A Critique on Logical and Scientific Reason

Science aims at nothing but making true and adequate statements about its object. The scientists only imposes two things, namely truth and sincerity, imposes them upon himself and upon other scientists.

(a) The Principle of Objectivation: Schrodinger wrote:

> Mind has erected the objective outside world of the natural philosopher out of its stuff. Mind could not cope up with this gigantic task otherwise than by the simplifying device of excluding itself – withdrawing from its conceptual creation. Hence the latter does not contain its creator.[9]

Schrodinger also quoted C.J. Jung and B. Spinoza who had identical views. Charles Sherrington expounded his observation in words:

> The material world has only been constructed at the price of taking the self, that is, mind out of it, removing it; mind is not part of it; obviously, therefore, it can neither act on it nor be acted on it by any of its parts.[10]

The time-space matrices (objective world) is at the successive consciousness. The subject of cognizance (true subject) is at the immanent quantum energy potential. There is no energy transfer between the object and the subject, hence there is no duality. The objective scientific mind is very factual, so the subject sees the object from fact to fact. It is a movement from the *known* to the *modified known*, which has higher sensitivity. The objective science constructs new models of known artifacts with a higher sensitivity. It leads to inventions, which flows to a new model of aeroplanes, frigidaires. It shows that the achievements of objective science are limited – it adds to human comfort, leisure and pleasure. It also contributes to competition, motivation, disorder and strife.

(b) The Principle of Understanding of Life Directly: The symbolic-dualistic frame of reference (psychological mode) of mind spectrum does not have the capacity to understanding of mind. The principle of understanding is embedded in the non-dual (timelessness) frame of Mind Continuum.

Many eminent scientists – including Albert Einstein, Werner Heisenberg, Erwin Schrodinger, Arthur Eddington, and so on – have stumbled upon another mode of perception (knowing), which is "understanding of life directly, instead of in the abstract, linear terms of representational thinking".[11] This implies "understanding the wholeness of life". Understanding expresses itself in the non-dual and non-conceptual frame of reference of Mind Continuum. *Understanding is a movement in the non-duality, a movement of the whole being.* Understanding comes only when the mind is very quiet, tranquil, silenced.

Understanding leads to a perception of the whole. Perception has nothing to do with knowledge (information). *The perceptions are born of non-duality and responses are born of spontaneity.* It awakens the awareness, the intelligence, which arises out of observation, self-knowing and compassion.

David Bohm, in a lecture in USA, in April 1977, said:

> The ultimate perception does not originate in the brain or any material structure, although a material structure is necessary to manifest it. The subtle mechanism of knowing the truth does not originate in the brain.[12]

It implies that ultimate perception happens in a trans-sensual dimension, which is timeless (beyond brain). As all knowledge is structured in Consciousness, so ultimate perception is a happening when Universal Consciousness bathes the brain cells. Perception bestows understanding the wholeness of life, which flowers to intuition, spontaneity and right action; it is a happening in the non-dual and non-conceptual frame of reference of Mind Continuum.

Observation is perception without psychological involvement, without any kind of reaction. Observation is a non-reactional attentiveness. Reaction implies evaluation, judgement, comparison. Observation is non-evaluatory perception. Observation without pseudo-observer is the state of wholeness of perception and wholeness of response. Observation flows to understanding, which happens in a non-dual and non-conceptual frame of reference. Understanding and learning blossoms to creativeness.

The "understanding, learning and spontaneous action" leads to inner silence. The energy born of silence – in the inner space, is not a part of inheritance. This energy is born of timelessness and is called Intelligence. Intelligence is an unconditioned energy born of the emptiness of silence. It has no past and no future. It resolves all psychological imbalances, ecological crisis, and other challenges. It is the highest hierarchical expression of Cosmic Consciousness.

(c) The Mystery of the Sensual Qualities of Time-Space-Causation Matrices: Erwin Schrodinger, Nobel laureate in Physics, 1931, wrote:

> Our world picture is colourless, cold, mute. Colour and sound, hot and cold are our immediate sensations; small wonder that they are lacking in a world model from which we have removed our own mental person .... The physical world picture lacks all the sensual that go to make up the Subject of Cognizance. The model is colourless, soundless and unpalpable.[13]

John Eccles pierces the sensory illusion with one startling but irrefutable assertion:

> I want you to realize that there is no colour in the natural world and no sound – nothing of this kind; no texture, no patterns, no beauty, no scent ....".[14]

In short, none of the objective facts upon which we usually base our reality is fundamentally valid. Erwin Schrodinger concluded (a) that all scientific knowledge is based on sense perception, and (b) that none the less the scientific views of natural processes formed in the way lack all sensual qualities and therefore cannot account for the latter".

4. Limitations of Education in Science

Maurice H. Wilkins, Nobel prize in Medicine, 1962 discussed how the scientist *ideally* works, in practice, in daily living. He stated:

(a) Science makes progress largely by a combination of two aspects: open-mindedness and the following of abstract concepts. Abstraction involves excluding and narrowing down, which is the very opposite of open-mindedness. Science always puts the most emphasis on this abstracting and narrowing down, on analytical procedures and on the role of the intellect as distinct from intuition and feelings.

(b) The logical and rational is a very important element in science. But emphasis on that tends to make people not notice the essential role of intuition. It is also a question of how science is developed through educational institutions and through institutionalized science. I agree with Einstein that the sort of scientific education we have now has produced a narrow-minded way of thinking amongst scientists, so that they give no proper attention to the moral and psychological dimensions.

(c) Science as it is today does not tell us very much about the real nature of human being. Scientists do not pay attention to the need of moral values in science. Science is a human activity and cannot be separated from all values.[15]

(d) We have to restructure the teaching of science in all educational institutions.

5. Open-Minded Enquiry of Scientists

(a) Albert Einstein wrote that we need "objective observing and understanding" as that leads to "peace and serenity". He added, "I see in nature a magnificent structure that must fill a thinking person with a feeling of humility".[16] The essential in the existence of a man is "how to think" and not "what to think"; and not what he does and suffers.

(b) Scientific Observation as a Total Action: Persistent and rigorous scientific observation implies to *see* an 'event/process/happening/object' without interfering in its flow, or to *see* without any transfer of energy to it, or to observe non-subjectively which means without any evaluatory comparison or judgement creeping in. It is to perceive without naming, identifying, without emotionally reacting.

(c) Attentivity: Observation without an observer is attentivity. The state of attentivity get energised in the whole being, where there is neither subject, nor object; or subject-object dualism ends. The sensitivity of the whole being is called attentivity.

(d) Spontaneity: A non-dual Mind perceives the truth spontaneously. It implies that there is no psychological time between expression of Life Field in the brain cells and its perception in the brain. Further there is no psychological time between pure perception and action. So the expression of Life Field in the brain cells, the perception and action are one dynamic flow. The whole Consciousness is a seamless, fluid web of process structures, it is one dynamic flow, and cannot be disassembled or split into simpler and simpler parts.

(e) Spirituality: It is an all-inclusive attentiveness to every aspect of life. Life in its wholeness is holy and sacred, it is the science of the wholeness of life. Spirituality is the science of the wholeness of life. In spiritual science, the emphasis is on the act of perception, the quality of the instruments employed for perception, the quality of consciousness behind the perception. Spirituality implies exploration of the maximum quantum energy potential of consciousness. Both science and spirituality have a non-authoritarian approach.

(f) The Right Kind of Education: Education is "objective observing (seeing), listening (hearing), understanding, learning and spontaneous action". It is the psychodynamics of the conditioned psyche and this imparts aliveness (a higher level of awareness) to the mind. It blossoms to true perception. It regenerates or rejuvenates to intelligence. This approach breaks through frontiers of particular cultures and establishes an entirely new set of values, which in turn can create a new civilisation and a new society.

6. Restructuring of Science

(a) Restructuring of Science: Karl Pribram wrote, "Brain science must deal with the awareness of awareness. It could no longer afford to shut out that part of the world which we call subjective".[17] Roger Sperry "spoke of the importance of a neglected area of science, namely the study of human subjective experience".[18] John Eccles also emphasized "to deal scientifically with the neglected area of inner conscious awareness".[19]

Roger Sperry wrote, "When the brain is whole, the unified consciousness of the left and the right hemispheres adds up to more than the individual properties of the separate hemispheres". So these studies raised the issue of consciousness in a new way. He adds, "When parts come together in a new whole, this new whole exhibits features – emergent properties – that can't be predicted as a rule from the parts, and cannot be explained in terms of the parts".[20]

The role of consciousness suggests to consider the objective and the subjective as one whole, or in other words, the outer (objective) and the inner (subjective) are not two separate movements but a unitary movement, one whole. The objective science must restructure itself to take in its ambient both the objective and the subjective dimensions of life. The energy exchange between the subject and the object takes less than a hundred microsecond. The restructured science deals with the wholeness of life.

(b) Ethics: It implies to take in its ambient the sensual and the trans-sensual dimensions of life.

(c) Health: All diseases are maladjustment of the human mind-brain-body system, or conditioned human psyche, or crisis of perception. Health is a complete physical, mental, psychological and social well-being; it is a dynamic state which is equipoised and balanced. A living in the *present*, the *now* heals; and living in the *past* and *duality* suppresses man's perceptive sensitivity and enhances disease. The curative functions of medicines have to be supplemented with understanding and living the "symbolic-dualistic frame of reference of mind spectrum (successive consciousness)" and the "non-dual and non-conceptual frame of reference of Mind Continuum (Simultaneous Consciousness)".

There is an urgency to restructure medical education which takes in its ambient the psychodynamics of the mind-brain-body system, or in other words, the wholeness of life. We have to jettison the illusion of the past, and to live in the present, the now. We have to restructure the biomedical model and adopt the holistic ecological perception. The restructured medical education takes in its ambient the wholeness of life.

(d) The "Ending of I-consciousness", "Ending of Thought-Time", and "Ending of the Fact" were discussed in the article entitled, ***"Symbiosis of Science, Education, Ethics and Health"*** .

(e) The role of consciousness suggests: (i) to consider the objective and the subjective as one whole, or in other words, the outer (objective) and the inner (subjective) are not two separate movements, but a unitary movement, one whole; (ii) it implies to take in its ambient the sensual and the trans-sensual dimensions of life; and (iii) health is a complete physical, mental, psychological and social well-being; it is dynamic state which is equipoised and balanced.[21,22] The subject of cognizance has to restructure itself, which may be written as:

[science-education-ethics-health]

(f) The observation of the objective fact without justification or condemnation releases tremendous energy. It is likely that at a certain critical quantum energy potential the fact makes a quantum jump across the "event horizon" and may move to the centre (subject). As it approaches the subject of cognizance, the fact is knocked off and what is left is pure Consciousness. When you conduct an enquiry, with each step there is a movement away from the false in our daily life, in our daily relationship.

In principle, it means that the subject of cognizance is at the highest quantum energy potential. Prigogine stated, "It is a physics that presupposes an observer situated within the observed world. Our dialogue with nature will be successful only if it is carried on from within nature".[23]

Galileo wrote, "He asked questions addressed to nature and succeeded in receiving answers directly from nature".[24] Nature responds to experimental interrogation if there is an inner non-verbal and non-rational urge for love, which is the spontaneous core and concern for fellow human beings which is in a sense of oneness of life.

7. A Creative Mind

Hideki Yukawa wrote, "Creativity implies discovering something that nobody has hitherto known about or inventing something new .... It is the discovery of a truth in nature unknown to man".[25]

All knowledge is structured in consciousness. This recognizes that consciousness alone can be totally self-referral, so the pure Mind or non-dual Mind can perceive the consciousness. When non-dual state of Mind is presented as an object to pure consciousness, then it reveals itself. In the non-dual Mind, the intelligence is born of the inner silence, of the inner space. The intelligence is uncluttered by the movement of thought. The dimension of intelligence, of unconditioned energy, of Fundamental Vibration, has no past, so it is free of all conditioning and divisive loyalties, etc. It bestows insight. Insight implies something which you perceive without any logic, without any motive. It is called immediate perception. *The perceptions are born of non-duality, of silence, and responses are born of spontaneity.* Perception has nothing to do with knowledge. Intelligence begins to use the human brain and help the human being to look at the world in quite a different way.

Intuition or insight or discovery is the timeless perception. It is insight alone that can decondition the brain. Insight is complete and from that completeness there can be logical, sane and rational action. Only then there can be flowering of the brain. Pure clear insight is perception without any doubt. *Understanding, learning and action are a cyclic process, which results in a spiral going upwards in the vertical dimension; it is a growth in the vertical dimension.* The higher the understanding, the higher the expression of the Consciousness in the human brain-body system.

Creativeness is a state of being (mind) in which conflicts, sorrows, psychological imbalances of the self are absent, a state in which the mind is not caught up in the demands and pursuits of desire. A creative mind resolves all human paradoxes.

Fritjof Capra wrote, "The rational part of research would, in fact, be useless if it were not complemented by the intuition that gives scientists new insights and makes them creative".[26] Constance Holden adds, "In high-level creativity the mind can do weeks of work in a short time".[27]

8. An Overview of Human Destiny

Fred Alan Wolf wrote, "Consciousness performs a dual role in the universe. In the world of the quantum, it is both the awareness and the creation of experience. It is the being and the knowing of experience. With the stroke of the twentieth-century quantum eraser, the dividing line between ontology (theory of being) and epistemology (theory of knowing) is rubbed out". The fusion of ontology, epistemology and psychology is beginning to emerge.[28]

Prigogine quoted (*From Being to Becoming*):

> For most of the founders of classical science – even for Einstein – science was an attempt to go beyond the world of appearances, to reach a timeless world of supreme rationality – the world of Spinoza. But perhaps there is a more subtle form of reality that involves both laws and games, time and eternity.[29]

Herman Weyl, prophetically wrote, "Another way [of approach to the phenomena of life] is of understanding from within is open to us …. I have a direct knowledge entirely different from the theoretical knowledge that represents the 'parallel' cerebral processes in symbols. This inner awareness of myself is the basis for the understanding of my fellow men". Prigogine and Isabelle Stengers agree completely with Herman Weyl. Man has to make a quantum jump from 'cerebral processes in symbols' to direct understanding of life.[30]

Voltaire wrote the following about human destiny:

> …. everything is governed by immutable laws …. everything is prearranged …. everything is a necessary effect …. we are both equally foolish, both toys in the hands of destiny.[31]

The human survival (destiny) is embedded in a quantum jump from time to timelessness, from knowledge to direct perception, from experience to perception, from reason to intuition (insight), and from intellect to intelligence.

Sri Aurobindo wrote, "The East alone has some knowledge of the truth, the East alone can teach the West, the East alone can save mankind". Aurobindo in a message on August 15, 1947, stated, "If this evolution (spiritual) is to take place, since it must come through the growth of the spirit and the inner consciousness, the initiative can come from India and although the scope must be universal, the central movement may be hers". The Divine Mother in 1963 wrote, "The only hope for the future is a change of man's consciousness and the change is bound to come". The message in 1967 read, "Men, countries, continents! The choice is imperative: Truth, or the abyss". The Mother had once observed, "Three-fourth of humanity is obsolete".[32]

9. Concluding Comment

Mankind has been engaged to end all chaos and conflict at least for the last 10,000 years, but all his efforts have proved to be sterile. We need a higher level of Consciousness to resolve all psychological imbalances and to live peacefully.

We have to educate our mind-brain-body system, to jettison our psychological mode and perceive direct perception. The psychological mode is invariably dualistic and constructs violence, strife, chemical pressure and nervous tension. The direct perception is in the non-dual and non-conceptual frame of reference of Mind Continuum. It is a quantum jump from the field of time to the realm of timelessness.

It is only in that silent state that the old neural paths are short-circuited and new neural paths are energized and it renews itself so that it is made fresh, young, and energetic. It is only a creative mind that can bestow discoveries. The creative state of mind is the noblest and most mysterious faculty of Mind.

We want a revolution, a psychological revolution in which the whole human race has to participate and every man is an equal partner. The revolution will regenerate or rejuvenate human Mind so that it perceives the new dimension of an integral and all-inclusive Consciousness, that can guarantee human survival along with all fauna and flora.

Modern physics tells us: (a) Either we grow together, or we mutually annihilate ourselves together, there is no third alternative; (b) To save us from the imminent catastrophes, we have to educate our mind-brain-body system and live in higher awareness; (c) Instead of the rigid centre of the "me", there is constant vibration of the perceptive sensitivity, instead of memory there is awareness, and instead of knowledge there is understanding.

The purification of perception is the beginning of a new dynamics of living. Nature invariably expresses itself as pure perception. Observation, understanding, learning, spontaneous action are states of reaction-free perception, or reaction-free attentiveness, or timeless perception.

The core of the Action Plan is to restructure our [science-education-ethics-health], which deals with the wholeness of life and resolves all our challenges – psychological imbalances, ecological crisis, and other challenges (terrorism, corruption, ethnic conflicts, poverty, and so on). It is governed by Cosmic Consciousness or Life Field. It bestows on us order, harmony, non-violence and peace. And if man escapes his responsibility, he does so at his own peril, and that of India and mankind.

References

1. Spencer Brown, *Laws of Form*, New York: Julian Press, 1972, pp. 104-105.
2. D.S. Kothari, *Some Thoughts on Science and Religion*, University of Delhi, 31.10.1977 and 3.11.1977, pp. 15, 28, 41.
3. Roger Sperry, in Erika Erdmann and David Stover, *Beyond A World Divided*, Shambhala, 1991, p. 42.
4. R. P. Kaushik, *Towards a New Consciousness*, Delhi: Darshan Yoga Sansthan, January 1974, p. 63.
5. J.W.N. Sullivan, *The Limitations of Science*, New York: Mentor Books, 1949, p. 140.
6. E.N. da C. Andrade, *An Approach to Modern Physics*, New York: Doubleday Anchor Books, 1957, p. 255.

7. Erwin Schrodinger, *What is Life*? and *Mind and Matter*, Cambridge University Press, 1967, p. 128.
8. Ilya Prigogine and Isabelle Stengers, *Order Out of Chaos*, Bantam Books, 1984, p. 5.
9. Erwin Schrodinger, *op. cit.*, p. 131.
10. Charles Sherrington, in Erwin Schrodinger, *ibid.*, p. 128.
11. Alan W. Watts, *The Ways of Zen*, New York: Vintage Books, 1957, p. 10.
12. David Bohm, in Gary Zukav, *The Dancing Wu Li Masters*, Fontana/Collins, 1982, p. 327.
13. Erwin Schrodinger, *op. cit.*, pp. 128, 148.
14. John Eccles, in Deepak Chopra, *Ageless Body, Timeless Mind*, New York: Three Rivers Press, 1993, p. 12.
15. Maurice H. Wilkins, in *Synthesis of Science and Religion*, Mumbai: The Bhaktivedanta Institute, June 1988, pp. 30-36.
16. Albert Einstein, *Nature*, 19 March 1979, Vol. 278, p. 214.
17. Karl Pribram, in Ken Wilber, *The Holographic Paradigm and Other Paradoxes*, Boulder: Shambhala, 1982, p. 17.
18. Roger Sperry, in *Synthesis of Science and Religion*, Mumbai: The Bhaktivedanta Institute, June 1988, p. 199.
19. John Eccles, in *Synthesis of Science and Religion*, *ibid.*, p. 199.
20. Roger Sperry, *Bulletin of the Theosophy Science Study Group*, India, February & April 1988, p. 23 and June & August 1988, p. 27.
21. Sampooran Singh and Dr Kanwaljit Kaur, *Science, Education, Health and Consciousness*, unpublished.
22. Ibid., *Symbiosis of Science and Spirituality*, unpublished.
23. Ilya Prigogine and Isabelle Stengers, *op. cit.*, p. 218.
24. Galileo, in Hideki Yukawa, *Creativity and Intuition*, Tokyo: Kodansha International Ltd., 1973, p. 111.
25. Hideki Yukawa, *ibid.*, pp. 123, 113.
26. Fritjof Capra, *The Tao of Physics*, New York: Bantam Books (Second Edition), 1988, p. 18.
27. Constance Holden, *Altered States of Consciousness: Mind Researchers Meet*, Science, 9 March, 1973, Vol. 179, No. 4077, p. 982.
28. Fred Alan Wolf, *Taking the Quantum Leap*, Harper & Row, 1981, p. 229.
29. Ilya Prigogine *et. al.* in *From Being to Becoming*, *op. cit.*, p. 310.
30. Herman Weyl, in Ilya Prigogine, *et. al.*, *op. cit.*, p. 311.
31. Voltaire, in *ibid.*, p. 257.
32. Sampooran Singh, Mrs (Dr.) Kanwaljit Kaur, Paramjit Singh, *The Masterplan Paradigm for Human Survival and Excellence*, Chandigarh: Faith Publishers, 1998, pp. 245-46.

21
Innovation in Science and in Man

Abstract

Man is an amphibian who lives simultaneously in two dimensions – the objective [the time-space-causation matrices, the outer] and the subjective [the timelessness, the inner]. All great discoveries have arisen from the realm of the timelessness, so scientific enquiry is not a cerebral activity and gathering knowledge [information] but an expression or a quest of the Timelessness, the Divinity, the Life Field. Innovation is to use the discovery to a commercially useful process/product. Discovery is fundamental and innovation is a derivative. Discovery is the timeless realm and innovation is in time field. Research is subjective, inspirational and often irrational; it is not a very structured activity.

This exploration deals with the miracle in science, with the miracle in human thinking during the last one year and that is capable of penetrating into inner workings of nature. President Dr. A.P.J. Abdul Kalam addressing a seminar at IIT, Delhi, on March 31, 2005, and said that India should "become a continuous innovator and creator of science and technology-intensive products". On December 31, 2004 at the University of Delhi, he talked about "creative leadership". Prof. R. Chidambaram also advocated "generation of innovations and also support systems". Prof. M.M. Sharma on July 14, 2004 said, "We as a nation will have to innovate to survive". We intuitively perceive that the survival and excellence of man is embedded in educating oneself for "discovery" and which itself guides for its "innovation".

1. Introduction

Science makes progress largely by a combination of two aspects : open-mindedness and the following of abstract concepts. Abstraction involves excluding and narrowing down, which is the very opposite of open-minded. Maurice H. Wilkins, Nobel prize in Medicine, 1962, wrote, "Science always puts the most emphasis on this abstracting and narrowing down, on analytical procedures and on the role of intellect as distinguished from intuition and feeling".[1]

Hideki Yukawa, a Japanese Nobel laureate in physics wrote, "Creativity implies discovering something that nobody has hitherto known about or inventing something new .... prior to creativity there has to be a struggle with one's own self .... it is a way of thinking in science which can be creative in the sense that it results in the discovery of a truth in nature hitherto unknown to man".[2] Discoveries come as intuitive flashes. Discoveries come in a non-dual and non-conceptual frame of reference, or it is timeless perception or true perception. Scientists observe a non-dual thought which expresses itself in a scientist's brain-body system, so it is not a brain or sensual activity, but nature uses the brain as a media. David Bohm, an eminent physicist, wrote:

> The ultimate perception does not originate in the brain or any material structure, although a material structure is necessary to manifest it. The subtle mechanism of knowing the truth does not originate in the brain.[3]

So creativity, discovery, intuition, insight, is in science (observation) and is hidden in man.

Einstein said once that the most incomprehensible fact is the fact that nature is comprehensible. *Creativity and Intuition* deals with this miracle – through which human thinking is capable of penetrating into the inner workings of nature. One aspect of the miracle is that nature is equally comprehensible to any scientifically trained human being, whatever his or her origin or cultural background.[4] Nature invariably responds to experimental interrogation.

There are 540 million youth below the age of 25 years in our population of a billion people. The nation needs young leaders, especially scientists, who can command the transformation of India into a developed nation embedded within an intuitive (not knowledge) society. We have to create a historical revolution and we must be in the driver's seat to contribute towards this societal change. We as a nation will have to innovate to survive.

This article explores a technology (not methodology) to have a new approach to the art of living. The last 10,000 years, or more, have demonstrated that the task of turning the age-old tide of human violence, aggression, ambition, motive is a stupendous task. Science also brought man at the control panel of human mind-brain-body system. We are deeply struck by man's concern about the large gap between the superhuman task (to live in peace) and the frailties of human endeavours. We intuitively observe that Nature is ready to help mankind. We observe that we have the possibility of gradually bringing a wider area of our humanity within the scope of intuitive living. We have to keep in mind that intuition is fundamental and rationality is a derivative.

2. Mind-Brain-Body Interactions

There is no problem more mysterious than the mind-brain-body interaction. Modern science has not made it less intractable. On the other hand, it has added a new urgency and also a new poignancy. Paradoxically, nothing is more commonplace than the reality of this interaction, yet our educational structure completely ignores this challenge. It is neither taught in the school nor at home. The professional scientists generally ignore it completely, as this is not measurable.

Roger Sperry, who won the Nobel prize in 1981, wrote, "Our values, beliefs and goals can directly affect our brain states and brain physiology".[5] In 1950, Roger Sperry wrote, "At the core of all metaphysical problems stands the mind-brain relationship, real understanding of which has vast influence on all the ultimate aims and values of mankind".[6] And obviously we cannot avoid exploring the mind-brain-body system.

The challenges to understand are how mind gathers impressions carried over by the sense organs, how they are stored into memory, how they are recalled when a new challenge comes, how certain chemicals are made, stored and their functioning, and how we interpret life around us according to the conditioning of our brain, is an open secret now.

3. Living in the False [Past] Universe

In modern physics, a very different attitude has now developed. Physicists have come to see that all their theories of natural phenomena, including the "laws" they describe are creations of the human mind, properties of our conceptual map of reality, rather than of reality itself. Relativity theory has shown that space is not three dimensional and time is not a separate entity. Both are intimately and inseparably connected and form a four dimensional continuum which is called space-time. Spencer Brown, in a most spectacular passage, wrote:

> The world we know is constructed in order to see itself. But in order to do so, evidently, it must first cut itself into at least one state which sees, and at least one another state which is seen. In this severed and mutilated condition, whatever it sees is *only partially itself*.[7]

The seer (subject) and the seen (object) are at the same quantum energy potential, hence there is energy transfer leading to mutilation of the both, and we are aware of mutilated object and the mutilated subject is recorded as modified memory. This introduces object-subject dualism. The error of dualism forms the root of intellection and is therefore next to impossible to uproot by intellection. To rectify this error of dualism, we have to resort to "objective observing and understanding", which demands a rigorous, consistent and persistent technology to meet this challenge.

Fred Alan Wolf wrote, "Consciousness performs a dual role in the universe. In the world of the quantum, it is both the awareness and the creation of experience. It is the *being* and the *knowing* of experience. With the stroke of the twentieth-century quantum eraser, the dividing line between ontology (theory of being) and epistemology (theory of knowing) is rubbed out",[8] and fusion of ontology, epistemology and psychology is beginning to emerge.

We are looking at ourselves and it is this process which we call the universe.

Every cell of the brain holds memory of million years of man. The present holds the totality of the past. The present mode of living – called the psychological mind or psychic pandemic – psychologically, we go through agonies, tortures, uncertainties, depression, a great deal of sorrow, chemical pressure and nervous tension. We live in past beliefs, concepts, ideals, verbalizing, identification, acquisition, name, fame and prestige, and so on. The mind-identified state is severally dysfunctional. It is a form of insanity. It is disharmonious and disordered. It lives in the past memory. All relations in this state are disturbances. It has no creativity, no newness, no originality. Every activity of the conditioned psyche leads to entropy, which is self-annihilation or self-destruction.

Mankind has started living in the *non-fact*, the *false*, the *past*, the *what should be*!, and has jettisoned the reality. The universe we live in is a peculiar state, destructive state. It is degenerating. There is violence, aggression, corruption, strife and so on. One is selfish, ambitious, greedy, possessive, success – these are parts of violence. All present efforts of man are leading to postponement, but not to eradicate violence. You cannot impose non-violence on a consciousness which is violent because that violence has deep-rooted causes in the psyche. Violence and so on can only be ended if there is mutation in the human psyche, or awareness of one's total psychological process. This regenerates intelligence, which dissolves all human challenges.

The human being has to learn to grow into that new dimension of consciousness, where direct communion, simple understanding, pure perception, becomes ingrained in him, then only there is a hope of creating a new society. Transcending the conditioned mind is a historical necessity for every man. The challenge is basically a psychological problem.

4. Living in the Objective Truth

Erwin Schrodinger, in about 1947, put forward "the principle of objectivation". He wrote:

> Without being aware of it and without being rigorously systematic about it, we exclude the Subject of Cognizance from the domain of nature that we endeavour to understand .... Mind has erected the objective outside world of the natural philosopher out of its own stuff. Mind could not cope with this gigantic task otherwise than by simplifying device of excluding itself – withdrawing from its conceptual creation. Hence the latter does not contain its creator.[9]

Charles Sherrington also explored the objective evidence of interaction between matter and mind.

> The material world has only been constructed at the price of taking the self, that is, mind, out of it, removing it; mind is not part of it; obviously, therefore, it can neither act on it nor be acted on by any of its parts.[10]

> Mind, for anything perception can compass, goes therefore, in our spatial world more ghostly than a ghost. Invisible, intangible, it is a thing not even of outline; it is not a 'thing'. It remains without sensual confirmation and remains without it forever.[10]

Schrodinger also quoted C.J. Jung and B. Spinoza in support of his principle of objectivation.

In objective science, the subject of cognizance is at the immanent quantum energy potential. The time-space matrices (objective world) is at the successive consciousness. There is no energy transfer between the object and the subject. The objective scientific mind is very factual, so the subject sees the object from *fact* to *fact*. It is a movement from the *known* to the modified *known*, which has higher sensitivity. The objective science constructs new models of known artifacts with a higher sensitivity. It leads to invention which flows to a new model of aeroplanes, air-conditioners, and so on. It shows that the achievements of objective science are limited – it adds to human comfort, leisure and pleasure. It also contributes to competition, motivation, disorder and strife.

The accelerative progress of science and technology, in the material field and the thrust of psychosocial evolution have led to the growth of globalisation, but global consciousness is not keeping pace with the globalisation.

Science and technology have made great breakthroughs in medicine, communication, industry and agriculture. Today, science and technology, if used with compassion, wisdom, and understanding, can abolish poverty, ignorance, illiteracy and malnutrition from the face of the earth.

5. Living in the Subjective World [Pure Perception]

(a) Restructuring of Science: Karl Pribram wrote, "Brain science must deal with the awareness of awareness. It could no longer afford to shut out that part of the world which we call subjective".[11] Roger Sperry "spoke of the importance of a neglected area of science, namely the study of human subjective experience".[12] John Eccles also emphasized "to deal scientifically with the neglected area of inner conscious awareness".[13]

Roger Sperry wrote, "When the brain is whole, the unified consciousness of the left and the right hemispheres adds up to more than the individual properties of the separate hemispheres". So these studies raised the issue of consciousness in a new way. He adds, "When parts come together in a new whole, this new whole exhibits features – emergent properties – that can't be predicted as a rule from the parts, and cannot be explained in terms of the parts".[14]

The role of consciousness suggests to consider the objective and the subjective as one whole, or in other words, the outer (objective) and the inner (subjective) are not two separate movements but a unitary movement, one whole. The objective science must restructure itself to take in its ambient both the objective and the subjective dimensions of life. The energy exchange between the subject and the object takes less than a hundred microsecond. The restructured science deals with the wholeness of life and it has the dimension:

[Science-Education-Ethics-Health]

The ethics imply to take in its ambient the sensual and the trans-sensual dimensions of life. A living in the present, the now heals and living in the past and duality suppresses man's perceptive sensitivity and enhances disease.

The observation of the objective fact without justification or condemnation releases tremendous energy.[15] It is likely that at a certain critical quantum energy potential the fact makes a quantum jump across the "event horizon" and may move to the centre (subject). As it approaches the subject of cognizance, the fact is knocked off and what is left is pure Consciousness. When you conduct an enquiry, with each step there is a movement away from the false in our daily life, in our daily relationship.

In principle, it means that the subject of cognizance is at the highest quantum energy potential. Prigogine stated, "It is a physics that presupposes an observer situated within the observed world. Our dialogue with nature will be successful only if it is carried on from within nature".[16]

Many eminent scientists – including Albert Einstein, Werner Heisenberg, Erwin Schrodinger, Arthur Eddington, Alan Watts – have stumbled upon another mode of knowing, called direct, or intimate, or intuition (insight), or "the understanding of life directly instead of in the abstract, linear terms of thinking".[17] Eddington wrote, "The intimate knowledge will not submit to codification and analysis, or rather, when we attempt to analyse it, the *intimacy is lost* and it is replaced by *symbolism*".[18]

As long as we are identified with an object that is bondage. As long as we think, act, live via an object, or as object: there is bondage. Subjectivity is never an object except in illusion, so subjectivity does not exist as an object "out there". Huang Po put it, "The perceived cannot perceive".[19] The pure Consciousness is always self-referral and cannot be perceived as an object. The highest quantum energy potential is *unity* which erases the distinction between observer and observed. In *unity*, everything you once thought to be "out there" is seen to be a part of yourself ("in here").

Alan Watts took no special attitude and found amazingly that this "letting go of expectation" was enough to free him. He wrote:

> For quite suddenly the weight of my own body disappeared. I felt that I owned nothing, not even a self, and that nothing owned me. The whole world became as transparent and unobstructed as my own mind. The problem of life simply ceased to exist.[20]

Einstein himself experienced episodes of complete liberation from space-time boundaries: "At such moments one imagines that one stands on some spot on a small planet gazing in amazement at the cold and yet profoundly moving beauty of the eternal, the unfathomable. Life and death flow into one, and there is neither evolution nor eternity, only Being".[21]

Einstein wrote to a friend, "I feel myself so much a part of all life that I am not in the least concerned with the beginning or the end of the concrete existence of any particular person in this unending stream".[22]

All knowledge is structured in consciousness. This recognizes that consciousness alone can be totally self-referral, so the pure Mind or non-dual Mind can perceive the consciousness. When non-dual state of Mind is presented as an object to pure consciousness, then it reveals itself. In the non-dual Mind, the intelligence is born of the inner silence, of the inner space. The intelligence is uncluttered by the movement of thought. The dimension of intelligence, of unconditioned energy, of Fundamental Vibration, has no past, so it is free of all conditioning and divisive loyalties, etc. It bestows insight. Insight implies something which you perceive without any logic, without any motive. It is called immediate perception. *The perceptions are born of non-duality, of silence, and responses are born of spontaneity.* Perception has nothing to do with knowledge. Intelligence begins to use the human brain and help the human being to look at the world in quite a different way.

Intuition or insight or discovery is the timeless perception. It is insight alone that can decondition the brain. Insight is complete and from that completeness there can be logical, sane and rational action. Only then there can be flowering of the brain. Pure clear insight is perception without any doubt. *Understanding, learning and action are a cyclic process, which results in a spiral going upwards in the vertical dimension; it is a growth in the vertical dimension.* The higher the understanding, the higher the expression of the Consciousness in the human brain-body system.

Creativeness is a state of being (mind) in which conflicts, sorrows, psychological imbalances of the self are absent, a state in which the mind is not caught up in the demands and pursuits of desire. A creative mind resolves all human paradoxes.

Fritjof Capra wrote, "The rational part of research would, in fact, be useless if it were not complemented by the intuition that gives scientists new insights and makes them creative".[23] Constance Holden adds, "In high-level creativity the mind can do weeks of work in a short time".[24]

6. The Art of Living

What is Life?: Life is a multi-dimensional phenomenon and human beings who are expressions of that phenomenon are themselves multi-dimensional creatures. Life is self-generated, self-controlled, self-regulated and self-sustained phenomenon. Life is Eternal, Infinite and Immortal. It expresses Itself at the highest quantum energy potential and is called Life Field or

Brahman. It expresses itself at the lowest quantum energy potential and is called successive consciousness or time-space-causation matrices.

Man is an amphibian, who lives simultaneously in two worlds – time and timelessness. The time dimension offers no guidance in regard to the more fundamental problems of relationship of man in the timeless realm.

The Inner Quality of Life: You have to educate your physical organism and psychological structure to move into non-action gracefully and voluntarily. It is a quantum jump which takes place through discontinuities and not through a linear path.

When thought travels in the body, it consumes nervous energy. When an emotion travels in the body, it affects the chemical system. In the timeless realm, the nervous tension and chemical pressure go to abeyance, and the whole system is relaxed. It is a balanced and equipoised state.

Silence is a state of cessation of mental activity, it is a state of no-mindedness or total relaxation of the psychological organism. When the mind is quiet, mutation happens. The highest quantum energy potential of Life Field is called love.

We have to learn the "objective observation and understanding" which blossoms to "peace and serenity".[25] The quest for truth has to begin with one's own inner life as lived day by day and as reflected in every relationship.

7. A Critique on Human Art of Living and Human Destiny

Today, after billions of years of prebiological evolution and metabiological evolution, mankind is at an evolutionary crossroads. We stand at a juncture wherein we are in supreme peril – not a peril from another species, not a peril from outer space, but a peril from ourselves – a man-made peril. From deep within the human psyche, there has developed this terrible power that threatens not only our generation but generations yet to be born, and not only our race, but all species on this planet.

The peril is not socio-economic-political structure, so it cannot be resolved by an act of will or Parliament's legislation. The crisis has been constructed by human mind over million and million years, which has resulted in violence and aggression, chaos and strife, chemical pressure and nervous tension in the human psyche. This can only be resolved by "objective observing (seeing), listening (hearing), understanding and living", which raises the awareness of the whole mind-brain-body system. This non-authoritative observing/listening bestows order, rhythm, nonviolence and peace. In other words, it activises Intelligence which resolves all human challenges.

8. An Open Letter to Mankind

Nature invariably responds to "experimental interrogation", when the whole system is governed by Intelligence. Galileo Galilei stated, "I asked questions addressed to nature and succeeded in receiving answers directly from nature".[26] It implies that when nature is questioned it provides answer as direct perception; and this is called "discovery". Discovery is in the *timeless* realm [non-dual and non-conceptual frame of reference of Mind Continnum] and innovation is in the field of time [dualistic frame of reference in the mind spectrum], so discovery flows to innovation.

Creative leadership emerges when man discovers new things hidden in nature. Man has to begin with intuition (insight), only then he can be creative. Intuition and discovery are synonymous. Intuition is fundamental and reason is a derivative. Reason emerges with the power of abstraction.

The subjective research is a dialogue with nature. There cannot be any innovation without first discovery. If "we as a nation have to innovate to survive" then the highest priority is to *understand* ourselves and live that understanding. There is an urgency to restructure the dimension of science – [Science-Education-Ethics-Health] – in mankind *here* and *now*. We cannot afford to waste even a single nanosecond.

Fred Alan Wolf wrote, "We are beginning a new age of awareness, the age of quantum consciousness, the age of the conscious atom. By looking within ourselves, we may be able to solve the problems facing us in the final frontier – the frontier of the human spirit".[27]

9. Concluding Comment

The humanity has taken a wrong turn at some early stage of evolution which has resulted in coherent superposition of the subject on the object and this created the chaos, strife in the world as psychological imbalances and so on. What needs to be done within the consciousness of man is "objective observing and understanding" to preserve our very existence and bring "peace and serenity" and a new culture.

The brain and mind are independent and shared by mankind. The successive consciousness and Simultaneous Consciousness are also shared by mankind. The restructured [Science-Education-Ethic-Health] deals with the wholeness of life, so the subject (true-subject) and the object are within nature; in other words, both are 'in-here'. The thought-structure coupled with volitions encounters the true-subject, which gets knocked off, or the thought-structure goes to abeyance. If this observation is continued for a long chronological time, the fact also goes to abeyance, and Cosmic Consciousness expresses Itself as order, harmony, nonviolence and peace.

The new vision of education is fundamentally a single dynamic flow of Universal Consciousness. The technologies are: (a) Epistemological Revolution Through Scientific Temper, (b) Pilgrimage to the Sacred Temple of Understanding and Learning, and (c) Symbiosis of Science and Spirituality. We have to work for: (a) Restructuring of Science, (b) Restructuring of Education, (c) Awareness of Human Values, and (d) Living in the Light of that

Understanding and Learning and Intelligence.

Many eminent scientists – Albert Einstein, Arthur Eddington, Alan Watts, Galileo Galiei have discussed the transcendence of space-time-causation matrices, perceived timeless perception and this blossomed to intuition and discovery. And the discovery has the potential to innovate itself. A growth rate of 8 to 10 percent flows, if the nation diligently and sincerely follows innovation *here* and *now*.

Physics tells us that: (a) the world extended in space and time is but our representation; (b) the objective scientific view of natural processes formed in this way lack all sensual qualities; (c) there is no energy exchange between 'objective fact' and 'subject of cognizance', so it moves from *fact* to *fact*; (d) many eminent scientists – Karl Pribram, Roger Sperry, John Eccles, Wilder Penfield – advocated that we consider the objective and the subjective as a single dynamic flow; and (e) if we follow a rigorous, consistent and persistent technology, then we have to restructure science – [Science-Education-Ethics-Health] – which deals with the wholeness of life and bestows on us the 'subjective truth', or 'pure perception', or 'timeless perception', or 'discovery', which flowers to order, harmony, nonviolence and peace.

References

1. Maurice H. Wilkins, in *Synthesis of Science and Religion*, Mumbai: The Bhaktivedanta Institute, June 1988, p. 30.
2. Hideki Yukawa, *Creativity and Intuition*, Tokyo: Kodansha International Ltd., 1973, pp. 123, 125, 113.
3. David Bohm, in Gary Zukav, *The Dancing Wu Li Masters*, Fontana/Collins, 1982, p. 327.
4. Hideki Yukawa, *op. cit.*, cover jacket.
5. Roger Sperry, in Erika Erdmann and David Stover, *Beyond A World Divided*, Shambhala, 1991, cover jacket.
6. Roger Sperry, *Neurology and the Mind-Brain Problems*, American Scientists, 1952, Vol. 40, p. 311; *ibid.*, in Erika Erdmann and David Stover, *op. cit.*, p. 42.
7. Spencer Brown, *Laws of Form*, New York: Julian Press, 1972, pp. 104-05.
8. Fred Alan Wolf, *Taking the Quantum Leap*, Harper & Row, 1981, p. 229.
9. Erwin Schrodinger, *What is Life? & Mind and Matter*, Cambridge University Press, 1967, pp. 127, 131.
10. Charles Sherrington, in Erwin Schrodinger, *ibid.*, pp. 128, 131.
11. Karl Pribram, in Ken Wilber, *The Holographic Paradigm and Other Paradoxes*, Boulder: Shambhala, 1982, p. 17.
12. Roger Sperry, in *Synthesis of Science and Religion*, Mumbai: The Bhaktivedanta Institute, June 1988, p. 199.
13. John Eccles, in *Synthesis of Science and Religion*, *ibid.*, p. 199.
14. Roger Sperry, *Bulletin of the Theosophy Science Study Group*, India, February & April 1988, p. 23 and June & August 1988, p. 27.
15. J. Krishnamurti, Krishnamurti on Education, Delhi: Orient Longman, 1974,

p. 90.
16. Ilya Prigogine and Isabelle Stengers, *Order Out of Chaos*, Bantam Books, 1984, p. 218.
17. Alan W. Watts, *The Ways of Zen*, New York: Vintage Books, 1957, p. 10.
18. Commins and Linscott, *Man and the Universe, The Philosophers of Science*, New York, Washington Square Press, 1969, p. 384.
19. Huang Po, in Ken Wilber, *The Spectrum of Consciousness*, USA: The Theosophical Publishing House, 1982, pp. 303, 273.
20. Alan W. Watts, in Deepak Chopra, *Ageless Body, Timeless Mind*, New York: Three Rivers Press, 1998, pp. 297-98.
21. Albert Einstein, in Deepak Chopra, *ibid.*, p. 280.
22. *Ibid.*, p. 303.
23. Fritjof Capra, *The Tao of Physics*, New York: Bantam Books (Second Edition), 1988, p. 18.
24. Constance Holden, *Altered States of Consciousness: Mind Researchers Meet*, Science, 9 March, 1973, Vol. 179, No. 4077, p. 982.
25. Albert Einstein, *Nature*, 19 March 1979, Vol. 278, p. 214.
26. Galileo Galilei, in Hideki Yukawa, *op. cit.*, p. 111.
27. Fred Alan Wolf, *op. cit.*, p. 248.

22

An Open Science: The Quantum Jump of Science to Spirituality

Abstract

Man is an amphibian who lives simultaneously in two dimensions – (a) the known, the time-space-causation matrices, and (b) the Unknown, the Timelessness. The field of time is leading man to conflict, violence and self-annihilation. All fundamental discoveries are expressions of timeless dimension, and not cold logic. All creativity is embedded in the timeless realm.

The occurrence of seeing or hearing, smelling or breathing is in the realm of non-duality and timeless. The observing, the understanding, the learning, the spontaneous action, the intuition (insight) are in the realm of the timeless.

An "open science" is the expression of Consciousness in the timeless realm or non-dual state of the mind-brain-body system. It is a pointer to tension-free, pressure-free, division-free, duality-free state. It rejuvenates Intelligence, which is energy of the Wholeness of Life. This journey within is often called "Symbiosis of Science and Spirituality". It is the consummation of human growth.

An open science leads to discovery. It flows to innovation which has the potential of 10-12 percent growth. It is a journey within from chaos and strife to nonviolence and peace.

The key to open science is Albert Einstein's words – "objective observing and understanding". This enhances the level of awareness of the whole system which brings mutation of the human psyche.

We enquire or explore on behalf of the human race an alternative way of living, an alternative culture, an alternative dynamics of human relationship.

It is the birth-right of every human being to grow into that dimensional transformation, an entirely new way of living. It bestows equipoise, a balance at the sensual level, and the invincible peace. This is the key to human survival and excellence.

1. Introduction

The progress of science and technology during the last four hundred years – especially the obsession of scientists with measurement and quantification – have accelerated the cultural decline, and critical level of insanity. J. Krishnamurti put it succinctly about India, "An ancient race, of three to five thousand years of certain culture has been wiped overnight".[1] In a dialogue between Krishnamurti and David Bohm, an eminent physicist, Krishnamurti said:

> I have a million years of experience, and it has given me a capacity. And I realize at the end of it all there is no relationship between me and truth.... Is it a shock to discover that your brain and your mind, your knowledge are valueless? All your examinations, all your struggles, all the things you have gathered through years and years, centuries are absolutely worthless.... It has given a tremendous shock to discover the truth, that all churches, prayers, books, have absolutely no meaning – except how we can build a better society, and so on.... So there is only one thing, and that is to discover that *all* that I have done is useless – ashes.[2]

Krishnamurti and David Bohm had a dialogue on: "*What is the future of humanity?*" David Bohm in the *Preface* wrote, "He (Krishnamurti) emphasizes that mind is essentially free of the distorting bias that is inherent in the conditioning of the brain, and through insight arising in proper undirected attention without a center, it can change the cells of the brain and remove the destructive conditioning".[3]

Roger Sperry, a Nobel prize in 1981, raised the most thoughtful questions: "What form religion and the teachings of Christ, Muhammed, Buddha, Confucious, and other founders might have taken, if Copernicus, Darwin, Einstein, and all the rest had come before their time instead of after?"[4]

Nature started expressing itself in the minds of eminent scientists during the last one century, and some case histories are as under:

(a) Many eminent scientists – including Albert Einstein, Werner Heisenberg, Erwin Schrodinger, Arthur Eddington – have stumbled upon another mode of knowing, called direct or intimate, or insight, or intuition, or the "understanding of life directly, instead of in the abstract, linear terms of representational thinking".[5]

(b) Many eminent scientists – Werner Heisenberg, Edwin Hubble, Max Planck, G.F. Chew, Bertrand Russell, Lincoln Barnett, Max Born and so on – have concluded that we cannot impart value education at reason or intellect level. The inculcating value education at the reason or intellect level is invariably limited and futile.[6]

(c) Recently many eminent scientists – Karl Pribram, Roger Sperry, John Eccles, Wilder Penfield – advocated that we consider the objective and the subjective as a single dynamic flow.[7]

To sum up, a closed system (conditioned psyche) evolves to disorder, and an open system (timeless) evolves to higher and higher order as there is flow of cosmic energy. An openness or receptivity implies that inwardly we are not preoccupied with any activity other than the action of perception, the action of audition, listening. An openness is a quantum jump from time to timeless, from science and spirituality. Modern science has brought man on the control panel of mutation or transformation of human psyche. Scientists have now discovered Life Field or Universal Consciousness, which is the substratum of all existences, and which is at a very high frequency (higher frequency means higher energy) throughout the cosmos. We are on the threshold of extending this energy into human dimensions.[8]

Science is an attempt to understand nature, to dig deeper into questions that have been asked from generation to generation – the relation between 'being and becoming', between 'permanence and change'. In a recent monograph [*From Being to Becoming*], one of the authors concluded in the following terms:

> For most of the founders of classical science – even for Einstein – science was an attempt to go beyond the world of appearances, to reach a timeless world of supreme rationality – the world of Spinoza. But perhaps there is a more subtle form of reality that involves both laws and games, time and eternity.[9]

Ontology is theory of being, and epistemology is theory of knowing.

This article explores a quantum jump from science to spirituality – it deals with time-bound experience to timeless awareness, from knowledge (information) to intuition (insight, pure perception), reason to intelligence, from *becoming* to *being*, from *change* to *permanence*. *Change* implies finite, transient, ephemeral, while *Permanence* identifies Eternal, Infinite, Immutable, Immortal. It is basically an extension of the recent paper.[7] We are living at one of the great turning points of human history.

2. The Evolutionary Rise of Man

(a) The Emergence of Man: Man had an evolutionary origin. The *Homo sapiens* emerged at about 2.5 million years ago. He had been observing, comparing and seeking to classify his observations. He recorded the partial experience and fragmentary knowledge (information) as memory or cultural belief structure. He started living in a symbolic world which was a construct of his mind-brain-body system. Our daily living is nothing but a chain of reactions to the compulsions that surround us – socially, culturally, economically, politically; and we look upon ourselves as victims of these pressures. This process is devoid of any creativity. This resulted in the exclusion of the absolute phase of reality. He made himself a prisoner of the ever changing phases of material constructs and his life lost stability. Man has emerged a violent species.

The psychological imbalances, ecological crisis, pollution challenges, a rising wave of violence and crime, terrorism, religious fundamentalism, ethnic conflicts, corruption, increasing unemployment, poverty, hunger, malnutrition, health crisis, are facing man inevitably to face the realities of human situation. For the last 10,000 years, or more, man has been trying to resolve these challenges, but all his efforts have proved to be sterile. All appeals to achieve full disarmament and permanent peace, all declarations by Nobel laureates; all appeals by Pugwash Conferences (since 1957), all appeals by International Physicians for the Prevention of Nuclear Weapons [IPPNW], by Secretaries-General [Dag Hammarskjold, U. Thant, Javier Perez de Cuellar] of the United Nations; all proclamations of a manifesto by eminent scientists have fallen on the deaf ears of mankind. George Wald, Nobel Prize in Medicine, 1967, wrote, "Our society has taken a wrong turn .... By now, the problem is not transcendence, but survival".[10] Jacques Monod, Nobel prize winning biologist, wrote in 1972, "What we see opening before us today is an abyss of darkness".[11] P. Krishna, an eminent physicist, said, "We are going towards self-annihilation and not much time is left now".[12]

(b) The Human Consciousness: Man is an amphibian who lives simultaneously in two dimensions – (a) the *known,* the outer world, the successive consciousness; and (b) the *Unknown*, the inner world, the Simultaneous Consciousness. The known offers no guidance in regard to the much more fundamental problems of the Unknown.[13] As a symbol, we may compare the *known* to a drop of water and the Unknown to a fathomless and boundless ocean. The known is in the field of time, so it is limited, finite, transitory and ephemeral. The Unknown is in the realm of the timelessness, which is Eternal, Infinite, Immutable and Immortal.

There is one reality – the Unmanifest (Passive Reality) and the Manifest (Active Reality). The active reality is a powerful, highly directional, monochromatic (same frequency) and coherent (the wave is in one phase) beam of energy; or a state of highest frequency (higher frequency means higher energy), or high quantum energy potential – this is called Simultaneous Consciousness or Life Field. The successive consciousness is at the lowest quantum energy potential of Consciousness.

Spencer Brown in a most spectacular passage wrote:

> In order to do so (see) it must cut itself up into at least one state which sees (subject) and at least one other state which is seen (object). In this severed and mutilated condition, whatever it sees is *only partially itself.*[14]

The mutilated object is perceived as mutilated perception, and the mutilated subject is recorded as modified memory. The subject, the observer is the past who has known what violence is. It is the past, it is knowledge, it is experience, it is all the stored up memories. This may be expressed as under:

Object + Subject → Mutilated Object + Mutilated Subject

The psychological mode of mind invariably gives mutilated perception, which is always non-fact, the false, the past, the what should be? It can never give the fact, the true, the present, the what is!

Observation: Observing is to see it objectively, not emotionally, not with prejudice, not with bias, but just observe it. Observing and reacting are incompatible. Observing and assessing, evaluating are incompatible. The I-consciousness does not know how to observe, it only knows how to react. Reaction means energy exchange or photon transfer between the subject and the object. This is directing mankind to disorder, disharmony, conflict and violence. Every activity of I-consciousness and thought-structure leads to enhancing entropy, which leads mankind to self-annihilation or self-destruction.

This bestows a realisation that the movement of thought, the movement of the 'I', the 'me', of knowing, carries with it the whole paraphernalia of the past, the inheritance, the human knowledge and the human experience.

Richard Feynman discussing the physical world uses concepts as, "Atoms with consciousness, matter with curiosity".[15] E.H. Walker speculates that "photons may be conscious!"[16] George Wald writes, "A physicist is an atom's way of knowing about atoms".[17] Every existence is a dance of atoms with consciousness. In the within, photons, the memory, the fact are conscious. The physical truth (atoms with consciousness) has its counterpart in the within (photons), the fact and the memory are conscious and the within has its verification in the without. It appears that every event/process in the without and the within are conscious. Everything in the cosmos is interpenetrated by consciousness and is dancing in a matrix of consciousness.[18]

3. Objective Science

Erwin Schrodinger quoted Charles Sherrington, B. Spinoza and C.J. Jung and showed that there is no energy transfer between the object and the subject of cognizance, so objective science deals from fact to fact. It is a movement from the known to the modified known. The objective science constructs new model of known artifacts with a higher sensitivity. It leads to invention. It contributes to competition, motivation, disorder and strife.

There is no objective world independent of the observer. Schrodinger wrote that our world picture is "colourless, cold and mute .... The objective world is colourless and soundless and unpalpable".[19] It is the awareness of the subject of cognizance that constructs the sensual qualities. In the new paradigm, the control of life belongs to the awareness. The biochemistry of the body is the product of awareness. Most of the body responses belong to our state of mind. Each solid particle in the universe turns out to be a ghostly bundles of energy vibrating in an immense field. This new world model accepts that we constantly make anew every second.[20]

The human mind-brain-body is composed of energy and information – the brain-body system is energy, and mind is information (consciousness or awareness), and these are coupled together. The brain-body system is objective while the mind is subjective. Every atom or every cell of the brain-body system is self-conscious.

4. The Subjective World

We have to restructure science so that it takes in its fold both the objective and the subjective dimensions of life. The restructured science has the dimensions :

[Science-Education-Ethics-Health]

The ethics imply to take in its ambient the sensual and trans-sensual dimensions of life. A living in the *present*, the *now* heals and living in the *past* and *duality* suppresses man's perceptive sensitivity and enhances disease.

An exploration of the timeless realm is as under:

(a) Albert Einstein stated the importance of "objective observing and understanding".[21] Both observing and understanding are embedded in the non-dual and non-conceptual frame of reference of Mind Continuum. These are happenings in the timeless realm. Observing is invariably non-subjective – to observe without any evaluatory comparison or judgement creeping in. Observing is reaction-free perception, unpolluted perception, pure perception. Observing, understanding, learning, spontaneous action emerge from the timeless realm, so it requires inner freedom.

(b) The occurrence of seeing or hearing, smelling or breathing is a movement in the sphere of non-duality or timeless realm. Seeing is direct, hearing is direct, smelling or breathing or touching is direct, but thinking is indirect.

(c) The movement of organic intelligence in the complexity of cosmic Life is a movement born of non-duality and conducted in non-duality. It has no association with volition. It is nonvoluntary.

(d) Intuition or insight is a happening in the timeless realm.

(e) Silence and emptiness leads to communion with 'what is', the present. The past, thought, words go to abeyance. It blossoms to non-verbal communion between what is outside of me and what is within me. It is a plunge into the Unknown. It is the non-cerebral and non-verbal exploration.

(f) Nature is not intrinsically random, but follows immutable laws. Nature's every activity is perfect. There is no scope for any error. An "Open Science" emerges out of immutable laws of nature.

(g) Life begins with breathing in and death begins with breathing out.

(h) Love is breathing out, hate is breathing in. Love is in the timeless realm hate is in the time field. It is the benediction of nature to gift the timeless realm to every human being.

Objective observing and understanding leads to higher level of awareness of mind that blossoms to penetrate a boundary that physicists call "the event horizon". On one side of the event horizon is anything that remains within the speed of light; on the other side is anything faster than the speed of light. The speed of light is absolute; it is like a wall that no object can crash through. As we approach the wall, time slows down, mass increases and space becomes curved.[22]

The observation of the objective fact without justification or condemnation releases tremendous energy.[23] It is likely that at a certain critical quantum energy potential the fact makes a quantum jump across the "event horizon" and may move to the centre (subject). As it approaches the subject of cognizance, the fact is knocked off and what is left is pure Consciousness. When you conduct an enquiry, with each step there is a movement away from the false in our daily life, in our daily relationship.

In principle, it means that the subject of cognizance is at the highest quantum energy potential. Prigogine stated, "It is a physics that presupposes an observer situated within the observed world. Our dialogue with nature will be successful only if it is carried on from within nature".[24]

A nervous system is just a machine for sensing photons. The boundaries of beliefs, concepts, ideals, acquisition, violence, fear are true "event horizons". The past is the enemy of wisdom. Any kind of linear thinking is doomed to remain trapped on the surface of life. The present is called the "eternal now". The timelessness is born when limitations of the sensual world are crossed with the help of awareness. Yoga is no less and no more objective than our quantum models.[25]

Wilder Penfield stated, "our minds and our brain are not at all the same.... It is the Mind that experiences and it is the brain which records the experience".[26] Observing and understanding are in the timeless realm, so this enhances the quantum energy potential of the Mind and this bestows insight. It is insight alone that can decondition the brain. Insight is complete and from that completeness there can be logical, sane and rational action. Only then there can be flowering of the brain. Only then you can have a complete relationship with the mind. Pure clear insight is perception without any doubt.

Nature invariably expresses itself as direct perception, so original (nature) and the mirror-image (direct perception) are identical. The world is given to me only once, not one existing and one perceived. Subject and object are only one. The barrier between them cannot be said to have broken down, as a result of recent experience in the physical sciences, for this barrier does not exist. David Bohm, an eminent physicist, wrote:

> The ultimate perception does not originate in the brain or any material structure, although a material structure is necessary to manifest it. The subtle mechanism of knowing of truth does not originate in the brain.[27]

Truth (or nature) expresses itself in a non-dual and non-conceptual frame of reference of Mind Continuum. The Mind Continuum is shared by the whole mankind.

Nature responds to "experimental interrogation" or "experimental dialogue", when a question is addressed in a non-dual frame of reference of Mind Continuum. Galileo Galiei wrote, "He asked questions addressed to nature and succeeded in receiving answers directly from nature".[28]

Herman Weyl, prophetically wrote, "Another way [of approach to the phenomena of life] is of understanding from within is open to us .... I have a direct knowledge entirely different from the theoretical knowledge that represents the 'parallel' cerebral processes in symbols. This inner awareness of myself is the basis for the understanding of my fellow men". Prigogine and Isabelle Stengers agree completely with Herman Weyl. Man has to make a quantum jump from 'cerebral processes in symbols' to direct understanding of life.[29]

Voltaire wrote the following about human destiny:

> everything is governed by immutable laws .... everything is prearranged .... everything is a necessary effect .... we are both equally foolish, both toys in the hands of destiny.[30]

5. Relationship

Life is relationship, which is expressed through contact with things, with people and with ideas. These relationships are based on the past, the psychological mode of mind, hence all relationships at the symbolic level lead to conflict and are false. It is the lack of understanding of relationship that causes conflict.

Relationship means communion without fear, freedom to understand each other, to perceive directly. Relationship expresses itself in a non-dual and non-conceptual frame of reference of Mind Continuum. Relationship expresses itself as love and compassion, and oneness of life.[31]

Love is not different from truth. Love is that state in which the thought process, as time, has completely ceased. Where love is there is transformation. Without love, revolution has no meaning.

Oneness of Life: Erwin Schrodinger wrote, "Consciousness is never experienced in the plural, only in the singular .... Consciousness is a singular of which the plural is unknown". He added, "Mind is by its very nature a *singulare tantum*. I should say: the overall number of minds is just one. I venture to call it indestructible since it has a peculiar time table, namely the mind is *now*. There is really no before and after for mind. There is only a *now* that includes memories and expectations".[32] There is oneness of Consciousness which is shared by every existence.

6. A Critique on Exploration of Science

(a) The objective science sees the "objective truth", the fact. There is no energy transfer between the object (fact) and the subject of cognizance, so objective science deals from fact to fact. It is a movement from the *known* to the modified *known*. The objective science constructs new models of known artifacts with a higher sensitivity.

(b) The observation of the objective fact without justification releases tremendous energy,[23] and this enhances the quantum energy potential of the fact. This energy enables the fact to cross the "event horizon" and move towards the subject (restructured science, the subjective world). The fact is annihilated and what is left behind is Pure Consciousness, which bestows on man pure perception.

(c) Science depends on intuition (insight) – where intuition is fundamental and reason is a derivative – and is one of the eternal truths. Intuition is direct perception and it is in the realm of the timeless. Science must encompass the field of time (objective dimension) and the realm of timeless (subjective dimension). It is an attempt to go beyond the world of appearances (time) and to understand the timeless world of supreme rationality – it is often called "an open science" and it deals with the wholeness of life. Yukawa wrote, "Man has to begin with intuition, and then he can proceed with the help of his power of abstraction .... The important point would appear to be the balance and cooperation between intuition and abstraction".[33]

(d) The famous formula Karl Popper expressed the quintessence of the art scientific research. This formula was modified and published. Due to short of space, we reproduce the formula in Fig. 1. It discusses the dynamic interplay of intuition and abstraction (intellection).

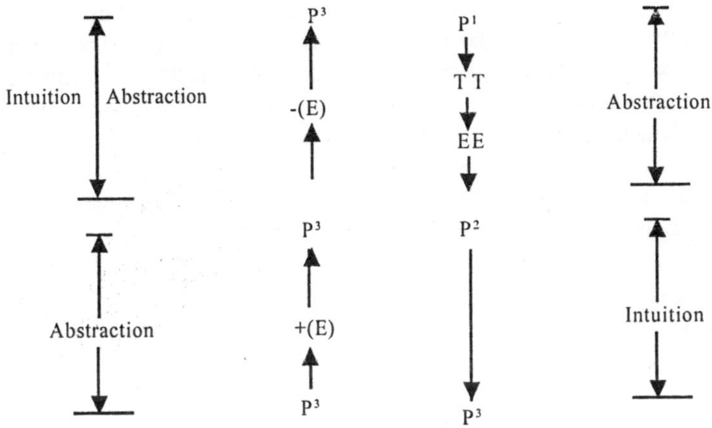

Fig. 1 An artist's view of the technology of gaining scientific knowledge. The intuitive non-verbal knowledge is represented as P^3. If some mistake occurs in expressing the knowledge P^3 in a formalised language, then we add the error (+ E) and knowledge is represented as P^4. The whole process of intuition-abstraction will have to be repeated to eliminate the error (- E) from P^4 to P^3, where P^3 and P^3 are the same.

7. A Critique on "National Knowledge Commission"

The Prime Minister Dr. Manmohan Singh has constituted the "National Knowledge Commission" on June 2, 2005 and launched the Commission on August 2, 2005. *All knowledge is rooted in the Universal Consciousness or Life Field, hence the foundation of National Knowledge Commission is higher awareness or consciousness at the National level.* Further the model[34] discussed above implies intuition-abstraction as the basis for any discovery, which flows to "creative leadership", "innovation" and higher "growth rate".

In essence, the National Knowledge Commission embraces the wholeness of life, which constitutes: (i) *Restructuring of Science:* which implies that objective science takes in its fold the subjective dimension of man; (ii)*Restructuring of Education:* to introduce dynamics of mind-brain-body system to understand the wholeness of life; (iii) *Awareness of Human Values*: and (iv) *to live in the light of that understanding*. This is not restricted to National level, but equally applicable at the International level throughout the globe. We quote some eminent scientists discussing that human fulfilment is embedded in the wholeness of life, the distilled "Intuitive Wisdom", the "Timeless Dimension of Life", "An Open Science".

(a) Maurice H. Wilkins, Nobel prize in medicine, 1962, wrote:

> Science makes progress largely by a combination of two aspects: open-mindedness and the following of abstract concepts .... Science always puts the most emphasis on this abstraction and narrowing down, on analytical procedures and on the role of the intellect as distinct from intuition and feeling .... The logical and rational is a very important element in science. But emphasis on that tends to make people not notice the essential role of intuition .... Science focusses attention on the material aspects of life and on technical solutions to human problem.[35]

It is emphasised that intuition depends on the "*direct perception*" of the wholeness of life.

(b) B.D. Josephson, Nobel prize in physics, 1973, wrote, "Science is biased in a materialistic direction".[36]

(c) Charles H. Towness, Nobel prize in physics, 1964, wrote, "Science gives great importance to truth. And a devotion to truth, in spite of what one's personal feelings and prejudices may be, takes us outside of ourselves".[37] Truth is "direct perception" of Divinity, Life Field, Timeless realm, so science gives great importance to the "pure perception" of the timeless.

(d) Fritjof Capra, an eminent physicist, wrote, "The rational part of research would, in fact, be useless if it were not complemented by the intuition that gives scientists new insights and makes them creative".[38] Constance Holden adds, "In high level creativity the mind can do weeks of work in a short time".[39]

(e) Ervin Laszlo, an eminent scientist, wrote:

> The whole .... is not externally related collection of diverse phenomena, but an internally integrated network that, despite its manifest diversity, forms an embracing and unitary totality. We are not just observing it, we are not just immersed in it, we are an integral part of it.[40]
>
> Thanks to the convergence of science with intuitive insight, we now possess the critical mass of essential concepts and ideas to fulfil the dream. We can be more than conscious parts in an evolving totality: we can be parts that truly fulfil their potentials and become conscious of the totality – and come to live in harmony with it.

(f) Man discovers new things hidden in the timeless realm of nature .... It is the discovery of a truth in nature hitherto unknown to man". Fundamentally "human truth" and "nature's truth" are one and the same – the original (Nature) and the mirror-image (man) in the timeless realm are identical. Many scientists have opined that "modern scientific work involves him and other scientists with society to an even greater extent than other people". Victor F. Weisskoff wrote, "Man divorced from nature could not be happy and man's power to resist nature is hopelessly puny".[41] The task of turning the age-old tide of human ambition and aggression can only be achieved if man makes a quantum jump from time to timeless, from experience to perception, from classical objective science to an "open science" (open to timeless perception). Hideki Yukawa wrote, "Science is a manifestation of man's efforts to open up the unknown world that lies before him, to discover new possibilities".[9] He added, "It is the discovery of a truth in nature hitherto unknown to man". Weisskoff wrote, "The mature Yukawa recognised that science is a part of human relations and that modern scientific work involves him and other scientists with society to an even greater extent than other people".[42]

(g) Galileo Galielei gave evidence of direct perception of nature.

(h) D.S. Kothari quoted Lord Buddha in words:

> What the Buddha proclaimed about the imperative need to *know* the mind, to *shape* the mind, and to *liberate* the mind has been more true than in the age of the Atom.[43]

One can intuitively perceive that highest state of Self is non-self. B.M. Khanna quoted Lord Buddha in words:

> Self is, Truth is. Where self is, Truth is not. Where Truth is, self is not.

And once 'self' dissolves, we have true freedom, freedom from time, freedom from all our suffering. This freedom alone will lead to a silent mind, a mind that no longer chatters, a mind that is deep and tranquil, a mind that does not live in the knowledge of yesterdays and tomorrows, a mind that lives in NOW, a mind that does not live in the knowledge of NOW but in the feeling of NOW.[44]

8. A Paradigm Shift or An Action Plan

(a) The aim of this research paper is "to initiate an open dialogue with an open mind", and call for responses of eminent scientists. It is suggested that this paper may please be printed (I will submit the floppy) on the highest priority, and copies are sent to other scientists of ISRO, DRDO, CSIR, DST, universities, Engineering and Medical Research Institutes. Perhaps such sharing may lead to a modified model of mutation of human psyche.

(b) The printed copies of the plan may please be circulated at the forthcoming session of the Indian Science Congress, January 2006.

(c) I suggest that the action plan may please be implemented in a certain time frame – say 4 months, because not much time is left *now*. We cannot play this game when our own survival is embedded in it.

(d) Fundamentally nature expresses itself as "direct perception" so the original (Nature) and the mirror-image (man) in the timeless realm are identical. The fundamental challenges facing man are mutation of the whole psyche or symbiosis of science and spirituality. And this rejuvenates Intelligence which resolves all challenges. This is the crucial transition to a higher consciousness. This guarantees human survival and excellence. This is the only vocation of mankind.

(e) It is the integral dimension of Consciousness that can resolve the challenges facing mankind. It is to understand the mind-brain-body system as one whole; which ensures order, harmony, security, nonviolence, peace and survival. The National Knowledge Commission may finalise its findings in maximum six months instead of thirty six months, as time is running out.

9. Concluding Comment

Man has to make a quantum jump from time-bound experience to timeless awareness, from knowledge (information) to intuition, from intellect to intelligence. The quantum energy potential of intelligence, which is totally independent of time, resolves all human challenges, and bestows order, harmony, nonviolence, peace and bliss. If we want to construct a sane world, we have to begin by changing ourselves – from mutilated experience to pure perception – the coded information embedded in energy. The secret of conquering the universal mystery lies not in our physical bodies, but in our Consciousness.

It is nature's benediction to bestow on man a hidden capacity of quantum jump from time to timelessness, from conflict and violence to "inner awareness". *Man discovered new things hidden in the timeless realm of nature.* The timeless perception bestows discoveries, which flowers to innovation and 10-12 per cent growth rate.

Every human activity (except that of the conditioned psyche) is governed by the expression of Consciousness as direct perception, or timeless perception. It is not an activity of the brain or the sensual world, although a material structure is necessary to manifest Consciousness. This rejuvenates Intelligence, which resolves all challenges. There is an urgency to share this with mankind.

We have to undertake an experiment in creating a "New Man" – a human being who lives in harmony with the inner and the outer, with himself and his environments, and who is free from all his ideologies and conditionings that now divide humanity. And if man escapes from his responsibility, it is at his own peril, and that of India and mankind.

We intuitively see that at this critical juncture nature is set to help mankind *here* and *now*. This culminates to peace and survival.

References

1. J. Krishnamurti, KFI Bulletin, 1983, No. 3, p. 27.
2. *Ibid., The Ending of Time, J. Krishnamurti and Dr. David Bohm*, San Francisco: Harper and Row, 1985, pp. 105, 107, 110.
3. J. Krishnamurti and David Bohm, *The Future of Humanity, Preface by David Bohm*, KFI 1987, pp. 2-3.
4. Roger Sperry, in Erika Erdmann and David Stover, Shambhala, 1991, pp. 177-78.

5. Alan W. Watts, *The Ways of Zen*, New York: Vintage Books, 1957, p. 10.
6. Sampooran Singh, Kanwaljit Kaur & Paramjit Singh, *Global Education and Human Survival*, Sanbun Publishers, March 2002, pp. 46, 35-36.
7. Sampooran Singh, *Innovation in Science and in Man*, unpublished.
8. *Ibid.*, *A Review on Symbiosis of Science and Spirituality*, Dagshai Cantt, The Atam Science Trust, October 7, 2002, p. 12.
9. *From Being to Becoming*, in Ilya Prigogine and Isabelle Stengers, *Order Out of Chaos*, Bantam Books, April 1984, p. 310.
10. George Wald, in Kishore Gandhi, *The Evolution of Consciousness,* New Delhi : National Publishing House, 1983, p. 14.
11. Jacques Monod, *Chance and Necessity*, New York: Vintage Books, 1972, p. 171; in *Synthesis of Science and Religion*, Mumbai: The Bhaktivedanta Institute, June 1988, p. 127.
12. P. Krishna, *Education, Science and Spirituality*, USA: The Institute for Higher Education, Research and Dialogue, August 1993, p. 12.
13. Aldous Huxley, in J. Krishnamurti, *The First and the Last Freedom*, London: Victor Gollancz Ltd., 1977, Foreword.
14. Spencer Brown, *Laws of Form*, New York: Julian Press, 1972, pp. 104-05.
15. Richard Feynman, *The Value of Science Project Physics Reader I,* Authorised Interim Version, 1968-69, pp. 3, 4.
16. E.H. Walker, *The Nature of Consciousness*, Mathematical Bio-sciences, 1970, pp. 175-76.
17. George Wald, in *Bulletin of the Theosophy Science Study Group*, India, December 1986, p. 71.
18. Sampooran Singh, *Science of Life*, Chandigarh: Faith Publishers, October 1994, p. 21.
19. Erwin Schrodinger, *What is Life*? and *Mind and Matter*, Cambridge University Press, 1967, pp. 128, 148.
20. Deepak Chopra, *Ageless Body, Timeless Mind*, New York: Three Rivers Press, 1998, pp. 36, 8.
21. Albert Einstein, *Nature*, 19 March 1979, Vol. 278, p. 214.
22. Deepak Chopra, *How to Know God*, Rider, 2000, pp. 270-71.
23. J. Krishnamurti, *Krishnamurti on Education*, Delhi : Orient Longman, 1974, p. 90.
24. Ilya Prigogine and Isabelle Stengers, *Order Out of Chaos*, Bantam Books, 1984, p. 218.
25. *Ibid.*, pp. 272, 296, 266, 291.
26. Wilder Penfield, in Deepak Chopra, *ibid.*, pp. 215, 217.
27. David Bohm, in Gary Zukav, *The Dancing Wu Li Masters*, Fontana/Collins, 1982, p. 327.
28. Galileo Galiei, in Hideki Yukawa, *Creativity and Intuition*, Tokyo: Kodansha International Ltd., 1973, p. 111.
29. Herman Weyl, in Ilya Prigogine and Isabelle Stengers, *op. cit.*, 1984, p. 311.
30. Voltaire, in *ibid.*, pp. 257-58.
31. J. Krishnamurti, *The First and Last Freedom*, London: Victor Gollancz Ltd., 1977, pp. 104-07, 177-81.
32. Erwin Schrodinger, *op. cit.*, pp. 94, 95, 145.
33. Hideki Yukawa, *op. cit.*, pp. 143, 119, 101-02.

34. Sampooran Singh, *The Holistic Education*, New Delhi: Siddharth Publications, 1987, pp. 149-56.
35. Maurice H. Wilkins, in *Synthesis of Science and Religion*, Mumbai: The Bhaktivedanta Institute, June 1988, pp. 30, 31, 36.
36. B.D. Josepheson, in *ibid.*, p. 40.
37. Charles H. Towness, in *ibid.*, p. 140.
38. Fritjof Capra, *The Tao of Physics*, New York: Bantam Books (Second Edition), 1988, p. 18.
39. Constance Holden, *Altered States of Consciousness: Mind Researchers Meet*, Science, 9 March 1973, Vol. 179, No. 4077, p. 982.
40. Ervin Laszlo, in *The Odyssey of Science, Culture and Consciousness*, New Delhi: Abhinav Publications, 1990, pp. 237, 238.
41. Victor F. Weisskoff, in Hideki Yukawa, *op. cit.*, Introduction, pp. 9, 12.
42. Hideki Yukawa, *ibid.*, pp. 205, 113, 112.
43. D.S. Kothari, *Atom and Self, The Evolution of Consciousness*, New Delhi: National Publishing House, 1983, p. 239.
44. B.M. Khanna, in *The Odyssey of Science, Culture and Consciousness*, *op. cit.*, p. 165.

23

The Art of Living in an "Open Science"

Abstract

Man has started living in the sensual world and he has divorced reality. He is aware of the non-fact, the past, the false, the what should be? The mind constructs chaos, strife and conflict; so he is living in nervous tension and chemical pressure. Every activity is a pointer towards self-annihilation.

Nature has bestowed man a hidden capacity to transcend thought-structure and imbibe non-dual frame of reference, or timelessness. This bestows attentivity of the objective world or objective truth. We call it as Silence because the I-consciousness, subconsciousness and unconsciousness have gone to abeyance and the fact (objective truth) expresses itself. It is a movement from the known to the modified known, and the intellect constructs artifacts with higher and higher sensitivity. It is called "limited objective state". It is responsible for invention (not discovery) in the materialistic field.

We have to restructure science so that it takes in its realm both the objective and the subjective dimensions of life. The dimensions of restructured science are: [Science-Education-Ethics-Health]. It deals with the highest quantum energy potential of the Universal Consciousness or Life Field. We call it an "open science", as it expresses the wholeness of life.

Meditation is a way of living that leads us through time (duality) to timeless (non-dual) state of our life. It transcends I-consciousness. It activises perceptive Intelligence, the wholeness of life. It bestows order, harmony, nonviolence and peace.

1. Introduction

For the past 10,000 years, or more, man has been trying to resolve the psychological imbalances, ecological crisis, violence, fear, poverty, hunger, malnutrition, and so on through different thought-structures, beliefs, concepts, ideals, but all these have proved to be sterile in transforming the human psyche. All appeals to achieve full disarmament and permanent peace; all declarations by Nobel laureates; all appeals by Pugwash Conferences (since 1957); all appeals by the International Physicians for the prevention of Nuclear

War (IPPNW); by Secretaries-General (Dag Hammarskjold, U. Thant, Javier Perez de Cuellar) of the United Nations; all proclamations of a manifesto by eminent scientists, philosophers, mystics have fallen on the deaf ears of mankind. George Wald, Nobel prize in Medicine (1967), wrote, "Our society has taken a wrong turn .... By now problem is not transcendence, but survival";[1] Jacques Monod, Nobel prize winning biologist, wrote in 1972, "What we see opening before us today is an abyss of darkness".[2] All conditionings are guiding mankind to self-destruction or self-annihilation.

There is no problem more mysterious than the mind-brain-body interaction. Modern science has not made it less intractable. On the other hand it has added a new urgency and also a new poignancy. There is no explanation, no clue, whatsoever, as to how the interaction of mind-brain-body occurs. Paradoxically, nothing is more commonplace than the reality of this interaction. My writing this sentence, and your reading it, is itself an expression of the mind-brain-body interaction, and yet nothing is more intractable, nothing a greater mystery.[3]

Man is an amphibian who lives simultaneously in two dimensions – (a) the known, the time-space-causation matrices, the objective; and (b) the *Unknown*, the timelessness, the subjective. Brain is a part of the objective world. It can be investigated objectively. On the other hand, the mind is subjective.

What we call mind (with a lower case 'm') is a conditioned energy, moving through the body, monitored by brain, operating through the senses. It is a movement of the structured framework of patterns and the dynamics of behaviour patterns are shared by all human beings. The framework of patterns (matter) is at the lowest quantum energy potential of consciousness and is called successive consciousness. The time-space-causation matrices represent the relative field of consciousness. It constructs pain, misery, suffering, chemical pressure and nervous tension. All knowledge, concepts, ideals intellectual approaches are sterile in transforming the psyche or to bring about a radical transformation in conditioned man.[4]

Thinking is a playback of thought coupled with feeling and response of the memory – cultural, racial, religious, communal – which is called willing. Willing flows to reaction. These conditionings have been organised and standardized. We are conditioned cerebrally, neurologically, chemically to react to situations in a specific way. To be aware of the dynamics, conditionings and limitations of the conditioned psyche is called self-knowledge.

What we call Mind (with upper case 'M') is the non-dual frame of reference of Mind Continuum. The Universal Consciousness expresses Itself in a non-dual frame as intuition (insight), spontaneity, love and compassion, order and harmony. It is the first-hand personal acquaintance of the quantum jump of the conditioned energy consciousness to Universal Consciousness;

or in other words, a quantum jump from fragmentary perception to the perception of the Wholeness.

Charles Sherrington wrote, "The mental is not available as a form of energy .... Our mental experience is not open to observation through any sense organ .... Mind for anything perception can compass, goes therefore in our spatial world more ghostly than a ghost. Invisible, intangible, it is a thing not even of outline, it is not a thing without sensual confirmation, and remains without for ever".[5]

Wilder Penfield concluded that "our minds and our brains are not at all the same .... It is the mind which experiences and it is the brain which records the experience".[6]

This article explores the common waking state or the psychological mode of mind; the objective science (the "principle of objectivation" in the words of Erwin Schrodinger), and the subjective world. It discusses a critique on "exploration of science" and on "National Knowledge Commission". An action plan is also presented.

This article basically is a response of Nature in the "intuitive-wisdom" realm, or its direct perception in a non-dual frame of reference in the Mind Continuum. It appears that Nature has released her truth to guide mankind for survival. The dimension of meditation is also discussed. There is an urgency "to open a dialogue with an open mind", and perhaps such sharing may lead to a modified field of mutation of human psyche. It is basically an extension of the earlier paper.[7]

2. Where is Psychosocial Evolution Leading Mankind?

Many eminent scientists, philosophers-mystics, political leaders have echoed that man has taken a perilous path and have suggested healing the diseased human psyche, or to end the crisis of perception. Some most spectacular passages are quoted as under :-

(a) The Committee on Science and Technology, USA, submitted in June 1981 a lengthy report to the House of Representatives, USA, which encouraged serious assessment of further research on "*the physics of consciousness*", and concluded that recent experiments:

> .... that there exists an 'interconnectiveness' of the human mind with other minds and with matter .... that the human mind may be able to obtain the information independent of geography and time.[8]

(b) The Union of Concerned Scientists (UCS) issued a "*World Scientists' Warning to Humanity*" in April 1993, signed by over 1670 scientists including 104 Nobel laureates, concluded that "Human beings and the natural world are on a collision course .... A new ethic is required – a new attitude towards discharging our responsibility for caring for ourselves and for the earth .... This ethic must motivate a great movement, convincing reluctant leaders and reluctant governments and reluctant peoples themselves to effect the needed changes".[9] As every event/existence is interrelated and interpenetrated, so the responsibility is not towards survival of one's self but for the human race.

(c) The 1993 Parliament of World's Religions, Chicago, concluded, "Peace eludes us …. the planet is being destroyed …. Neighbours live in fear …. women and men are estranged from each other …. children die!". The Parliament also suggested, "Earth cannot be changed for the better unless we achieve a transformation in the consciousness of individuals and in public life …. This transformation must also be achieved in the area of ethics and values!".[10]

(d) Einstein – Out of His Later Years – wrote, "The confidence in the stability, even the very basis of existence, of human society has largely vanished. One senses not only a threat to man's cultural heritage but also that a lower value is placed upon all that one will like to see defended at all costs …. Life in human society meant dangers of all sorts …. even standards of virtue, aspiration and practical truth were taken for granted as an inviolable heritage, common to all civilised humanity …. The sanctity of life vanished …. The lie was raised to the dignity of a political instrument. It was a manifestation of weakness of character and lack of responsibility on the part of the leaders in the affected countries". This presents evidence that the human psyche during the first five decades of twentieth century got heavily conditioned leading to lowering of its sensitivity.

We have brought the human race to an explosive situation and our life is poised on razor's edge. Einstein wrote, *"Man changes but little …. though current trends like the present bring him unimaginable sorrow. Nothing of all that will remain but few pitiable pages in history books, briefly picturing to the youth of the future generations the follies of its ancestors".*[11] Perhaps, the benign turn of destiny, who have faith in the mysterious will be saved.

(e) Recent rise of terrorism, religious fundamentalism, drug trafficking indicates an enhancement of the conditioning of the psyche. To sum up, the twentieth century has added to conditionings of the psyche and is leading to self-annihilation, or self-destruction. And this is now threatening the extinction of all life-supporting system within a foreseeable period of time.

3. The Psychological State of Mind

R.P. Kaushik, based on Patanjal Yoga, wrote, "The first stage prior to recognition and naming is a state of non-verbal or thought-free perception (*nirvikalpa jnana* or pure perception). From this pure perception we proceed to thought, recognition and naming (*savikalpa jnana*).[12] In the psychological field, both these processes of identification and alienation are one and the same. Vimala Thakar, an eminent philosopher-mystic, wrote, "The superstructure of psychological time of a concept (is) grafted on timeless reality, and we live in this superstructure…. We live in these enclosures but the enclosure is not the totality of life. Psychological time, created by man, used for the convenience of the brain for the security of the brain, is not the reality.[13] J. Krishnamurti wrote, "In the seeing of any fact there is no "me". Me is non-seeing. The "me" cannot see, cannot be aware .... There is no me in the present".[14] Obviously seeing the fact is in the timeless realm.

George Wald, Nobel prize in medicine, 1967, spent most of his scientific life studying mechanisms of vision. He prophetically wrote:

> One can put together everything that we have learned, and add to it everything that workers in this area hope to learn; and none of it comes anywhere near, or even aims in the direction of *what it means to see* .... Seeing – the event in consciousness – seems to lie in another universe, unapproachable by science .... Science offers no way of testing for the mere presence or absence of consciousness, let alone its content.[15]

Wald's observation that the event in consciousness seems to lie in another universe, unapproachable by science" implies the event in consciousness is in the timeless realm.

Spencer Brown, in a most spectacular passage, explained:

> In order to see itself, it (the mind) must cut itself up to at least one state which sees, and at least one other state which is seen. In this severed and mutilated condition, whatever it sees is only partially itself.[16]

The subject and the object are at the same quantum energy potential, so there is energy exchange, and both are mutilated. The mutilated subject gets recorded as modified memory and the object is perceived as mutilated object. The mutilated observer is invariably aware of the *non-fact*, the *false*, the *past*, the *what should be*!

In the psychological field, the identification and alienation are one and the same. In either case there is no integration of the observer and the observed. Hence there is no observation but only distortion.

The psychological mode of mind is that: (i) the Unmanifest expresses itself as manifest (direct perception); (ii) the perception recalls the past memory; (iii) there is direct superposition of different webs (psychological structures); (iv) this results in mutilated perception and modified memory; (v) the modified memory is recorded and we are aware of the mutilated perception; (vi) the mutilated perception is the *non-fact*, the *false*, the *past*, the *what should be?* This may be expressed as under:

Unmanifest (Subject) + Recalled Memory →
Mutilated Perception + Modified Memory

Without being aware of it and without being systematic about it, the whole mankind is caught in an unprecedented crisis of perception. The conditionings are related to the process of becoming which is the foundation of modern civilization and culture, which is comparative, competitive and aggressive.

The space between the observer and the observed in which self-projection takes place is the psychological space. In the psychological space, the mutilated observer is constantly projecting itself as desires, aversions, fear, violence and knowledge, so it is not observing the observed and hence not understanding what the observed actually is.

Knowledge: It is always an indirect activity, it is in the field of duality. The conditioned I-consciousness has voluntarily accepted the role of an observer – so there is the knower, the knowing and the knowledge. Knowledge is stored as memory.

4. Emergence of I-consciousness, the Ego

For the past 2,000 to 10,000 years, or more, man had been observing, comparing and seeking to classify his observations as subjective dimension (inner) and objective dimension (outer). Thinking, knowing, memorizing, drawing conclusions from the information fed into us – all these are involuntary processes, with which we are born – and these are called inheritance, or I-consciousness, or ego. These are conceptual processes and do not have any factual reality. So thinking is in the dualistic field or indirect perception. This is also called "I" or the "me", or "ego". The partial experience and fragmentary knowledge is always indirect, and therefore partial and limited.

The whole mankind is heavily conditioned over uncountable aeons and the mind responds due to these conditionings. The past is the "me". There is no "me" in the present. As long as the mind is operating in the past, there is the "me"; and the mind is the past, the mind is the "me".

The observer is the past who has known what violence is. It has taken up the role of I-consciousness voluntarily, so it is always *non-fact*, the *false*, the *past*, the *what should be*! It is never aware of the *fact*, the *true*, the *present*, the *what is*? We are products of not only conditioned energy but cultivated, moulded, directed channelised energies. The "I", the "me", the "self", the "ego" is constituted of partial experience, fragmentary knowledge, past memory, and constructs chaos and conflict as nervous tension and chemical pressure.

To sum up, I-consciousness has no factual reality at all. It has a conceptual reality. It has an ideational reality, and it has relevance to our lives. The concepts are grafted on the perception leading to distorted perception. Education requires developing a new way of living – economically, politically, socially, which is supportive to the inner balance – the inner and the outer have to be harmonized.

5. Transcendence of I-Consciousness

The ancient sages/seers saw that the first step is pure cognition, or pure perception, or direct perception; then there is recall of past memory and there is coherent superpositions of different webs which result into distorted perception. If we observe the flow of thought structure, we see that the movement of thought structure is within me, it is independent of my volition, it is independent of the ego. The discovery has a dual effect – (a) the pride and vanity that I think – that these are my thoughts – goes to abeyance, so there is humility; and (b) the mental movement has a momentum independent of me, and I act upon them. I gratified the needs of thought which means that I do not evaluate them and I do not judge them. I just observe the thought without interfering in its flow. This flows to abeyance of I-consciousness which implies transcendence of desire, sentiments, ambition, and so on.[17,18]

To sum up, "When it is seen that there is a racial consciousness and racial momentum of knowledge, experience and thought, and there is nothing like an individual 'ego', the 'self', the 'me', there is sense of great freedom within".

6. The Art of Scientific Observation

Biologist T.H. Huxley once wrote, "Sit down before fact like a little child and be prepared to give up every preconceived notion, follow humbly wherever and to whatever abyss Nature leads, or you will learn nothing".[19] Fritjof Capra observed that "all the difficult economic, environmental, social, political and human problems are really different facets of one and the same crisis, and that crisis is essentially a crisis in perception".[20] He underlined the quintessential importance of "the art of seeing" as a basic prerequisite to effective problem solving. Lehmann wrote, "The demand for more and more observation and experiment implies a warning to both physicists and biologists: do not trust pure imagination; do not lose yourself in wild speculations; in meta-physics and meta-biology which tempt you to transcend the limits of possible experience".[21] Gilbert Blane, in 1819, wrote, "Observation and more of observation, on the actual working of the Conservative and Restorative Principle in health and disease, is the foundation of medical science".[22] E.F. Schumacher said, "Our task is to look at the world and see it whole".

Prigogine stated that "the behaviour of a reality (is) unrelated to our beliefs, our ambitions, or our hopes".[23] Erwin Schrodinger stated, "a moderately satisfying picture of the world has only been reached at the high price of taking ourselves out of the picture, stepping back into the role of a non-concerned observer".[24] David Bohm concluded, that "to observe nature with a certain kind of objectivity could be called the germ of scientific attitude",[25] or scientific temper.

Observation is perception without psychological involvement, without any kind of reaction. Reaction implies evaluation, judgement, condemnation, comparison. Observation is non-evaluatory perception. Observation without the pseudo-observer is the state of wholeness of perception and wholeness of responses.

7. Objective Science

About 1600 Kepler and Galileo simultaneously and independently formulated the principle that the laws of nature are to be discovered by measurement. All propositions were to be confined to that which was objectively measurable and verifiable. Eddington put it succinctly, "Nature thus provides that knowledge of one-half of the world will ensure ignorance of the other half".

Sullivan put it, "We cannot observe the course of nature without disturbing it",[26] or Andrade, "Observation means interference with what we are observing .... observation disturbs reality".[27] George Wald wrote, "The very crux of twentieth-century physics is the recognition that the observer cannot be excluded from his observations; he is an intrinsic participant in them".[28] It was abundantly clear to those physicists that objective measurement and verification could no longer be the mark of absolute reality.

About 1947, Erwin Schrodinger proposed the principle of "objectivation". He wrote:

> Mind has erected the objective outside world of the natural philosopher out of its own stuff. Mind could not cope with this gigantic task otherwise than by the simplifying device of excluding itself – withdrawing from its conceptual creation. Hence the latter does not contain its creator.[29]

He quoted Charles Sherrington in words:

> The material world has only been constructed at the price of taking the self, that is, mind, out of it; obviously, therefore, it can neither act on it nor be acted on by any of its parts.

He also quoted C.J. Jung and B. Spinoza to support his view.

> The material world has only been constructed at the price of taking the self, that is, mind, out of it; obviously, therefore, it can neither act on it nor be acted on by any of its parts.[30]

Schrodinger also quoted C.J. Jung and B. Spinoza to support his view.

The scientific mind is very factual. Discovery is its mission, its perception. It sees things through a microscope, through a telescope; everything is to be seen actually as it is; from that perception, science draws conclusions, builds up theories. Such a mind moves from *fact* to *fact*. The spirit of science has thing to draw from individual conditions, with nationalism, with race, with prejudice.

The physicists call the boundary between objective world and the inner world as "the event horizon". In the outer world, everything that remains is within the speed of light; on the other side is anything faster than the speed of light. The speed of light is absolute; it is like a wall that no object can cross through. As we approach the wall, time slows down, mass increases and space becomes curved.[31] The photons cannot cross the event horizon, so there is no energy transfer between the object and the subject. There is awareness of the fact, the objective content, the objective truth. The objective awareness responds and provides answer to the objective challenge.

The objective science deals from fact to fact. The objective world is invariably partial, so it invariably provides partial answers to the challenge which is called 'invention' in the field of materiality; it cannot provide answer to the wholeness of life.

It is a movement from the known to the modified known. The objective science constructs new model of known artifacts with a higher sensitivity. It leads to invention. It contributes to competition, motivation, disorder and strife.

There is no objective world independent of the observer. Schrodinger wrote that our world picture is "colourless, cold and mute .... The objective world is colourless and soundless and unpalpable".[32] It is the awareness of the subject of cognizance that constructs the sensual qualities. In the new paradigm, the control of life belongs to the awareness. The biochemistry of the body is the product of awareness. Most of the body responses belong to our state of mind. Each solid particle in the universe turns out to be a ghostly bundles of energy vibrating in an immense field. This new world model accepts that we constantly make anew every second.[33]

The human mind-brain-body is composed of energy and information – the brain-body system is energy, and mind is information (consciousness or awareness), and these are coupled together. The brain-body system is objective while the mind is subjective. Every atom or every cell of the brain-body system is self-conscious.

8. Subjective World

We have to restructure science so that it takes in its fold both the objective and the subjective dimensions of life. The restructured science has the dimensions :

[Science-Education-Ethics-Health]

The ethics imply to take in its ambient the sensual and trans-sensual dimensions of life. A living in the *present*, the *now* heals and living in the *past* and *duality* suppresses man's perceptive sensitivity and enhances disease.

An open science is born when nature expresses itself as "direct perception". It is a manifestation of the highest quantum energy potential of Universal/Cosmic Consciousness or Life Field. It is the expression of pure Consciousness in the timeless realm or non-dual state of mind-brain-body system. An open science deals with the wholeness of life, or it is manifestation of the immutable laws of nature. An open science flowers to all fundamental discoveries, which are expressions of timeless dimension, and not cold logic. All creativity is embedded in the timeless realm. It flows to innovation which has the potential of 10-12 per cent growth rate.

9. Meditation

(a) Meditation is a dimension in which one can live and from which one can move and function. You have to be aware from the beginning to the end, not become inattentive in the middle. The new quality of awareness is attention. There cannot be attention if you are not sensitive to the structure and the nature of these man-made traps. The purpose of human life is to grow into a state of inner equipoise, which is not disturbed by the movement of relationships. It has the capacity to function in a balanced way on the sensual level, in any situation. Meditation is to allow the total conditioned mind to go into abeyance and relax completely, and to explore the unconditioned energy in the timeless realm, the non-dual field. It is a trans-psychological need for the human existence. There is urge for freedom and for peace.

(b) The State of Observation: In the state of observation, there is non-reactional attentiveness, so observation is in the realm of non-duality, the timeless state. The doer and the experiencer go to abeyance, the comparison and judgement, the acceptance and rejection, the evaluation and naming also go to abeyance; and the state of observation is activised. The beauty of defencelessness comes about. Innocency is utter defencelessness.

(c) Intuition (Insight): All knowledge is structured in Consciousness. This recognizes that consciousness alone can be totally self-referral, so the pure Mind or non-dual Mind can perceive the consciousness. When non-dual state of Mind is presented as an object to pure consciousness, then it reveals itself. In the non-dual Mind, the intelligence is born of the inner silence, of the inner space. The intelligence is uncluttered by the movement of thought. The dimension of intelligence, of unconditioned energy, of Fundamental Vibration, has no past, so it is free of all conditioning and divisive loyalties, etc. It bestows insight. Insight implies something which you perceive without any logic, without any motive. It is called immediate perception. *The perceptions are born of non-duality, of silence, and responses are born of spontaneity.* Perception has nothing to do with knowledge. Intelligence begins to use the human brain and help the human being to look at the world in quite a different way.

Intuition is direct understanding, non-comparative expression, non-evaluatory perception. The acquisitive culture and psychological aggression benumbs the sensitivity. It is insight alone that can uncondition the brain, not the effort of thought, which is time-bound. Thought is at a lower quantum energy potential of consciousness than insight. Where there is ending of thought-time, there is insight. Only then there can be flowering of the brain. Only then can you have a complete relationship with the Mind.

When there is clear insight into violence, for instance, that very insight banishes all violence. Intelligence itself expresses itself as insight in the brain. It uses the brain for its expression. Insight is not due to recall of knowledge, so it brings mutation in the psyche. That insight is complete and from that completeness there can be logical, sane, rational action.

(d) Spontaneous Action: A non-dual Mind perceives the truth spontaneously. It implies that there is no psychological time between expression of Life Field in the brain cells and its perception in the brain. Further there is no psychological time between pure perception and action. So the expression of Life Field in the brain cells, the perception and action are one dynamic flow. The whole Consciousness is a seamless, fluid web of process structures, it is one dynamic flow, and cannot be disassembled or split into simpler and simpler parts.

10. Dimensional Transformation

(a) The restructured science leads to inner transformation in the quality of consciousness. It is mutation. It is a dimensional transformation. Mutation is not the result of an action and meditation is not the product of human activity.

When the highest quantum energy potential of Cosmic Consciousness expresses itself in a non-dual frame of reference of Mind Continuum, dimensional transformation or mutation happens. It is the highest level of awareness in a non-dual frame of reference that bestows dimensional transformation.

(b) We need a genuine urge to find out the meaning of life, the nature of truth, the nature of reality, faith in life and fearlessness to carry out and what you understand. The whole universe, the life, the divinity is with us, that power of intelligence is with us and we shall be helped. The intelligence provides the help when it is really needed.

(c) Silence: Meditation is the only way to set oneself free of the wheel of thought and karma and destiny. So what remains is an ocean of Silence. The "I" and the "me", and its subconscious and unconscious go to abeyance, so the contents of the silence is the total discontinuation of the mental activities.

The discontinuity of the ego takes us to a different dimension of consciousness, when knowing and experiencing may not happen at all, but perception, understanding and action get activised. This happens in a non-dual and non-conceptual frame of reference. This refers to a non-cerebral, non-verbal exploration .

In silence the subject of cognizance moves to higher quantum energy potential and it splits up to the objective world, the *fact*, the *true*, the *present*, the *what is*! The world of the fact cannot cross the "event horizon", so there is no exchange of energy between the subject and the object, and we are aware of the objective truth.

Erwin Schrodinger defined the "Principle of Objectivation" in words, "Without being aware of it and without being rigorously systematic about it, we exclude the Subject of Cognizance from the domain of nature that we endeavour to understand". There is no verbalization of the subject of cognizance, but the objective truth is verbalized. The object moves from fact to fact. It is a movement from the known to the modified known. The objective science constructs a new model of known artifacts with a higher sensitivity.

(d) Total Silence: In common terminology, it is called the subjective world. The quantum energy potential of the subject of cognizance rises further and the fact also goes to abeyance. The space is full of cosmic movement, or Universal Intelligence, or Life Field. The vibration of the silence also stops and there is total emptiness. It is free of verbalization and free of conditioning.

The energy of life is a Perceptive Intelligence. The Cosmic, the Universal Intelligence sees the wholeness of life, understands everything. There is no ignorance contained in the Intelligence. The wholeness is very holy.

The Cosmic Intelligence is not a faculty of the cerebral organ, it is not related to sensual faculty, the cerebral and the sensual becomes a channel. The Cosmic Intelligence uses the brain-body for its expression. The Universal Consciousness uses the brain-body instrument for the welfare of all mankind. The person is reduced to nothingness inside, which the Universal Consciousness uses for the service of all people.

(e) Awareness: It is an energy born of the emptiness of Silence. This energy is released by the non-action of knowledge, of conditioned consciousness. It is an energy that permeates the whole being, vibrates in the whole being, that becomes the source of perception and response. It has no relation with the volition. It shows that Life is one, that Life is holistic.

Understanding and learning are timeless. The timeless present is the emanation of Eternity; it is flowering of Infinity. This present moment contains all time and is therefore itself timeless, and hence this *timeless present is Eternity itself* – a moment without date or duration, extension or succession, past or future, before or after, "having the whole of its existence simultaneously, which is the nature of Eternity".[34]

(f) Life: The holistic life is a totality. It is a homogeneous organic wholeness in which everything is related to every thing else. The inter-relatedness of everything that lives and moves is the content of complexity. Life is always Whole, the Wholeness of Life can never be fragmented. It is non-divisible, non-fragmentable Wholeness. Knowledge cannot transform the inner core of our being. Knowledge is sterile. It sees only a part.

Character is to live according to the laws of nature. There is really a crisis in character – individually and collectively. The sense of values – which have their base in the highest quantum energy potential of consciousness – is lost. Character building activities are not the functions of the government. Characters are not built up through legislation.

Meditation is not the activity of the mind or the body. It is a state where the conditioned mind, the movement of the past in us, the movement of the subconscious and unconscious along with the conscious mind voluntarily discontinues, stops and there is a total silence of the conditioned mind. It is a state of motionlessness as far as the psycho-physical part of our being is concerned. It is a state born of non-duality and sustained by non-duality, it is a state of timelessness. So from activity to non-action, from speech to verbal silence, from movement to non-motion, we become voluntarily silent. Meditation is defined as the art of living in an "open science".

Meditation is awareness of the time and timeless dimension, of motion and motionlessness, of form and formlessness, of sound and silence, of darkness and light, of birth and death – the two together constitute the wholeness of life. "An open science" introduces us to the motionlessness, formlessness. Meditation tells us that man is an amphibian who simultaneously lives in two dimensions – time and timelessness, sound and silence, and so on. In philosophical language, meditation introduces us to our pure ISNESS (Cosmic Consciousness, Life Field) which has never been conditioned and shall never be conditioned.

11. Understanding

Everybody is aware of his own thoughts, so he is self-conscious. Thinking is indirect perception, so it is repetitive and mechanistic; it is the past; so thinking can never give the understanding. Right thinking comes with self-knowledge. Without understanding yourself, you have no basis for thought. Without self-knowledge, what you think is not true.

Understanding in Objective Science: Arthur Eddington wrote, "The stuff of the world is mind-stuff .... The mind-stuff is not spread in space and time .... Recognizing this the physical world is entirely abstract and without "actuality" apart from its linkage to consciousness, we restore consciousness to the fundamental position".[35] Obviously, the material world "can neither act on it nor be acted by any of its parts". In other words, in objective science there is no exchange of energy between the objective world and the subject of cognizance. The *limited* objective science gives us a modified picture of the object with higher sensitivity, so it leads to artifacts with higher sensitivity. It cannot give us the subjective truth. The world view generated by objective science is limited because it focuses attention on the material aspects of life and on technical solutions to human problems. The logical and rational is a very important element in objective science.

Understanding is not an intellectual affair. There is understanding only when the mind and the heart are in perfect harmony. Understanding means love and the sanity of the mind. And love cannot be practiced or taught. The sanity of the mind can only come where there is clear perception, or direct perception, which flows from a non-dual mind or timeless frame of reference. Schrodinger wrote, "The mind is always *now*. There is really no before or after for the mind. There is only a *now* that includes memories and expectations".[36] Understanding flows from the *now*, the *present*, the *non-dual* frame, the *timeless* frame.

There is an art of listening. To be able to listen one should abandon or put aside all prejudices, pre-formulations and daily activities. We listen really to our own noise, to our own sound, not to what is being said.

The objective science gives us knowledge of the objective truth, so it is partial. It has no capacity to bestow us the perception of the wholeness of life.

Direct Understanding of the Wholeness of Life: Many eminent scientists – including Albert Einstein, Werner Heisenberg, Erwin Schrodinger, Arthur Eddington, and so on – have stumbled upon the culminating insight of 400 years of persistent and consistent scientific research of *another mode of* *knowing* Reality. Arthur Eddington wrote:

We have two kinds of knowledge which I call symbolic knowledge and intimate knowledge .... [The] more customary forms of reasoning have been developed for symbolic knowledge only. The intimate knowledge will not submit to codification and analysis; or, rather, when we attempt to analyze it the *intimacy is lost and it is replaced by symbolism*.[37]

Eddington call the second mode of knowing "intimate" because the subject and the object are intimately united in operation. Alan W. Watts wrote that "unconventional knowledge (aimed at) the understanding of life directly, instead of in the abstract, linear terms of representational thinking".[38] We have three basic modes of knowing – (a) symbolic, or map, or inferential, or dualistic knowledge; (b) objective truth; and (c) subjective truth. The subjective truth is synergistic to objective truth. The subjective truth is also called intimate, or direct, or intuitive (insight), or non-dual knowledge.

Truth cannot be given to you by somebody. You have to discover it; and to discover there must be a state of mind in which there is direct perception. Understanding comes through being aware of *What is*. We can have understanding of *what is*! when we recognize it without condemnation, without any comparison, without justification, without identification.

Resolution of Human Paradox: The answer to human paradox, in the words of Albert Einstein, is "Objective Observing and Understanding" the inner world. The observing is reaction-free perception. A non-subjective perception or direct perception is the foundation of scientific life or true life.

Unconditional and total relaxation means that there is no pressure on the body chemistry, no tension on the neurological system. This is the substance of relaxation or silence. It is in the state of silence with unconditional emptiness that the other, the Unnamable, the Immeasurable, the Infinite, the Eternal gets energized. It resolves all challenges, and ushers a new era of order, harmony, nonviolence, peace and bliss.

12. A New Way of Living

(a) Personal discovery of truth is the essence of religion, and the religious person is one who has fearlessness to live the truth, come what may. One lives fearlessly with faith in life and trust in one's own understanding, with a sense of fellowship with all the species – human and non-human – one walks through the corridors of pleasure and pain, honour and humiliation, because life is for living.

By offering our life for that psychic mutation to take place, we might be creating a living cell of a new life style, a new way of living. We might be setting into the orbit of human consciousness a spark of love and compassion. We embark on this enquiry on behalf of the whole human race.

(b) ***Individual and Relationship:*** The relationship between "individual man" and nature is very holy. We have to learn to read the book of the past memory with clarity without any interpretation, without any evaluation. We have to educate our physical organism and psychological structure to move into non-action gracefully and voluntarily. It is a quantum jump which takes place through discontinuities and not through a linear path.

Our brain, thought structure and memory does not belong to individual man. J. Krishnamurti wrote, "I am the conclusion of all the experiences of mankind. The story of mankind is me .... I am humanity, not the collective". He added, "Our brain is a kind of machine and it grows old because we live with friction, with conflict".[39] We have to learn to read the book without any interpretation.

The objective science gives a partial picture of the objective world of materialism. It constructs artifacts with higher sensitivity. The subjective realm expresses itself as "direct perception". It expresses itself as perceptive "Intelligence", which bestows order, harmony, nonviolence and peace.

Man is an amphibian who lives simultaneously in two dimensions – (i) the *known*, the time-space-causation matrices, and (ii) the Unknown, the timelessness. Nature, out of its benediction, has bestowed two hidden dimensions – the understanding and love – to every man. Depending on his faith (and education), he can regenerate or rejuvenate any, or both, potentials for his fulfillment.

(c) Individual and Self-Identity: Eugene Wigner, Nobel prize in Physics, 1963, stated, "Self-identity demands consciousness. You would not have identity without consciousness. Consciousness creates identity. I know that I am I. And it is my consciousness which tells me that".[40] Fritjof Capra wrote, "By consciousness I mean the state of mind characterized by self-awareness, where one is aware that he is aware".[41] Fred Alan Wolf wrote that consciousness "performs a dual role in the universe. In the world of the quantum, it is both the awareness and the creation of experience. With the stroke of the twentieth-century quantum eraser, the dividing line between ontology (theory of being) and epistemology (theory of knowing) is rubbed out".[42] He added, "We are beginning a new age of awareness, the age of quantum consciousness, the age of the conscious atom. By looking within ourselves, we may be able to solve the problems facing us on the final frontier – the frontier of the human spirit".

To sum up, man at individual level, does not have any original activity at the waking state of consciousness. But nature, due to its benediction has embedded two potentials (Understanding and Love) in man which he can rejuvenate at the "highest level of consciousness"; or in Albert Einstein's words, "Objective Observing and Understanding". This bestows new relationship between man and nature, and between man and man.

13. Concluding Comment

(a) The common waking state or psychological frame of reference is repetitive and mechanistic. It is the past, and there is no understanding or creativity.

(b) The "Objective Observing and Understanding" bestow on us a capacity for higher awareness which enable us to rise above our terrestrial limitations, and finally to approach the possibility of transcending the throbbing abyss of space and time itself. This is the first quantum jump from dualistic state to non-dual state.

(c) The objective science is aware of the objective fact, an objective reality. It is partial and does not deal with the wholeness of life. It has a limited understanding of the objective world. In objective science, there is no energy transfer between the subject and the object, so there is awareness of the objective truth.

(d) We have to restructure science so that it takes in its realm both the objective and the subjective dimensions of life, and has the following dimensions:

[Science-Education-Ethics-Health]

The person is chemically poised and nervously relaxed, and the energy moves with the highest quantum energy potential. It is a movement in the non-duality, a movement of the whole being. Energy is sensitivity; energy is intelligence. The person then is an effortless, choiceless awareness of the infinite movement of life. It deals with the wholeness of life. It bestows discoveries. It is called an "open science".

(e) Meditation is a revolutionary dimension in which it is possible to grow and live, and whenever it is necessary to move from that into human relationships. It is not the end, it is not the destination, it is rebirth. We move from activity to non-action, from speech to verbal silence, from movement to non-motion, we become voluntarily silent.

(f) We have to understand that to be religious or spiritualist, we have to meet every relationship with the awareness of the presence of the Divine. If we can feel the presence of the immeasurable, the unnamable totality around us – even when we are busy with the particulars – then the very complexion of our behaviour will change.

(g) There is an urgency for inner revolution and there are two parallel roads – one is understanding and the other is love. Both understanding and love are born of non-duality and sustained by non-duality. Love is that state in which the thought process, as time, has completely ceased. Where love is, there is transformation.

(h) Man has to grow into the condition in which a society will be based on love, friendship and cooperation where social, economic and political order will be free from exploitation, corruption and violence. This is the art of living in an "Open Science"

References

1. George Wald, in Kishore Gandhi, *The Evolution of Consciousness*, New Delhi: National Publishing House, 1983, p. 14.

2. Jacques Monod, in *Chance and Necessity*, New York: Vintage Books, 1972, p. 171; in *Synthesis of Science and Religion*, Mumbai: Bhaktivedanta Institute, June 1988, p. 127.
3. D.S. Kothari, *Some Thoughts on Science and Religion*, The University of Delhi: Oct. 31 & Nov. 3, 1977, pp. 41, 28.
4. Sampooran Singh, *A Review of Symbiosis of Science and Spirituality*, Dagshai: The Atam Science Trust, October 7, 2002, p. 2.
5. Charles Sherrington, in D.S. Kothari, *Some Thoughts on Science and Religion*, *op. cit.*, p. 46.
6. Wilder Penfield, in Deepak Chopra, *How to Know God*, London: Rider, 2000, pp. 215, 217.
7. Sampooran Singh, *An Open Science: The Quantum Jump of Science to Spirituality*, unpublished.
8. *Report to the House of Representatives, USA: Committee on Science and Technology*, June 1981, in R.G. Jahn, February 1982, Proceedings of the IEEE, Vol. 70, No. 2, p. 165.
9. *Union of Concerned Scientists, World's Scientists' Warning to Humanity*, USA: 26 Church Street, Cambridge, MA 02238, April 1993.
10. *World Parliament of World's Religions*, USA: Chicago, 1993.
11. Albert Einstein, *I Believe: Nineteen Personal Philosophies*, London: Unwin Books, Postscript, 1962, pp. 29-31.
12. R.P. Kaushik, *Towards a New Consciousness*, Delhi: Darshan Yoga Sansthan, January 1974, pp. 63-71.
13. Vimala Thakar, *Contact with Vimala Thakar*, Holland: Nr. 19, April 1986, p. 23.
14. J. Krishnamurti, *The Urgency of Change*, London: Victor Gollancz Ltd., 1979, pp. 10, 11.
15. George Wald, in Kishore Gandhi, *The Evolution of Consciousness*, *op. cit.*, pp. 6, 7.
16. Brown Spencer, *Laws of Form*, New York: Julian Press, 1972, pp. 104-05.
17. Vimala Thakar, *Silence*, Blaricum: Holland, 1989, p. 6.
18. *ibid.*, *Through Silence to Meditation*, Ahmedabad: Vimal Prakashan Trust, 1998, pp. 22-27.
19. T.H. Huxley, in Ken Wilber, *The Holographic Paradigm and Other Paradoxes*, Shambhala, 1982, pp. 15-16.
20. Fritjof Capra, *Uncommon Wisdom: Conversation with Remarkable People*, New York: Simon & Schuster, 1988, p. 232.
21. F.M. Lehmann, *Two Kinds of Natural Science*, The Scientific and Medical Network Newsletter, August 1993, No. 52, p. 17.
22. Gilbert Blane, in F.M. Lehmann, *ibid.*.
23. Ilya Prigogine and Isabelle Stengers, *Order Out of Chaos*, New York: Bantam Books, 1984, p. 5.
24. Erwin Schrodinger, *What is Life*? and *Mind and Matter*, Cambridge University Press, Ed. 1967, p. 128, 131, 127.
25. David Bohm, *New Scientist*, November 11, 1982, p. 364.
26. J.W.N. Sullivan, *The Limitations of Science*, New York: Mentor Books, 1949, p. 140.
27. E.N. da C. Andrade, *An Approach to Modern Physics*, New York: Doubleday Anchor Books, 1957, p. 255.

28. George Wald, in *Synthesis of Science and Religion*, *op. cit.*, p. 19.
29. Erwin Schrodinger, *op. cit*, p. 131.
30. Charles Sherrington, in Erwin Schrodinger, *ibid.*, p. 128.
31. Deepak Chopra, *How to Know God*, *op. cit.*, pp. 271-72.
32. Erwin Schrodinger, *op. cit.*, pp. 128, 148.
33. Deepak Chopra, *Ageless Body, Timeless Mind*, New York: Three Rivers Press, 1998, pp. 36, 8.
34. Ken Wilber, *The Spectrum of Consciousness*, USA: The Theosophical Publishing House, 1982, p. 98.
35. Arthur Eddington, *Nature of the Physical World*, Cambridge University Press, pp. 276-77.
36. Erwin Schrodinger, *op. cit.*, p. 145.
37. A. Commins and R.N. Linscott, ed., *Man and the Universe, The Philosophers of Science*, New York: Washington Square Press, 1969, p. 453.
38. Alan W. Watts, *The Ways of Zen*, New York: Vintage Books, 1957, p. 10.
39. J. Krishnamurti, *KFI*, 1982, No. 1, p. 40; and 1983, No. 3, pp. 27-29.
40. Eugene Wigner, *Synthesis of Science and Religion*, *op. cit.*, pp. 257-64.
41. Fritjof Capra, in *ibid.*, pp. 272-76.
42. Fred Alan Wolf, *Taking the Quantum Leap*, San Francisco: Harper.

24
The First and Last Letter to Mankind

Abstract

The whole mankind is caught in a crisis of perception, of conditionings. These are leading mankind to self-annihilation.

The answer to this paradox is embedded in Albert Einstein's prophetic words, "Objective Observing and Understanding"; or in Max Planck's words – a single point of our "individual awareness"; or in Edwin Hubble words – there is a "world of pure values"; which flows from insight (intuition).

The Prime Minister has constituted a "National Knowledge Commission" on June 13, 2005. Our "intuitive-insight" for human survival is: (a) regeneration of an "open, pliable and creative Mind" which is basically time-free, concept-free and measure-free. A creative mind is alert, attentive and sensitive; and (b) it is the growth of a person into a higher level of Consciousness, or quantum energy potential of Consciousness.

Humanity needs to have direct perception and to live in that higher awareness. The awareness energises the intelligence, which resolves all challenges and bestows order, harmony, nonviolence, peace and bliss.

1. In modern physics, a very different attitude has now developed. Physicists have come to see that: (a) all their theories of natural phenomena, including the 'laws' they describe, are creations of the human mind; properties of our conceptual map of reality, rather than of reality itself; (b) the model is colourless, soundless and unpalpable, which implies that the physical world picture lacks all the sensual qualities that go to make up the subject of cognizance; (c) the objective outside world excludes the subject of all cognizance or adequate understanding of the subject of cognizance, and (d) as all human perennial values are embedded in the subjective dimension, so in the objective world the values are missing and quite particularly meaning and end are missing. Erwin Schrodinger, Nobel laureate in physics, called this as "Principle of Objectivation", or in other words, this procedure leads to objective world, and there is no room for the subject of cognizance and its immediate structures.

2. Intellect gives us objective truth so it ignores the subjective world. Intellect is invariably dualistic, so it has no capacity to gather *true* knowledge,

and it is aware of *distorted* or *mutilated* knowledge. It leads to invention in the technological field. It is an improvement in the field of the known to the known. It suggests improvement in the field of higher sensitivity of matter, and thus science can construct better aeroplane, cars, frigidaires, air-conditioners, more efficient use of war machines, and designs and constructs artifacts with a greater sensitivity. Science deals with man-made social cosmos or synthetic construct of the mind-brain-body system.

3. Science cannot resolve the psychological imbalances, ecological crisis and other human challenges. Albert Einstein wrote:

(a) The significant problem we face cannot be resolved at the same level of thinking we used when we created them.

(b) Humanity needed to reach a higher level of awareness to solve the problems of nuclear war.

4. Sri Aurobindo, the greatest philosopher-mystic, wrote:

(a) In 1911, Sri Aurobindo in the *Song of Humanity* wrote:
A time will come when the Indian mind will shake off the darkness that has fallen upon it, cease to think or hold opinions at second and third hand and reassert its right to judge and enquire with perfect freedom into the meaning of its own culture and tradition.

(b) The evolution of mind working upon life has developed an organization and use of matter which can no longer be supported by human capacity without an inner change. The problem is fundamental and in putting its evolutionary Nature is confronting man with a critical choice which must one day be solved in the true sense if the race is to arrive or even to survive.

(c) Sri Aurobindo in the message on August 15, 1947 stated, "A catastrophe may intervene and interrupt and destroy what is being done, but even then the final result is sure .... A new spirit of oneness will take hold of the human race".

(d) The Divine Mother in 1963 wrote, "The only hope for the future is a change of man's consciousness and the change is bound to come". The message in 1967 read, "Men, countries, continents! The choice is imperative: Truth, or the abyss". The Mother had once observed, "Three-fourth of humanity is obsolete".

Science deals with gathering of fragmentary experience and partial knowledge (information), so it has no potential to resolve human challenges. We have to restructure science to a new dimension – science-education-health – of the subject of cognizance, and this energises Intelligence/Wisdom.

5. Albert Einstein, in prophetic words, indicated the conditioning of the psyche in the first five decades of the twentieth century, in words:

(a) Einstein – **Out of His Later Years** – wrote, "The confidence in the stability, even the very basis of existence, of human society has largely vanished. One senses not only a threat to man's cultural heritage but also that a lower value is placed upon all that one will like to see defended at all costs .... Life in human society meant dangers of all sorts .... even standards of virtue, aspiration and practical truth were taken for granted as an inviolable heritage, common to all civilised humanity .... The sanctity of life vanished .... The lie was raised to the dignity of a political instrument. It was a manifestation of weakness of character and lack of responsibility on the part of the leaders in the affected countries". This presents evidence that the human psyche during the first five decades of twentieth century got heavily conditioned leading to lowering of its sensitivity.

(b) Einstein – Out of His Later Years – wrote, "Man changes but little .... though current trends like the present bring him unimaginable sorrow. Nothing of all that will remain but few pitiable pages in history books, briefly picturing to the youth of the future generations the follies of its ancestors". Perhaps, the benign turn of destiny, who have faith in the mysterious will be saved.

(c) The accelerative conditioning of the psyche continued from 1950 to date. The recent challenges are terrorism, organised crime, religious fundamentalism, ethnic conflicts, corruption and so on. The conditionings have gone from the conscious to the sub-conscious level. We have reached a critical stage in conditioning the psyche.

Einstein painted a very grim picture of the future of mankind based on his perception of the Reality. And man has to face this situation. Many eminent scientists – George Wald, Nobel laureate in Medicine, 1967; Hideki Yukawa, a Japanese NL in Physics; Jacques Monod, NL in Biology, also talked that the world is heading towards self-destruction or self-annihilation. The fundamental cause as to why science and technology has failed mankind to lead to 'peace and serenity' is that he started living in the external world of time-space-causation (objective world) and rejected the Reality (the subjective dimension).

This implies that knowledge (information) proliferates while wisdom atrophies. This is a serious crisis. This crisis has enhanced between 1950 to date. At present, at all levels, there is a growing realisation that the existing models of education in India and in the rest of the world have failed and there is a total lack of relevance between the human being and the complex contemporary society. The increasing human challenges are forcing man inevitably to face the realities of the human situation. Many eminent philosophers-mystics have pointed out that the challenge is not only addressed to the structure of education, but to the nature and quality of man's mind, his way of living, consciousness and life.

6. Pupul Jayakar, in an essay entitled "*Changing Horizons, 1962-1977*" wrote, "A general destructive violence, a fragmentation at all levels, and a growing lack of sensitivity was permeating the landscape in India. J.

Krishnamurti said, "One observes this (violence) .... happening around us socially, what is happening at the periphery of life, but also from what is happening inwardly".

J. Krishnamurti met Jawaharlal Nehru and Acharya Kripalani in 1931 and said, "If they ignored inner regeneration in the struggle for independence, India would lose its way". It appeared that the Indian mind was degenerating and this is continuing at an accelerative pace during the last 74 years and it appears that now we have reached an abyss. India must wake up from deep slumber to regenerate or rejuvenate the Indian mind.

J. Krishnamurti said, "India has lived all these centuries on ideas. She has to move from a life based on these ideas, which are non-facts, to seeing of the fact only". P. Krishna, an eminent physicist, in a talk at California on April 3, 1993 said, "We are going towards self-annihilation and not too much time is left now".

The accelerative growth of violence inwardly is happening but no steps appear to have been taken to end the challenge of violence, fear, acquisition, psychological imbalances. The answer to this paradox is mutation of the psyche – *mutation is possible in the brain through insight, not through knowledge, not through accumulation. That insight is the ability to see a truth directly, a direct perception of the truth, and education must foster that.*

7. A Scientific Enquiry: A scientific theory is always provisional. It is a guide to further development, to its own replacement. It deals with the object and ignores the subject. All great discoveries have arisen from the realm of the timelessness, so scientific enquiry is not a cerebral activity. D.S. Kothari wrote, "The new theorems, unless trivial, are of course first discovered and then proved". Discovery is a flash of insight. It is not a step by step process, but a sort of quantum jump. A rational account of discovery, the proof, follows and not precedes the discovery. Fundamental discoveries which go deep into the nature of things are the result of uncommon intuition and faith, not cold logic. Albert Einstein stated, "The most beautiful experience (perception) we can have is the mysterious. It is the fundamental emotion which stands at the cradle of true art and true science". All creativity in science flows from a *deep faith* in the mysterious, the Cosmic Consciousness, the Life Field. It is also called your own *inner quantum energy potential*, or subjective dimension.

The advancements made in material science has involved a progression from simple materials to more complex form of technology and engineering, and we are today at the convergence of Nano, Bio and Information technologies. The material technology leads to conditioning and lowering of sensitivity of mind, and has no correlation with the emergence of creative scientists. *Mankind needs a direct perception, that lay beyond the conditioned mind, beyond brain, memory and all responses of consciousness.*

A state where observing is whole, inclusive, unfragmented. This is the emergence of a new human being, a creative being, who lives beyond conflict, action of will and assertiveness.

The Resolution of the Human Paradox

8. The fundamentals of creativity in science and technology [or "generation of innovations and support system to multiple application] is as under:

(a) When one is aware of the "objective truth", and ignorant of the "subjective world", the mind is engaged in "invention" (materialistic approach). If one is simultaneously aware of the subjective world (inner) and the objective world (outer), then the mind is engaged in discovery. Discoveries come as intuitive flashes.

(b) All knowledge is structured in Consciousness, and Consciousness bestows creativity. It is the expression of pure consciousness in a pure Mind. Creativeness happens when one lives in communion with nature. Creativeness, discovery, direct understanding of life, intuition, insight, are synonyms. Intuition over a long chronological time flowers to Intelligence or Wisdom.

(c) Understanding comes into being from the *now*, the *present*, which is always timeless. It comes only when the mind is very quiet and still. In stillness there is no projection of thought. We can have understanding of '*what is*' when we recognise it without condemnation, without justification, without identification. Einstein stated that we need "objective observing and understanding", which flows to "peace and serenity".

The above attributes of creativity are identical to the Einstein's perception, who defined that all creativity in science flows from the mysterious, the Life Field.

9. Scientists are now talking of "inner space labs". Houston said, "In high-level creativity, the mind can do weeks of work in a short time". Fritjof Capra wrote, "The rational part of research would, in fact, be useless if it were not complemented by the intuition that gives scientists new insights and makes them creative". A.N. Whitehead wrote, "It is no exaggeration to say that the future of our civilisation depends on the degree to which we can balance the forces of science and religion. Bryce-Smith wrote, "We face an evolutionary challenge that requires a moral response: we shall survive only if we can show that we deserve to survive".

10. It seems to be imperative that "to educate the educator", or "bestowing awareness to scientists" is the first priority. Discovery is the beginning of creativeness; and without creativeness, do what we may, there can be no peace or happiness for man. Creativeness comes into being when there is constant awareness of the ways of the mind, and of the hindrances it has built for itself.

11. The "Creativity in Science" that blossoms to "Symbiosis of Science and Spirituality". This results in higher level of awareness, or higher hierarchical expression of Consciousness, or regeneration of Intelligence. It is the harbinger of new era of freedom, peace and bliss. This guarantees human survival and progress.

12. A Cosmic Plan

Some physicists are deeply inspired by the remarkable harmony, order and unity of nature that recent advances have uncovered. Fred Hoyle, in a pithy phrase, says, "The universe is a put-up job". Many eminent philosophers/mystics discuss the concrete actuality of the "territory" and the essence of their perception is as under:

(a) The Cosmic Consciousness or the Fundamental Vibration is not yours or mine. It is the consciousness of all mankind and all existence. David Bohm says, "Deep down the consciousness of mankind is One". Every existence and event exist in the matrix of consciousness and are permeated by consciousness.

(b) What we call mind, or the successive consciousness, or the scattered vibrations is a conditioned energy moving through the body, monitored by brain, operating through the senses. The self-conscious energy is a structured movement, it is a movement in the framework of patterns and structures, and are shared by all human beings.

(c) All human beings suffer, go through great agonies, shed tears having the sense of loneliness, pain, anxiety, depression, uncertainty. The content of our consciousness is our beliefs, tendencies, desires, anxieties, loneliness and so on. Each one is the conclusion of all the experiences of mankind, the story of mankind. He is neither individual nor collective. He is humanity.

(d) The cultural belief structure or memory is shared by mankind. Thought and its meanings are shared by mankind. Through thought, the total human past gets expressed through an individual, but it does not belong to that person. They are not "your" thoughts though they get expressed through you with a tremendous momentum. Thinking is never original.

(e) The survival of mankind is embedded in man's learning and living the harmonious whole movement of seeing, of hearing, of acting, of learning, and of meditating. There is one harmonious whole movement. There is one movement of consciousness at a certain hierarchical level or quantum energy potential.

(f) Albert Einstein stated, "The most beautiful experience we can have is the mysterious. It is the fundamental emotion which stands at the cradle of true art and true science". These flow from "pure perception".

(g) Nature invariably expresses itself as "*direct perception*". David Bohm, an eminent physicist wrote, "The ultimate perception does not originate in the brain or any material structure, although a material structure is necessary to manifest it. The subtle mechanism of knowing the truth does not originate in the brain. Similarly man has to learn to express himself as '*direct perception*', as man's truth is fundamentally the same as that of nature, which implies that man has to learn and live according to the immutable laws of nature. This only is the social, psychological and even the political revolution. This is the only vocation of mankind.

(h) The Mind is "open, pliable and creative", and it expresses itself in a non-dual and non-conceptual frame of reference of Mind continuum, which is basically time-free, concept-free and measure-free. A creative mind is alert, attentive and sensitive. It is reaction-free attentiveness. It expresses itself as the wholeness and oneness of life, so it is holy and sacred.

The cosmic plan is that every existence and every event are vibrations of consciousness at different hierarchical levels or different quantum energy potentials. There is Oneness that governs the whole cosmos and beyond cosmos. It is a benediction to be alive. It is a blessing to be alive and have the opportunity to relate ourselves to the life within and around us. It is a privilege to be alive in a human form where self-knowing and self-discovery is possible, where relationships are possible.

Concluding Comment

13. We intuit that it is still time to introduce remedial measures for "Creativity of Science and Technology" and "Symbiosis of Science and Spirituality". In Einstein's words, the solution is embedded in "objective observing and understanding". It evidently implies that we will have to draw an Action Plan to achieve "Creativity in Science?", or to be "a continuous innovator and creator of science and technology". This guarantees our survival.

14. Einstein stated, "Who would have thought around 1900 that in fifty years time we would know so much more and understand so much less .... It is in fact, nothing short of a miracle that the modern methods of instruction have not yet entirely strangled the holy curiosity of inquiry; for this delicate plant, aside from stimulation, stands mainly in need of freedom; without this it goes to wrack and ruin without fail".

15. There is an urgency "to initiate an open dialogue with an open mind" with other eminent scientists of India. We have to initiate "objective observing (seeing), objective listening (hearing) and understanding" of the wholeness of life. It is emphasised that "objective observing and understanding" rejects the psychological symbolic-dualistic world, and is embedded in the non-dual and non-conceptual frame of reference of Mind Continuum. And naturally, the first step is a quantum jump from symbolic-dualistic frame of reference to non-dual frame, or from intellect to intuition, or from reason to quantum logic.

16. During the last 10,000 years, or more, man has been trying to end all violence, but he has utterly failed. We have built a society which is violent. Man has emerged as a violent species. The challenge now is to transcend violence and live in order, harmony, nonviolence, peace and bliss.

17. There is an urgency to understand the mind-brain-body system – we

need psychological revolution, in which the whole human race has to participate and every man is an equal partner. This revolution has the potential to regenerate and rejuvenate human conditioned psyche, so that it perceives the new dimension of an integral and all-inclusive Consciousness, that can guarantee human survival, human excellence, human fulfilment.

18. The symbolic-dualistic frame of reference of mind spectrum constructs time-space-causation matrices. Einstein wrote that "his thoughts and feelings are something separated from the rest – a kind of optical delusion of his consciousness. This delusion is a kind of prison for us .... Our task must be to free ourselves from this prison", or in other words to transcend his thoughts and I-consciousness.

19. The destructive conditionings can be ended through the art of objective observing and understanding. The individual holds the key to averting disaster. The truth can be discovered through self-observation and the mirror of relationships.

20. It is the Cosmic Consciousness that governs the mutations of the mind-brain-body system. It activises the Intelligence which expresses as order, harmony, nonviolence and peace.

21. The first epoch of the 21st century grants India the last chance to regenerate her lost heritage and to reinstate itself as the pioneer for ushering a new golden age of mind-brain-body research. Einstein wrote, "The chief motivating force behind all technical effort ought to be a supreme concern for the destiny of the human race". It is a unique service to mankind. It has the potential to steer the ship of humanity from confrontation to sharing, from annihilation to survival.

22. And if man escapes his responsibility to bring mutation of his psyche, it will be at his own peril and that of India and mankind.

Part-IV
A New Vision on Earthquakes and Tsunamis

The word "tsunami" is from the Japanese words *tsu* (harbor) and *nami* (waves). A tsunami is a wave or series of waves in the ocean that can be hundreds of miles long and have been known to reach heights up to 34 ft (10.5 m). These "walls of water" travel as fast or faster than a commercial jet. The massive December 26, 2004 tsunami traveled 375 miles (600 km) in 75 minutes. That is 300 mph (480 kph). These walls of water are capable inflicting massive damage along coastal lands.

On December 26, 2004, a massive underwater earthquake off the coast of Indonesia's Sumatra island rattled the Earth in its orbit. The quake measuring 9.0 on the Richter scale is the largest one since 1964. Dozens of aftershocks with magnitude of 5.0 or higher occurred in the following days. The death toll has reached higher than 2,83,000, and many communities suffered devastating property damage.

Some scientists believe that Earthquake/Tsunami are frightening weaponry unsheathed by Nature. These devour countless lives and leave a trail of misery and suffering. The concept of 'fury', or 'wrath' of nature, or 'revenge' of nature contradicts all known scientific and philosophical-mystics knowledge/insights, which suggests Nature is benevolent and its direct perception is "order and harmony", "non-violence and peace", and "love and compassion". There is no explanation in literature regarding the cause of suffering – misery, disease, loss of life and property – happening on a gigantic scale in a few moments.

Billions of men are constructing violence at every instant and throwing it in the external environment. We know nothing how this violent energy, when it becomes critical in the environment, gets destroyed or dissipated. This is governed by the immutable laws of nature.

We investigate "a new vision on earthquake/tsunami". The vision, in brief, is that the human conditioned mind constructs violence, which is thrown into the environment, and this violence, in some mysterious way gets dissipated as earthquake/tsunami. We have not been able to give any experimental evidence to prove our hypothesis or to end or lessen the hazards; but one important fact is that the hypothesis does help to ameliorate the human suffering.

The hypothesis is supported by eminent scientists – say Albert Einstein – and eminent saints/sages – say, Mahatma Gandhi. Sri Aurobindo, Vimala Thakar, Lama Govinda, John Selby, Tolly Eckhart are also quoted. Lord Buddha's proclamation is also presented. Every word uttered by these saints/ sages is in the timeless realm. Nature's every expression is in the timeless realm. It suggests strongly to me that there is no scope for a great calamity, or catastrophe, or a tsunami, except human failure to deal with the crisis of transcending violence.

Violence is an indication of psychological immaturity. Physics concludes that violence is a construct by symbolic-dualistic frame of reference of mind spectrum; and non-violence is a direct percept in the non-dual and non-conceptual frame of reference of the Mind Continuum. The non-dual frame is synergistic or at higher quantum energy potential to symbolic-dualistic frame. We have to understand and learn a quantum jump from violence to non-violence, from symbolic-dualistic frame to non-dual frame, from reason to intuition. And in this is embedded our survival.

President A.P.J. Abdul Kalam visited St. Petersburg [The Tribume, May 26, 2005] and called for "pooling of scientific knowledge and resources between the countries to enhance capabilities for the predictim of earth quakes, tsunamis and other national disasters". A dozen joint research projects have been formulated.

25
Tsunami: Its Moral Implications for Humanity

Abstract

The physical evolution and biological evolution over billions and billions of years have shown that there is harmony and order in all events/existences. There is love and compassion and nature's benediction flows everywhere. Many scientists and philosophers call the earthquakes and tsunamis as due to nature's fury or wrath; it causes disharmony, disorder and chaos; it gives pain, misery and suffering to mankind; but this contradicts all known scientific and ethical/spiritual knowledge. We have to undertake an exploration of the cause and the nature of calamities, catastrophes, earthquakes and tsunamis.

A true scientist, or a deep religious man, living in non-dual and non-conceptual frame of reference of Mind Continuum perceives that the physical calamities are not constructs by Nature (Cosmic Consciousness, Life Field), but are constructs of the conditioned mind-brain-body system, or inheritance of man. The catastrophes are signals of Nature to warn man that his living is antagonistic to the Laws of Nature, and he must purify his psyche.

Observing (seeing), listening (hearing), understanding and learning, action is direct; while thinking, and psychological inheritance is indirect. Nature's expression is invariably direct. All inanimate, physical, animate, psychological structures suggest exceedingly ingenious design, a purpose, behind every event/existence. Modern physics suggests that the purpose includes us.

The conditionings of the psyche during the last 100 years have increased manifold. The conditionings have now travelled from the conscious level to sub-conscious level. The conditionings express themselves as violence, fear, psychological imbalances, ecological crisis, and other human challenges. The conditionings cannot be resolved through analysis or intellectual introspection, but can end through understanding and learning.

The mutation in Consciousness happens when the "restructured science-education-health" observes the conditioned psyche and the Unconditioned Energy, or in other words, the wholeness of life. This resolves the age-old tide of I-consciousness, thought-structures, psychological imbalances, ecological crisis, and other human challenges. This guarantees human survival and excellence.

We have to open an institute to explore "Science with Creativity" and "Symbiosis of Science and Spirituality". The words of the "Union of Concerned Scientists (April 1993) are true today that "This ethic must motivate a great movement, convincing reluctant leaders and reluctant Governments and reluctant peoples themselves to effect the needed changes".

Modern science has dramatically shown that man (*Homo sapiens*) is an amphibian who lives simultaneously in two worlds – (a) the *known*, the successive consciousness; and (b) the *Unknown*, the Simultaneous (Cosmic) Consciousness. The successive consciousness constructs the time-space-causation matrices, so man has divorced the Reality, the Timelessness, the Life Field; and this expresses itself as conflict, pain and misery. The objective science and technology bestows on us materialistic benefits, but this has not contributed to ending of psychological imbalances, and bestows on us joy, non-violence and peace. In the twentieth century, an attempt has been made to jettison all the conflicts, but we have miserably failed.

Mahatma M.K. Gandhi obtained political independence in India by non-violent means. Through his life and teachings, he bears testimony to the values of Non-violence and Truth – which have moral/spiritual implications. A review of his teachings was published in February 1998.[1] A great calamity – known as Bihar Earthquake – which descended on India in January 15, 1934. Nearly 25,000 people died in the earthquake. Many more thousand suffered from injuries sustained and still more were homeless. Gandhiji worked to ameliorate the suffering. He wrote extensively in the Harijan and called Bihar earthquake as "The Great Calamity". He wrote, "Now why should this calamity come upon us? I request you to think with me".[2] In other words, he wanted free India to embark on research on the instrument of non-violence. It is sad that during the last 57 years, India has not started any exploration on the instrument of non-violence.

When Nature unsheaths its frightening weaponry with full fury, there is pretty little that mankind can do despite its many technical advances. That sort of helplessness and vulnerability was witnessed on December 26, 2004 when huge tidal waves generated by a strong earthquake hit coastal areas of India, Sri Lanka, Indonesia and Thailand and flattened everything that came in the way. This whole episode is called *Tsunami's* terror. The death toll crosses 2,83,000 with many missing, injured and homeless.

This article presents Gandhiji's teachings on the Bihar earthquake of 1934. It discusses the recent Tsunami which happened on December 26, 2004 and its moral and spiritual implications. It explores the relevance of Gandhiji's teachings to mankind facing the massive rehabilitation of men who are sick, teachings to mankind facing the massive rehabilitation of men who are sick, and are now seeking shelter. Perhaps such an exploration, and more so in the light of recent discoveries of science to avert such a great catastrophe are of utmost urgency. Further, mankind has inherited a conditioned psyche, and every activity of the psyche is leading him to increase of entropy, which is death or self-destruction; and perhaps this exploration may lead to human survival and excellence.

Mahatma Gandhi's Teachings in response to Bihar Earthquake

Gandhiji wrote, "Behind even this calamity there is a divine purpose that works for the good of all humanity .... this earthquake is a divine chastisement sent by God for our sins .... that not a blade of grass moves but by the divine will .... For me, there is vital connection between the Bihar calamity and the untouchability campaign".[3]

He further wrote, "Let us derive the lesson from this calamity that this earthly existence is no more permanent than that of the moths we see every night dancing round lights for a few minutes and then being destroyed".[4] He added, "This absolutely unthinkable affliction in Bihar your sins and my sins are responsible.... this affliction has come upon us because of this atrocious sin of untouchability". He advocated that "we have to open the book of life. The root meaning of education is drawing out".

A Critique on Gandhiji's Teachings in the Light of Modern Physics: Let us explore the words "sin" and "untouchability". These are words of a language. The whole language depends on time. Language is the creation of the mind, time is also a creation of the mind. Modern physics has confirmed that the concepts, ideals, language are limited, that these are not features of reality, but creations of the mind – parts of the map, not of the territory. Space and time are nothing but names, forms of thought, words of common usage; these are now reduced to the subjective role of the elements of the language a particular observer uses for his or her description of natural phenomena, each observer will describe the phenomena in a different way.[5]

What Gandhiji expressed in his writings was in the timeless realm and true relationship of man to the Divine. Every word Gandhiji uttered regarding Bihar earthquake is the deepest truth. Gandhiji expressed two relationships – the "*sin*" and "*untouchability*" – which are in time. Both sin and untouchability add violence to the human mind. In the non-dualistic and non-conceptual frame of Mind Continuum, man has to jettison thought-time (concepts) if he wants his survival.

In the modern physics language, the words "*sin*" and "*untouchability*" are replaced by the word "*conditioning*" of the psyche; it implies to store the partial experience and fragmentary knowledge in the psyche and called the cultural belief structure or memory. The cultural belief structure is metastable, so it spontaneously disintegrates and the product is called thought-time or thought-structure. And thought is invariably coupled with volition. This flows to I-consciousness, which does not have any physical reality except conceptual reality or ideational reality. So modern physics advocates jettisoning of I-consciousness and thought-structure. [In Gandhiji's words, the sin and untouchability are considered as conditionings or violence in the psyche]

Enhancing of Conditionings (Violence) in the Human Psyche

In 1974, Pupul Jayakar wrote, "A general destructive violence, a fragmentation at all levels, and a growing lack of sensitivity was permeating the landscape of India".[6] Krishnamurti added, "One sees disintegration, overpopulation, callousness, a growing indifference to man and environment". J. Krishnamurti stated, "India has lived all these centuries on ideas. She has to move from a life based on ideas, which are non-facts, to seeing the fact only".[7] Osho wrote, "The spiritual traditions of the past have made a deep split within the individual, and this is reflected in all the instruments of society .... Modern man is so burdened with the traditions from the past and the anxieties of modern-day living, that he must go through a deep cleansing process before he can begin to discover the thought-free, relaxed state of meditation".[8]

Sri Aurobindo wrote, "A catastrophe may intervene and interrupt and destroy what is being done, but even then the final result is sure .... A new spirit of oneness will take hold of the human race".[9] The message in 1967 of the Divine Mother (Madame Mirra Alfasa) read, "Men, countries, continents! The choice is imperative: Truth or the abyss". Mother had once observed, "Three-fourth of the humanity is obsolete".[10] All this suggest that (a) since 1947, the conditioning or violence has increased tremendously; (b) man has started living in a conceptual world, a world of traditions; (c) lowering the level of sensitivity; and (d) this is leading to self-destruction and self-annihilation.

Sri Aurobindo wrote, "The evolution of mind working upon life has developed an organization and use of matter which can no longer be supported by human capacity without an inner change. The problem is fundamental and in putting its evolutionary Nature is confronting man with a critical choice which must one day be solved in the true sense if the race is to arrive or even to survive".[11]

Krishnamurti met Jawaharlal Nehru in 1931 at which Acharya Kripalani, Achut and Pupul Jayakar were present. The subject for discussion was that the Indian mind was degenerating. Achut said, "Nehru and Kripalani" felt that it was essential for India to be politically free before a regeneration in the Indian mind could take place. Krishnamurti had said that "if they ignored inner regeneration in the struggle for Independence, India would lose its way".[12] The events after Independence clearly indicate that the Indian mind continued its deterioration.

wiped out in the twinkling of an eye. Therefore while we have yet breathing time, let us get rid of the distinctions of high and low, purify our hearts and be ready to face our Maker (Cosmic Consciousness) when an earthquake or some natural calamity or death in the ordinary course overtakes us".[13] Nature is benevolent, its every activity is in harmony, in order and coherence, so "calamity is a blessing in disguise, so there is a Divine purpose that works for the good of all humanity".[14] A natural calamity happens when our mode of living is antagonistic to the immutable laws of nature, so natural calamity is an indirect warning to us to heal our conditioned psyche. It is a reminder to man to undertake radical transformation of the human psyche.

Nature's Trading Account

Violence is an indication of psychological immaturity. It is a construct by symbolic-dualistic frame of reference, and nonviolence is a percept in the non-dual and non-conceptual frame of reference of the Mind Continuum. The non-violent frame is synergistic or at higher quantum energy potential to symbolic-dualistic frame of consciousness.

The Second Law of Thermodynamics states that disorder can never spontaneously give rise to order. More precisely, this law legislates nature's trading account in a quantity called entropy, which is, roughly the degree of disorder in a physical system. When ordered energy becomes disordered, the entropy rises. The device by which an organism maintains itself stationary at a fairly high level of orderliness (= fairly low level of entropy) really consists in continually sucking orderliness from its environment. It is the higher level of attentiveness or awareness that leads transformation from violence to non-violence. Man has to understand and learn to remain at the highest level of attentiveness. It is the moral/spiritual duty of man to keep fairly low level of entropy.

A Cosmic Plan?

Paul Davies, an eminent physicist, proposed a unified theory of all existence. He wrote, "Few physicists have been deeply inspired by the remarkable harmony, order and unity of nature that recent advances have uncovered .... Every advance in fundamental physics seems to uncover yet another facet of order .... There is unexpected harmony and coherence of nature .... We cannot fail to be struck by the ingenuity of the natural world? Nature is astonishingly clever in the way it gets thing to work .... The laws which enable the universe to come into being spontaneously seem themselves to be the product of exceedingly ingenious design .... The universe must have a purpose".[15]

There has been no collision between any two stars in the cosmos. We have to accept perfect harmony, order, symmetry and coherence in Nature. It suggests strongly to me that there is no scope for a great calamity, or catastrophe, or a Tsunami except human failure to deal with the crisis of transcending violence. The moral/spiritual implications for humanity is to transcending violence. The moral/spiritual implications for humanity is to understand the cause of conditioning which leads to violence and to transcend the conditionings in the human psyche. It is felt that billions of men throw the energy of violence which gets locked up in the Earth and after uncountable aeons may erupt as great calamity. We have to learn to live in the fact, not the invented opposite. We initiate Institute for Ending of Violence, or Science with Creativity, or Symbiosis of Science and Spirituality.

Bertrand Russell wrote, "Thanks to science and technology, mankind is united in evil but not yet united for good; that people have learnt the technique of mutual destruction, but not the more desirable technique of worldwide cooperation". Russell believed that "wisdom becomes increasingly necessary with every advance in knowledge and technology, and that although our age has surpassed all others in knowledge, it has not enjoyed a correlative increase in wisdom"; and he called for a "new moral outlook".[16]

Mutation or Radical Transformation of the Psyche

(a) Restructuring of Science-Education-Health: A galaxy of scientists – Karl Pribram, Roger Sperry, John Eccles – have opined, "Brain science could no longer afford to shut out that part of the world which we call subjective". We have to "restructure science-education-health", which takes in its ambient both the objective and the subjective dimension of life. Similarly the dimension of value education is embedded in the trans-sensual dimension of human personality, the Life Field. Again a living in the *present*, the *now* heals and living in the past, the duality suppresses man's sensitivity and enhances disease. We have to restructure science-education-health, which deals with the wholeness of life, and the subject of cognizance is called true-subject.[17]

(b) Ending of I-consciousness: We transfer what is happening at the biological level and transfer it to the psychological level; and this we call ego, or I-consciousness, so the 'I' is not an entity, but a concept. I-consciousness has no factual reality. It has a conceptual reality. It has an ideational reality. We graft the subjective on the perceptual level and transform it to conceptual level.

(c) Ending of Thought-Time: When thought-time comes in the realm of true-subject, the photons meet an 'impregnable wall', or "event-horizon", or "frequency-barrier", and are dissipated, which implies that thought-time goes to abeyance and what is left behind is the *fact*, the *true*, the *present*, the *what is*!

(d) Ending of the Fact: If we continue observing or listening to the *fact*, without justification, or condemnation, or motivation, we observe that there is only the *fact* and nothing else. The fact itself is the operator and not you operating on the fact. If you leave the fact alone, it has no time. The fact is free from fear. The fact grows, which implies that there is release of tremendous energy and the fact moves to higher quantum energy potential. This is expressed as Pure Consciousness → Fact.

(e) Ending of Violence: The violent society is based on laws which are antagonistic to the immutable laws of nature. It is based on callousness and opposition, and not on reverence of life, not on love or tenderness. It is based on acquisition, competition, comparison, exploitation and violence. The violence ends when I-consciousness and thought-structure go to abeyance. This regenerates Intelligence, which perceives life in its wholeness. This guarantees human survival and excellence.

The Future Tsunami, the Great Calamity

Some British and US academics have predicted that volcanic eruption in Spain's Canary Islands could unleash a "mega-tsunami" larger than any in recorded history and may hit the US.[18,19] These predictions have alerted mankind to the great calamity. This has added fear to human mind and thus enhance the conditionings.

The fundamental challenge facing mankind is the conditioned psyche. The conditionings have to be rooted out from our midst in its entirety, for at the bottom of this ending of conditionings, there is realisation of the ending of this calamity and bestow order, harmony, nonviolence and peace. We urgently need the purification of our mind-brain-body system.

Dissipation of the Energy of Thought in the Waking and Dream State

In order to see, the mind first cuts itself up into at least one state (subject) which sees, and at least one other state (object) which is seen. The subject is in the inner realm, and the object in the outer field. Apparently it is one of the most difficult things for a human being to look at things directly, not through images, opinions, conclusions, which are all symbols. This state of mind is called the common waking state. The energy of thought is dissipated in the environment.

Everyone is aware of three common states – waking, deep sleep and dream state. If due to some cause, the deep sleep state releases energy, then to dissipate this energy, nature has designed the dream state. Dream state is a device of the brain to protect the psyche; this is the whole process of thought and I-consciousness. The 'me', the ego, the I-consciousness is a symbol, not an actuality.

There are cases of dream when it happens for a minute or two, but the dreamer sees it as a month, or a year, or whole life. It means that the velocity of thought in the dream state enhances. This implies increase of momentum in the dream state. The volition is sometimes carried with the thought. The energy of the waking state and the dream state, which is violence, is dissipated in the environment. Scientists postulate that energy cannot be destroyed, so it stores itself in some way, which is a miracle in itself.

During our waking hours, the subject and the object are different, similarly during the dream state, the dreamer and the dream are different, so we are engaged in interpretation which constructs conflict. This is an illusory process. In principle, the subject and the object in the waking state (and the dreamer and the dream) are one entity, so there is no question of understanding or interpretation. Therefore your search for a meaning in the waking state or dream state is the action of conflict.

Consciousness is the whole of man and does not belong to a particular man. When there is the consciousness of one particular man, there is the complex problem of fragmentation, contradiction, conflict and war. Man has to understand, learn and live in the wholeness of life.

Dissipation of the Energy of Photons in a Black Hole

Black holes and quasers are the nearest things to windows on the edge of infinity. Black holes are the remnants of old stars that collapsed onto themselves when they ran out of fuel. As light and energy get sucked into a black hole, they disappear from our cosmos. If a photon of light tries to go round a black hole, it will start to curve in the hole's direction until it falls in. The photon falls into the black hole forever, frozen in time. Inside the black hole, the photon has already been devoured, in less than a hundred microseconds.[20] This implies that they are going somewhere. We conjecture that the energy of light (photons) gets converted into material of black hole, which at some appropriate time gets emanated as gamma ray particles – it appears that these are governed by some laws of nature, but we are unaware of it.

Transcendence of Thought-Time

There is an impregnable wall between the time field and the timeless realm, which the physicists call "the event horizon" (or frequency barrier or time barrier). No photon can penetrate this boundary – in the 'time-field', everything is within the speed of light, and in the 'timeless realm, there is anything faster than the speed of light. The speed of light is absolute; it is like a wall that no object can crash through. *As we approach the wall, time slows down, mass increases and space becomes curved.*[20]

The boundaries of psychological imbalances, ecological crisis, fear, violence are true event horizons, because the conditioned mind cannot go beyond them. If violence encounters true-subject (restructured science, education and health), the violence goes to abeyance. The symbolic-dualistic frame makes a quantum jump to non-dual and non-conceptual frame of Mind Continuum. If thought-time encounters true-subject, there is flow of spiritual energy which increases the potential of the object and this flowers to annihilation of thought. The energy of thought gets transformed to the energy of the being which flows to increase of potential of the Being. So thought energy gets transformed to the energy of the Being, which expresses itself as order, harmony, non-violence and peace.

A Quantum Jump from Violence to Nonviolence

Wilder Penfield wrote about "the separate existence of mind". He added, "it is the mind which experiences and it is brain which records the experience .... our minds and our brains are not at all the same".[21] Brain is a part of the objective world. It can be investigated objectively. Mind is subjective, it is vibrational self-conscious energy. The energy transfer from subjective to objective happens in less than five thousandths of a second, so we treat the subjective and the objective as one dynamic movement. This understanding is responsible for keeping one attentive, which means a quantum jump from violence to nonviolence. It is the higher potential of consciousness that can resolve the human challenges.

Osho wrote, "History records only violence. History does not record silence – it cannot record it. All records are of disturbance. Whenever someone becomes really silent, he disappears from all records".[22] All records of earthquakes or other calamities are records of disturbance. Where there is silence, all violence ends, and with the ending of violence, all disturbance ends. Modern man with all his scientific knowledge is not aware of the power of the silence. It is imperative that man undertakes an experiment in the laboratory of his mind to make a quantum jump from violence to nonviolence.

Jonas Salk wrote, "Violence may be wholly pathological in origin, it may also be a manifestation of resistance and an attempt to correct unhealthy extremes of excess or of greed inimical to the health of the species and therefore of individuals as well".[23] Violence may be release of cerebral or chemical tensions in the human psyche. Similarly an earthquake or a great calamity is a release of stored violence in the earth.

A Critique on Origin of Earthquake or Tsunami

We discuss the origin of earthquakes (or catastrophes, or calamity) based on the exploration of the inherited conditioned mind-brain-body system. The conditioned mind-brain-body system constructs energy of violence throughout the day. Millions and millions of people are throwing energy of violence throughout the day in the atmosphere. Niels Bohr stated that the world was "fundamentally a discontinuous and quantum jumping world .... To him, discontinuity was a fundamental truth".[24] The energy of violence is fundamentally discontinuous, so between any two consecutive quantum energy packets, there is the Unmanifest – it is basically "violence-Unmanifest-violence" packet. Nature due to its ingenuity permits the energy from the Unmanifest to flow to the manifest. When a conditioned mind-brain-body system throws such a packet of energy in the atmosphere, it is stored up in the atmosphere and from there it goes deep down the earth and gets stored there. Physics tells us that energy cannot be destroyed, so it is locked up deep in the earth frozen in time. If the stored up energy cannot be sustained, it gets unlocked and gets dissipated as a earthquake or volcano eruption. When the energy of violence attains a critical mass, it explodes as a earthquake, but this criticality is not amenable to any measurement. If there is any 'fault' in the earth's crust it appears as earthquake which has an elemental dynamics of its own. *The earthquake is defined as an explosion when the energy of violence has a critical mass and coupled with the elemental dynamics of the fault in the earth.*

Gandhiji, in prophetic words, addressed his message to all mankind:

> To me the earthquake was no caprice of God nor a result of a meeting of mere blind forces.... He and His Law are one. The Law is God. Anything attributed to Him is not a mere attribute. He is the Attribute. He is Truth, Love, Law, and a million things that human ingenuity can name .... For God is the Law. But I submit that we do not know the Law or the laws fully, and what appear to us as catastrophes are so only because we do not know the universal laws sufficiently.... Visitations like droughts, floods, earthquakes and the like, though they seem to have only physical origins, are, for me, somehow connected with man's morals.... With me the connection between cosmic phenomena and human behaviour is a living faith that draws me nearer to my God, humbles me and makes me readier for facing Him.[25]

To a deep religious man, these quotes give evidence that earthquake, Tsunami are not constructs by Nature (Cosmic Consciousness, Life Field), but are constructs of the conditioned mind-brain-body system, or inheritance of man.

A true scientist living in a non-dual and non-conceptual frame of Mind Continuum suggests strongly ingenuity, harmony, order, coherence, purpose, nonviolence, peace and intelligence in life. So all droughts, earthquakes, floods, Tsunamis are signals of Nature to tell man that his living is not in accordance to the immutable laws of Nature. We have to learn a moral lesson from these catastrophes and learn to live in harmony with the Nature's laws. It is a message addressed to all mankind.

To be more precise, the highest hierarchical expression of the Cosmic Consciousness is harmony, order, nonviolence and peace. The lowest manifestation of successive consciousness is conflict, violence, which expresses itself as a calamity, a catastrophe, an earthquake, a Tsunami. It is the Cosmic Consciousness, the Life Field, the Divine, the Absolute, the *Brahman*, that governs every event/existence in the cosmos and beyond.

Underground Observatories: The Sudbury Neutrino Observation (SNO) is 6800 feet underground. It is intended to detect neutrinos from astronomical sources. There have been several neutrino observatories built in the world. We cannot rule out packets of "violence-Unmanifest-violence" energy travelling deep down underground. Even if we reject all insights of religious men, we cannot explain how harmonious, ordered and benevolent nature acts as a tyrant in case of earthquakes/tsunamis. This exploration is only possible by a scientist who lives in a non-dual and non-conceptual frame of reference of Mind Continuum – we need intuitive wisdom.

The Future of "Man and His Destiny"

Jacques Monod, Nobel-prize biologist wrote, "What we see opening before us today is an abyss of darkness".[26] In prophetic words, George Wald, Nobel Prize in Medicine, 1967, wrote:

> We who know where we are going, are now heading rapidly toward self-extinction, toward removing humankind from the cosmic scene, and for the most trivial of reasons: profit, the rivalry of the United States and the Soviet Union, the momentum of the military-industrial complex in both nations. The cosmic drama will go on elsewhere, ignorant of our demise, caring nothing; but we will have no further part in it .... Does that matter? It matters to us.[27]

Maurice H. Wilkins, Nobel Prize in Medicine, 1962, wrote:

> The whole question of moral, spiritual, and other dimensions is normally pushed out of science. So, something in the nature of science often does lead people towards destruction. Abstraction can be regarded as a form of violence and, as various poets have remarked, science often does violence to nature.[28]

P. Krishna, an eminent physicist and educationist, in a talk in USA, said, "We are going towards self-annihilation and not too much time is left now".[29]

The irony of human destiny is that while man is spending billions and billions of dollars on rehabilitation of Tsunami victims, but he is not prepared to spare few coppers for "Science with Creativity", and "Symbiosis of Science and Spirituality". We intuit that man's peace, survival and excellence is embedded in the proposed project. I have been trying this for the last 25 years, but I have miserably failed as my limited words have fallen on the deaf ears of mankind.

Concluding Comment

The physical evolution and biological evolution over billions and billions of years have shown that there is harmony and order in all events/existences. There is love and compassion and nature's benediction flows everywhere. The entire saga is an example of ingenious mechanism and governed by Cosmic Consciousness or Supreme Intelligence. The earthquakes, tsunamis add to human suffering, pain, misery and chaos, so these cannot be explained at an intellectual or scientific level. We have to establish new relationships connecting the calamities with the human way of living.

The giant calamity, called Tsunami, is a by-product of conditioned human mind-brain-body system over a long chronological time. The total responsibility lies with 'I' and 'you', 'we' and 'they', and the whole mankind. As long as there is violence or aggression in the human psyche, the fear of a catastrophe, in some parts of the world, will remain. A sane and harmonious world is only possible if Govt. of India approves exploration of the conditioned psyche.

The conditionings of the psyche during the last 100 years have increased manifold. The conditionings have now travelled from the conscious level to sub-conscious level. The conditionings express themselves as violence, fear, psychological imbalances, ecological crisis, and other human challenges. The conditionings cannot be resolved through analysis or intellectual introspection, but can end through understanding and learning.

At all levels of life, the components of living systems are interlinked in network fashion. It appears that the physical calamities and human behaviour are interlinked in some complex network fashion, but we are not aware of the laws of the universe. The exploration is only possible by a non-dual and non-conceptual frame of reference of Mind Continuum. We need some major breakthroughs.

The moral-spiritual implications of Tsunami is that it is a man-made calamity over a long chronological time, which is basically due to his inherited violence of the mind-brain-body system. Millions of people have lived a life which is antagonistic to the immutable laws of nature over a long chronological time. It is as a result of the humanity and the collective conflict and chaos in which we have lived. We must now better appreciate this fact and develop the wisdom to direct the process towards his fulfilment rather than towards his destruction, annihilation. For underlying the new ideas, including those of modern physics, is a unifying order; but it is not causality; it is purpose, and not the purpose *of* man and *of* the universe, but the purpose *in* the universe and *in* man.[30]

Let us initiate "a global dialogue with an open mind" where every human being is an equal partner. The Cosmic Consciousness (Life Field) will govern the whole globe (and the whole cosmos) and make the globe a paradise. This provides mankind the strongest faith in the Order, Harmony, Nonviolence and Peace; this bestows Intelligence (Wisdom), and survival of mankind.

Buddha's central theme was desire is the cause of all pain. What the Buddha proclaimed about the imperative need to ***know*** the mind, to ***shape*** the mind, and to ***liberate*** the mind – has never been more true than in the age of the atom".[31] Buddha said, "Self is, Truth is. Where self is, Truth is not. Where Truth is, self is not". And once 'self' dissolves, we have true freedom, freedom from time, freedom from all our suffering. This freedom alone will lead to a silent mind, a mind that no longer chatters, a mind that is deep and tranquil, a mind that does not live in the knowledge of yesterdays and tomorrows, a mind that lives in NOW, a mind that does not live in the knowledge of NOW but in the feeling of NOW".[32]

We want a revolution, psychological revolution, in which the whole human race has to participate and every man is an equal partner. This revolution will regenerate or rejuvenate human mind so that it perceives the new dimension of an integral and all-inclusive consciousness, that can guarantee human survival along with all fauna and flora.

The mutation in Consciousness happens when the "restructured science-education-health" observes the conditioned psyche and the Unconditioned Energy, or in other words, the wholeness of life. This resolves the age-old tide of I-consciousness, thought-structures, psychological imbalances, ecological crisis, and other human challenges. This guarantees human survival and excellence.

An exploration of the mutation in consciousness is the only consummation of life. This is the only vocation of mankind. If we escape our responsibility, it will jeopardize our future, the future of India and of mankind.

References

1. Sampooran Singh, Dr (Mrs) Kanwaljit Kaur and Paramjit Singh, *The Masterplan Paradigm for Human Survival and Excellence*, Chandigarh: Faith Publishers, February 1998, pp. 169-204.
2. Mahatma Gandhi, *Speech at Public Meeting, Rajapalayam*, 25.01.1934, *The Hindu*, 26.01.1934.
3. *Harijan* 02.02.1934; *The Hindu*, 24.01.1934; and *The Hindustan Times*, 25.01.1934.
4. *Harijan* 02.02.1934; *The Hindu*, 26.01.1934.
5. Fritjof Capra, *The Tao of Physics*, Bantam Books (Second Edition), 1988, pp. 147, 153, 277.
6. Pupul Jayakar, *Krishnamurti: A Biography*, Harper and Row, 1986, p. 314
7. J. Krishnamurti, in Pupul Jayakar, *ibid.*, p. 379.
8. Osho, *The Book of Nothing*, Pune: Rebel Publishing House, 1975, p. 265.
9. Sri Aurobindo, *A Message on August 15, 1947*, in The Tribune, August 15, 1997.
10. The Divine Mother (Madame Mirra Alfasa), in *CIRHU*, Centre of International Research in Human Unity, Auroville, Paper 4, February 1997, p. 55.
11. Sri Aurobindo, in *The Odyssey of Science, Culture and Consciousness*, New Delhi: Abhinav Publications, 1990, p. 41.
12. Jawaharlal Nehru, in Pupul Jayakar, *op. cit.*, p. 397.
13. *Harijan*, February 2, 1934 (address at Madura, 26.01.1934).
14. *Speech at Public Meeting*, 24.01.1934; *Harijan*, 02.02.1934.
15. Paul Davies, *Superforce*, New York: Simon and Schuster, 1984, pp. 224, 223, 232, 243.
16. Bertrand Russell, in *The Courier*, May 1988, p. 4.
17. Sampooran Singh, *A Scientific Vision on Inter-Faith Harmony and Social Cohesion*, unpublished.
18. *The Tribune*, Editorial, December 28, 2004.
19. *The Times of India*, New Delhi, January 6, 2005.
20. Deepak Chopra, *How to Know God, op. cit.*, pp. 270-71.
21. Wilder Penfield, in Deepak Chopra, *ibid.*, pp. 217, 214.
22. Osho, *The Book of Nothing*, Pune: Rebel Publishing House Pvt. Ltd., 1975, pp. 15-17.

23. Jonas Salk, *Man Unfolding*, New Delhi: Allied Publishers Pvt Ltd., 1972, p. 117.
24. Niels Bohr, in Fred Alan Wolf, *Taking the Quantum Leap*, Harper and Row, 1981, pp. 85, 21.
25. *Harijan*, 16.02.1934; *The Collected Works of Mahatma Gandhi*, Govt. of India: The Publication Division, No. LVII, pp. 164-165.
26. Jacques Monod, *Chance and Necessity*, New York: Vintage Books, 1972, p. 171; & in *Synthesis of Science and Religion*, Mumbai: The Bhaktivedanta Institute, June 1988, p. 127.
27. George Wald, *op. cit.*, p. 13.
28. Maurice H. Wilkins, in *Synthesis of Science and Religion*, *op. cit.*, p. 39.
29. P. Krishna, *Education, Science and Spirituality*, USA: The Institute of Higher Education, Research and Dialogue, 1993, p. 12.
30. Jonas Salk, *Man Unfolding*, *op. cit.*, pp. xi, xii.
31. Buddha, in D.S. Kothari, in *The Evolution of Consciousness*, *op. cit.*, pp. 35, 239.
32. Buddha, in B.M. Khanna, *The Odyssey of Science, Culture and Consciousness*, New Delhi: Abhinav Publications, 1990, p. 165.

26

Tsunami : Its Implications on the Fundamental Dimensions of Man

Abstract

Physicists have concluded that "Everything we know about nature is in accord with the idea that the fundamental processes of nature lie outside space-time but generate events that can be located in space-time". Some physicists proposed "superluminal (faster-than-light) transfer of negentropy (Negentropy is another term for order) without signals".

It appears that when a packet of energy (violence-Unmanifest-violence) stored in the bowels of the earth reaches a critical level, it explodes (information transfer) in the realm outside space-time and this is governed by superluminal channel; but this release is located in space-time matrices, which appears as earthquakes. This suggests that all earthquakes and tsunamis are not due to nature's wrath but are constructs of the conditioned mind-brain-body system over uncountable aeons.

If all "physical" calamities are constructs of the conditioned mind-brain-body system, then we have to culture a good (sensitive) mind capable of direct perception at every moment – a non-dual and non-conceptual frame of reference of Mind Continuum. It is imperative that we have to restructure our educational system. It discusses the implications on the fundamental dimensions of man, that is, to be concerned with the whole development of man.

Man has emerged as a violent species. He has been trying to end conflict, violence, aggression, war in the psyche for the last 10,000 years or so; but all intellectual approaches, thought-structures, beliefs, concepts have proved to be sterile in transforming the human psyche, rather these have added to the conditionings. We have brought the human race to an explosive situation and our life is poised on a razor's edge.

Man has an evolutionary origin. He emerged on this planet at about 2.5 million years ago. Every event/existence is governed by the hierarchical expression of Cosmic Consciousness, or in other words, quantum energy potential of Life Field. The successive consciousness is violent, aggressive, greedy, envious, jealousy, hate, fear, power, prestige, name, success, and so on – these are called psychological imbalances. Psychologically the individual has not changed at all and the structure of the society throughout the world has been created by the individuals. The outward social structure is the result of the inner psychological structure of our human relationships. Each one of us is the storehouse of the past – he is not collective but humanity. The individual is the human who is all mankind. The whole history of mankind is written in ourselves. It implies that the hierarchical level of consciousness has not changed during the last 2.5 million years.

Man has made accelerative progress in the material field – agriculture, medical, travel and innumerable gadgets – which have added to our comfort, luxury, but this has not contributed to our happiness. Man has an inbuilt or hidden capacity to make a quantum jump to higher level of consciousness. Many eminent scientists and non-institutionalized religious men have written that at this critical juncture nature is trying to help man to make a quantum jump from psychosocial evolution to spiritual evolution. The characteristics of the non-dual and non-conceptual frame of reference of Mind Continuum is harmony, order, nonviolence and peace. And this is the fundamental dimension of man.

Ilya Prigogine wrote, "In the twentieth century, an attempt has been made .... to seek behind all the conflicts a more universal conception of human dignity that is less subject to the constraints of history. In this respect too, the twentieth century has marked an irreversible turning point in the history of mankind – a turning point that heralds a new age, a bifurcation".[1] All these attempts have been in the field of time-space-causation matrices. He is basically working on a conceptual framework that integrates life's biological, sensual world (cognitive faculty) and social-economic-political dimensions. But here we have also failed to develop a coherent, systemic approach to some of the crucial issues of our time.

The human brain is objective and the mind is subjective. The mind has an independent existence from the brain. In order to see, the mind cuts itself up into at least one state that sees, and at least one other state that is seen. There is an energy interaction between the seer (subject) and the seen (object), and this happens in less than a hundred microseconds. This leads to distortion of perception and modified memory. It is basically coherent superpositions of concepts on the fact leading to an illusionary conceptual model. The conceptual model disintegrates to thought-structures coupled with volitions, which are called conditionings. This gives chemical pressure and nervous tension, which constructs all psychological imbalances, ecological crisis and other challenges.

This article explores the cause of physical catastrophes and concludes that the physical calamities are not constructs by Nature (Cosmic

Consciousness, Life Field), but are constructs of the conditioned mind-brain-body system. The catastrophes, when expressed in the space-time matrices; and are signals of Nature to warn man that his living is antagonistic to the immutable laws of nature, and he must work on priority to rectify himself. It also discusses the relationship of violence and tsunami. It also presents a technology to make a quantum jump from psychosocial evolution (false dimensions of man) to spiritual evolution (true dimensions of man). It leads to an ending of all violence, which blossoms to ending of calamities. The art of living is also discussed. This is the only technology to relieve man of the pain and sufferings, and to regenerate "the fundamental dimensions of man", which are free gifts of nature.

A Cosmic Plan

The physical evolution and biological evolution over billions and billions of years have shown that there is harmony and order in all events/existences. There is ingenuity and intelligence in nature. Paul Davies, an eminent physicist wrote, "The laws which enable the universe to come into being spontaneously seem themselves to be the product of exceedingly ingenious design .... The universe must have a purpose, and the evidence of modern physics suggests strongly to me that the purpose includes us".[2] Ruth Nanda Anshen wrote, "Modern physics is a unifying order, but it is not causality, it is purpose, and not the purpose of the universe and of man, but the purpose *in* the universe and *in* man .... No civilization has previously had to face the challenge of scientific specialization, and our response must be new .... The spiritual and moral needs of man as a human being and the scientific and intellectual resources at his command for life may be brought into a productive, meaningful and creative harmony".[3]

The Great Calamities

Many scientists and philosophers call the droughts, floods, earthquakes and tsunamis are due to nature's fury or wrath; these cause disharmony, disorder and chaos; it gives pain, misery and suffering to mankind; and this contradicts all known scientific knowledge. We have to undertake an exploration of the cause and the nature of calamities, catastrophes, earthquakes and tsunamis.

A great calamity – known as Bihar Earthquake descended on India on January 15, 1934. Nearly 25,000 people were swallowed up in the bowels of the earth in the twinkling of an eye, and tens of thousands had been rendered homeless.

Fritjof Capra, an eminent physicist, discussed 'natural' disasters in 1998 that are caused in part by human-induced global climate change. He stated that Hurricane Mitch took 10,000 lives, the catastrophic flood of the Yangtze River in China caused more than 4,000 deaths, and Bangladesh suffered floods killing 1,400 people. Capra added that "the rising tide of natural catastrophes over the past decade is a clear indication that the climatic instability caused by human action is increasing".[4] Janet Abramovitz, World Watch Institute, wrote, "The 'unnatural disasters' were made more frequent or more severe due to human action".[5] Capra concluded that, "the principal lesson to be learned is that the cause of most of our environmental and social problems are deeply embedded in our economic systems". We opine that man must undertake through observation healing his conditioned psyche. Man has to initiate exploring "Science with Creativity" and "Symbiosis of Science and Spirituality".

On December 26, 2004, huge tidal waves generated by a strong earthquake hit coastal areas of India, Sri Lanka, Indonesia and Thailand and flattened everything that came in its way. The death toll crosses 2,83,000 with many missing, injured and homeless. A massive relief operation is in progress to ameliorate human suffering.

Mahatma Gandhi's Response to Bihar Earthquake

The human intellect considers earthquakes as a Divine wrath. It implies that we have lost our relationship with nature, because in nature there is harmony and order; and we inevitably lose our relationship with human beings, so with human being we behave as disharmonious and disordered. This has been emphasised by Gandhiji in words:

> Even in this indescribable calamity there is a divine purpose that works for the good of humanity.[6]
>
> To me the earthquake was no caprice of God nor a result of a meeting of mere blind forces .... Visitation like droughts, floods, earthquakes and the like, though they seem to have only physical origins, are, for me, something connected with man's morals.[7]
>
> And since God is love, we can say definitely that even the physical catastrophes that He sends now and then must be a blessing in disguise.[8]

We have to understand, learn and live in harmony with nature and harmony with all human beings.

Relationship of Violence and Tsunami

In the world there is violence, extraordinary turmoil, crisis after crisis. There are wars, division of nationalities, religious differences, racial and communal differences, one set of systematized concepts against another. The outer (objective) is a mirror-image of the inner (subjective); or to be more precise, there is energy exchange between the outer and the inner, and this takes less than a hundred microseconds. One is responsible, totally, for the way one lives one's daily life.

Violence is anger, hatred, conformity, imitation, obedience. The denial of all that is the opposite of that. We have not changed fundamentally during the last uncounted aeons. We have emerged as a violent species.

Billions and billions of people are throwing the energy of violence throughout the waking state and the dream state of mind in the environment, and the energy of violence gets stored in the environments. When it reaches criticality, nature, out of its benediction, causes the energy of violence (violence-Unmanifest-violence packets) to go deep in the bowels in the earth and it gets stored there. If energy of violence reaches a critical mass in the earth, it explodes as an earthquake (which may take the shape of tsunami if faults in the earth are present). This dissipates all violence stored in the earth.

It may take man to build violence packets for uncountable aeons, but the release of this energy as a earthquakes takes fraction of a second. As an analogy, we recall that a dream state may take from 1-2 minutes, but the dream at times appears to have covered many years or whole life. Dream state is a release of excess of violent energy in the brain. Similarly nature takes many years or uncountable aeons, to store violent energy, but release as a earthquake may take few seconds. This only suggests that we have no awareness of the immutable laws of nature.

Observing (seeing), listening (hearing), understanding and learning, action are direct; while thinking and psychological inheritance are indirect. The physical evolution and biological evolution suggest remarkable harmony, order and unity of nature that recent advances of science have uncovered. Nature invariably expresses itself as fact in a non-dual and non-conceptual frame of reference of Mind Continuum. It encodes a message on the superluminal channel (faster-than-light). Henry Stapp wrote, "Everything we know about nature is in accord with the idea that the fundamental processes of nature lie outside space-time but generate events that can be located in space-time".[9] In 1975, Jack Sarfatti, a physicist, postulated that faster-than-light connections exist in nature. It is superluminal transfer of negentropy (another term for order) without signals.[10]

Physicists have shown rationally that "our rational ideas about the world in which we live are profoundly deficient".[11] It appears that something very exciting is happening. When the packet of energy (violence-Unmanifest-violence) reaches a critical level, it explodes (information transfer) in the realm outside space-time and this is governed by superluminal channel; but this release is located in the space-time matrices. This release of energy in the space-time matrices appears as an earthquake.

To sum up, *all droughts, floods, catastrophes, calamities, earthquakes, tsunamis are not due to nature's wrath against mankind, but are constructs of the conditioned mind-brain-body system or inheritance of man over uncountable aeons*. An understanding gives you the feeling that man divorced from nature could not be happy and man's power to resist nature is hopelessly puny. This attitude is different from the tenets of scientific civilization to conquer nature. It reminds man that the age-old tide of human ambition and aggression must be jettisoned. And naturally these can be ended when the human mind-brain-body system is purified. An understanding of relationship between man and nature enables you to see God (Life Field, Cosmic Consciousness) face to face. No one can have true relationship without being 'intuitively' convinced that God alone is real and that all else is an illusion.

In Love, or Truth, or Beauty, or Reality, there is superluminal transfer of negentropy. In nature everything happens simultaneously and everything is interrelated. It is integrated and self-emerging system, a totality that exists. Life is mysterious wholeness. It is non-fragmentary, indivisible whole. It is homogeneous, undifferentiated and dimensionless. It is timeless, eternal and infinite. Life is immeasurable and unnamable.

The Art of Living

The whole process of living must emerge from the understanding, learning and action, which are based on direct perception, or communion with nature. The different phases of the art of living, which are basically one dynamic flow of life are as under:

(a) Phase I: The psychosocial evolution over uncountable aeons has led us to living of coherent superpositions of recalled past or remembrances over the present, and this has led to mutilated perception or crisis of perception. What one calls living is that which has happened and gone. That which has gone is dead. This is not living, this may be called existence or vegetating like a plant. This has led to jettisoning the Reality, the Life Field. This is leading him to pain and conflict.

(b) Phase II: About 1600, Kepler and Galileo simultaneously and independently formulated the principle that laws of nature are to be discovered by measurements. This has led to accelerative progress in the material field. Man accumulated partial experience and fragmentary knowledge, and not whole. This has led to comfort and luxury in the material field, but has not added to human happiness and peace.

(c) Phase III: Eminent scientists – Karl Pribram, Roger Sperry, John Eccles – have shown that the outer (objective) and the inner (subjective) are not two separate movements but one whole. We have to integrate the objective and the subjective dimensions of life.

(d) Phase IV: Recently we proposed restructuring of science (science-education-health) as the subject of cognizance, and then observe or listen to everything happening in the mind-brain-body system. To be psychologically free and original happens when you are aware of your own inward activities. This leads to pure perception which bestows freedom, order, nonviolence and peace.

(e) Phase V: We have to understand deeply that there is harmony, order and oneness of life. Nature's benediction is flowing everywhere. All droughts, floods, earthquakes, tsunamis are not due to nature's wrath but are constructs of the conditioned mind-brain-body system over uncountable aeons.

(f) Phase VI: There must be freedom from knowledge to understand that which is eternal, which is timeless. We must understand freedom, dignity and beauty of life. This bestows intelligence.

(g) Phase VII: Observation over a long time bestows intuition or insight. Intuition is fundamental and reason is a derivative. Insight is direct perception of the Unknown. We base ourselves on intuition or insight and then, explain intuition by reason. We have to learn to live in communion with nature.

(h) Phase VIII: To grasp this whole movement of life requires intelligence, not the intelligence of thought, or books, or knowledge; but the intelligence of love and compassion with its sensitivity. We have to understand the movement of life and death. We have to work for 'psychological death' and regenerate 'psychological living'.

(i) Phase IX: J. Krishnamurti wrote, "If you are in harmony with nature, with all the things around you, then you are in harmony with all human beings. If you have lost your relationship with nature you will inevitably lose your relationship with human beings".[12] We have to reject our intellectual approach that all calamities are due to Divine wrath and understand that nature is benevolent and we have to live in harmony and order with nature, and that will bestow harmony with human beings.

(j) Phase X: The political leaders and people must initiate the above phases (III to IX) simultaneously and on war footing. The different phases are not watertight compartments, but a single dynamic flow of life. This is a joint venture of the whole mankind; and in this is embedded human survival and excellence. It is the greatest vocation of the world.

To sum up, *the art of living is a harmonious movement of understanding, learning, acting – this bestows direct perception or communion with nature. Meditation is not a state; it is dynamic movement of unconditioned energy without any conflict or observer, or meditator.*

Concluding Comment

The arts of observation, listening, understanding, learning, action are direct, while thinking and psychological inheritance are indirect. Nature's expression is invariably direct. Living in direct perception, or in communion with nature ends all psychological imbalances, ecological crisis, challenges of poverty and hunger, and so on. If the relationship with nature is in harmony and order, all droughts, floods, earthquakes, tsunamis will vanish. It is the highest quantum energy potential of Cosmic Consciousness or Life Field that can end all pain, sorrow, suffering, and bestow harmony, order, nonviolence and peace.

All harmonious relationships of man and nature, man and man is nonviolence; and disharmonious relationship is violence. Ilya Prigogine stated, "There is a need for new relations between man and nature, and between man and man".[13] Maurice H. Wilkins wrote, "Something in the nature of man often lead people towards destruction. Abstraction can be regarded as a form of violence".[14] *As nature's expression is invariably 'direct perception', so it is invariably nonviolence. It strongly suggests that a 'direct perception' is nonviolence, and a 'mutilated perception' is violence.* In other words, *a symbolic-dualistic frame of reference of mind spectrum expresses itself as violence, disharmony and disorder and a non-dual and non-conceptual frame of reference of Mind Continuum expresses itself as nonviolence, harmony and order.*

It is imperative that mankind must understand and learn to make a quantum jump from symbolic-dualistic frame to non-dual and non-conceptual frame. Man must learn to live in the intuitive or insight mode of mind and jettison the intellectual frame of reference. It is based on the law that insight is fundamental and reason is a derivative. We use the intellect only, if it is essential to explain the insight.

Mahatma Gandhi wrote, "Nonviolence is the law of our species as violence is the law of the brute".[15] He added, "I hoped the fragrance of nonviolence of India would permeate the whole world".[16] It is law of harmony and order that will govern the world. It is evident that Gandhiji's insight is the same as the recent advances of modern science.

Brain is objective and mind is subjective. Mind is independent of brain. We have to educate and culture a good mind – a mind which is capable of dealing with the many issues of life as a whole, and which does not try to escape from them. And it is essential for the mind to be aware of the conditioning of the mind-brain-body system, its own motives and pursuits. To be psychologically free and original can only come about when you are aware of your own inward activities, watch what you are thinking and never let one thought escape without observing the nature of it, the source of it. The need of the hour is to rejuvenate the mind from successive consciousness to Simultaneous Consciousness, Life Field. We have to jettison the old dimensions (time, past, violence) of mind and regenerate the new dimensions (timelessness, harmony, nonviolence, peace) of mind. We are concerned with the whole development of each human being, helping him to realize his own highest and fullest capacity.

Nature is full of love and compassion and its benediction flows everywhere. J. Krishnamurti concluded, "The benediction of the timeless is waiting, the ground is ready. There is a sense of benediction waiting, and we are not moving towards it".[17] The essence of the 'timeless wisdom' is that 'you' and 'I', 'we' and 'they', and all humanity have to bring in radical transformation or mutation in the human consciousness; or a quantum jump from psychosocial evolution to spiritual evolution. This jettisons the lower dimensions of successive consciousness, and imbibe the higher dimensions of Simultaneous Consciousness, or Life Field. This is a pilgrimage to the higher dimensions of insight or intuition, which the intellect can never fully grasp.

References

1. Ilya Prigogine, in *The Courier*, May 1988, p. 13.
2. Paul Davies, *Superforce*, New York: Simon and Schuster, 1984, p. 243.
3. Ruth Nanda Anshen, in Jonas Salk, *Man Unfolding*, New Delhi: Allied Publishers Private Ltd., 1972.
4. Fritjof Capra, *The Hidden Connections*, New Delhi: Harper Collins Publishers, India, 2003, pp. 183-84.
5. Janet Abramovitz, in Fritjof Capra, *ibid.*, p. 184.
6. *Harijan*, 02.02.1934.
7. *Ibid.*, 16.02.1934.
8. *Ibid.*, 20.04.1934.
9. Henry Stapp, in Gary Zukav, *The Dancing Wu Li Masters*, Fontana/Collins, 1982, p. 310.
10. Jack Sarfatti, in Gary Zukav, *ibid.*, pp. 310-11.
11. Gary Zukav, *ibid.*, p. 309.
12. J. Krishnamurti, *Krishnamurti to Himself (His Last Journal)*, London: Victor Gollancz Ltd., 1987, p. 107.
13. Ilya Prigogine and Isabelle Stengers, *op. cit.*, p. 312.
14. Maurice H. Wilkins, in *Synthesis of Science and Religion*, Mumbai: Bhaktivedanta Institute, June 1988, p. 30.
15. M. K. Gandhi, in Prabhu & Rao, *The Mind of Mahatma Gandhi*, Ahmedabad: Navjivan Publishing House, February 1967, p. 112; *Young India*, 11.08.1920, p. 3.
16. *Ibid.*, *The Leader*, p. 44; *Young India*, 31.12.1931, p. 428.
17. J. Krishnamurti, in Pupul Jayakar, *Krishnamurti: A Biography*, San Francisco: Harper and Row, 1986, pp. 380, 384.

27

New Relation Between Earthquake/Tsunami and Man and Nature

Abstract

Nature unsheathed its frightening weaponry – earthquake and tsunami – with full fury on December 26, 2004 and devoured 2,83,000 human lives and left unprecedented trail of misery and suffering. The concept of 'fury', or 'wrath' of nature contradicts all known scientific and philosophical-mystics knowledge/insights, which suggests that nature is benevolent and its direct perception is "harmony and order", "non-violence and peace", and "love and compassion".

An exploration indicates that: (a) the truth of man and truth of nature emerge from the Cosmic Consciousness or Life Field; (b) the catastrophes are constructs of the conditioned psyche over a long chronological time; and (c) this suggests a new relation between earthquakes/tsunamis and man and nature. It is basically an intuitive insight and rejects intellectual approach. Fundamental discoveries which go deep into the nature of things are the result of uncommon intuition and faith, not cold logic. Intuition is fundamental and intellect is a derivative. It shows the unity of Life Field, or Oneness of the Cosmic Consciousness.

If the thought-structure coupled with volitions encounters the restructured science (science-education-health), the thought-structure is annihilated. The mind makes a quantum jump to non-dual realm and is regenerated to its pristine effulgence. It regenerates the supreme Intelligence which resolves all challenges. It is the benign turn of destiny that enables us to stave off disasters – psychological imbalances, earthquakes and tsunamis. The time to act is Now or Present. Now or never.

Some physicists have been deeply inspired by "the remarkable harmony, order and unity of nature". Paul Davies, an eminent physicist wrote, "The laws which enable the universe to come into being spontaneously seem themselves to be the product of exceedingly ingenious design .... The universe must have a purpose, and the evidence of modern physics suggests strongly to me that the purpose includes us".[1] It appears that all nature's events/

existences are utterly perfect. Nature invariably expresses itself as 'direct perception'. Nature is astonishingly clever in the way it gets the things to work. There is ingenuity and intelligence in nature. Many physicists are compelled to believe that there is something behind it all; in a pithy phrase of Fred Hoyle, "The universe is a put-up job".[2] Ruth Nanda Anshen wrote, "Modern physics is a unifying order, but it is not causality, it is purpose, and not the purpose of the universe and of man, but the purpose *in* the universe and *in* man".[3]

In nature, all biological systems are open, they are running up, becoming more, not less, complex and organised; these are moving to higher and higher levels of order and harmony; or higher and higher hierarchical expression of consciousness. Nature invariably expresses itself as direct perception; so each step of nature contributes to higher harmony, higher order and higher Consciousness; and rejects moving to disharmony and disorder, chaos and so on. In nature, spontaneous higher harmony, order arise out of lower harmony and order. Similarly, we find harmony and order in trans-psychological dimension of Consciousness in man.

Observation, listening, understanding, learning, action and nature's expressions are 'direct', so these are in communion with nature's laws. The evidence of modern physics suggests strongly that the universe is the product of ingenious design and the principle of 'harmony and order' is incorporated in the wholeness of life and the physical universe. The intellectual approach of objective science leads us to a blind alley. Weisskopf wrote that "this attitude is different from the tenets of scientific civilization which pretends to have conquered our impotence in the face of nature in the raw".[4]

Life is relationship, which is expressed through contact with things, with people and ideas. In understanding relationship, we must meet life fully, adequately. Relationship means communion without fear, freedom to understand each other, to communicate directly. Relationship is possible in a non-dual and non-conceptual frame of reference of Mind Continuum. It happens when there is direct perception. Ilya Prigogine and Isabelle Stengers prophetically wrote:

> It is quite remarkable that we are at a moment both of profound change in the scientific concept of nature and of the structure of human society as a result of the demographic explosion. As a result, there is a need for new relations between man and nature and between man and man.[5]

This article explores the relationship between physical calamities (earthquakes, tsunamis) and man and nature. It discusses the relationship by Saints-Seers. An exploration between man and man, and man and nature has been presented. We observe that the human truth and nature's truth emerge from the Cosmic Consciousness or Life Field.

If thought-structure coupled with volition encounters the restructured science (science-education-health) dimension, the thought-structure and I-consciousness go to abeyance, and all violence ends. Mind lives in the non-dual realm and is regenerated to its pristine effulgence. This energises Supreme Intelligence which ends all earthquakes and tsunamis.

Exploration by Saints-Seers

The responses by saints-seers, philosophers-mystics of East and West are as under:

(a) Albert Einstein: He wrote, "God is not malicious". He added, "What is truly valuable in our battle with life is not the nation. I should say that the creative and impersonable individuality, the personality – he who produces the noble and sublime while the common herd remains dull in thought and insensible in feeling". This implies that man has the capacity to generate noble and sublime thoughts which transcends all pain, misery and suffering, and bestows "peace and serenity". This deeply suggests that earthquake and tsunami are not due to nature's wrath or fury, but due to some deeper laws of nature, which we do not know.

Further, a development of human personality suggests that the conditioned psyche produces an energy of violence. And some mysterious laws of nature, the energy of violence in the atmosphere get dissipated.

(b) Mahatma Gandhi: Gandhiji, the apostle of nonviolence and peace, wrote very extensively on Bihar Earthquake which happened on January 15, 1934. Nearly 25,000 lives were swallowed up in the bowels of the earth in the twinkling of an eye. Tens of thousands had been rendered absolutely homeless. Gandhiji wrote:

> Even this indescribable calamity there is a divine purpose that works for the good of humanity.[6]
>
> To me the earthquake was no caprice of God nor a result of a meeting of mere blind forces .... Visitation like droughts, floods, earthquakes and the like, though they seem to have only physical origins, are, for me, something connected with man's morals.[7]
>
> And since God is love, we can say definitely that even the physical catastrophes that He sends now and then must be a blessing in disguise.[8]

We have to understand, learn and live in harmony with nature and harmony with all human beings. Gandhiji, at the highest intuitive level, stated that the Bihar Earthquake was basically 'good for humanity' and a 'blessing in disguise' because this reminded man to give up living which is antagonistic to laws of nature and to live according to the immutable laws of nature. Further, the physical calamities are intertwined with man's morals.

(c) Sri Aurobindo: In a message on August 15, 1947, Sri Aurobindo wrote, "A catastrophe may intervene and interrupt and destroy which is being done, but even then the final result is sure .... A new spirit of oneness will take hold of the human race".[9]

(d) Vimala Thakar: Life is immense, it is infinite, it is immeasurable. It is homogeneous and harmonious whole. Vimala Thakar, the greatest living seer, wrote, "If the human race is behaving in such a way that it disturbs the harmony of the Universe, perhaps the universal forces will check the stupidity of man and bring the human race back to the awareness how harmony is the fundamental principle required for existence".[10] We opine that the 'universal healing forces' of harmonious living have initiated the Tsunami of December 26, 2004, which has devoured 2,83,000 human lives in the twinkling of an eye and innumerables are wounded and without shelter.

(e) Lama Govinda: If the words of Paul Davies[1] are true that the purpose includes us, then it is evident that "*the human truth becomes the truth of nature*". This is what a Tibetan Buddhist, Lama Govinda speaks.[11]

(f) John Selby: John Selby wrote, "Bliss lives in the presence of love, compassion, acceptance, trust. Bliss is the flow of pure spirit in our lives .... When there is no past, there is the bliss of the present moment".[12] Selby quoted Krishnamurti, "In that act of seeing, there is all beauty. And with beauty there is love – and when there is love, you have nothing more to do. Where you are, you have bliss, you have heaven. Bliss and passion lie on the energetic continuum. Bliss is the passive dimension and passion is the active dimension".[12]

(g) Tolle Eckhart: In his latest book, entitled *Stillness Speaks,* Tolle Eckhart wrote, "We depend on nature not only for our physical survival, we also need nature to show us the way home, the way out of the prison of our own mind .... When you perceive nature, let there be spaces of no thought, no mind. Be aware of life within the form – the sacred mystery, and perceive its all-encompassing consciousness".[13]

(h) Lord Buddha: He proclaimed about the imperative need "to *know* the mind, to *shape* the mind and to *liberate* the mind".[14] He added, "Self is, Truth is. Where self is, Truth is not. Where Truth is, self is not".[15] To *know*, to *shape*, to *liberate* the mind depends on understanding the conditioned psyche. To end the self and to know the truth depends on observing, listening, understanding, learning, action and living, because all these are ***direct***, and communion with nature. The wholeness of life is the only direction. The attentivity, the sensitivity, the receptivity is the only energy to be exercised. This flows to radical transformation of the psyche, which implies a quantum jump from symbolic-dualistic frame of reference to non-dual and non-conceptual frame of Mind Continuum. This flows to a quiet mind, or "no-mind state".

To sum up, the above philosophers-mystics have stated that man is a miniaturised cosmos so all cosmic phenomena and human behaviour seems to be intertwined and interrelated. Modern science seems to support this gospel. Victor F. Weisskopf, quoted Laotse and Changtse that "man divorced from nature could not be happy and man's power to resist nature is hopelessly puny".

The physical evolution, biological evolution and trans-psychological revolution of man, and exploration of saints-seers confirms that the truth of man and the truth of nature have emerged from Cosmic Consciousness, the Life Field. We understand the truth of nature if we are stabilised in the truth of man. The physical calamities can only be ended through understanding the truth of man.

New Relations Between Man and Man

Man (*Homo sapiens*) is an amphibian who lives simultaneously in two worlds – (a) the *known*, or the time-space-causation matrices, or the successive consciousness; and (b) the *Unknown*, or the timelessness, or the Simultaneous Consciousness, or the ultimate quantum energy potential of Cosmic Consciousness, or the Life Field. The known or cultural belief structure is projected as time-space-causation matrices; it is the past, the thought; it is limited and transient; hence it cannot give any relation between man and man. The Unknown expresses itself in a non-dual frame of reference of Mind Continuum; it is eternal, infinite and immutable; and it gives true relationship between man and man in the timeless realm.

When thought-structure coupled with its volitions comes face to face with the true-subject (restructured science-education-health), the thought-structure goes to abeyance and what is left behind is pure Consciousness observing Itself. This bestows true relationship between man and man. The fundamentals of true relationship is without fear, violence and with perfect peace.

(a) The Art of Observation: R.P. Kaushik quoted Patanjal Yoga, and wrote, "The first stage prior to recognition and naming is a state of non-verbal or thought-free perception (*nirvikalpa jnana* or pure perception). From this pure perception we proceed to thought, recognition and naming (*savikalpa jnana*)".[16]

The concepts are grafted on the perception, which reduces a perception to a concept. We are not aware of direct perception, but are aware of mutilated perception or conceptual model. This is known as the crisis of perception. The whole mankind, barring a few persons, has lost the faculty of direct observation. This implies that we are not aware of the present moment, but aware of the past.

We have to observe non-subjectively, without any evaluatory comparisons or judgements. Observation without an observer is attentivity. The whole being gets tuned into a state of attentivity, where there is neither subject nor object. It is a subject-free and object-free attentivity. A quiet mind is exceedingly sensitive and aware and spontaneously responsive. A quiet mind is at a higher quantum energy potential of consciousness than a chattering mind. Such a mind acts totally, not fragmented, because it acts out of complete stillness. In order to observe actually 'what is', in oneself and in the world a quiet, very still mind is necessary.

Observation leads to a quantum jump from a 'cognitive mode' to perceptual mode. It is the direct perception of 'what is'. It leads to a harmonious mind-brain-body system. Observation leads to without analysis, without the analyzer. It implies that the observer and the observed are one.

Observation is the pure and simple Law of Equanimity; neither dislikes or hatred nor likes or undue attachments, pure awareness of the conditioning as it arises, pure awareness of thoughts arising as a rsult of this conditioning, simple pure awareness, no denial or suppression of desire, no joining in the resulting thought process, simply seeing the basic conditioning as it arises, as it is.[17]

(b) The Art of Listening: The moment we hear a word, we recall its past memory and graft it on the word; this leads to distortion of what we hear. The human mind at once loses its equanimity and responds by resistance or identification, and psychological pleasure and pain come into operation. This disturbance distorts listening. This resistance creates a chain reaction and ultimately destroys our peace and freedom.

We have to learn to listen directly, without any interpreting or qualifying words, which implies that the mind is peaceful and quiet. If we listen with complete attention, all images, all conditionings are shattered, the mind is rejuvenated, made innocent and fresh to learn and live. Thus the arts of observing and listening will bring about a tremendous understanding – not only of self but of the whole world.

(c) The Art of Understanding: We understand that when thought goes to abeyance, there is integration of the observer and the observed. You observe yourself without the eye of the past, watch yourself in action, which is relationship, without any movement of thought. When there is no thought, there is the bliss of the present moment. Meditation means the awareness of both the outer world and the inner world. It is to observe without any choice, without any distortion, without any thought. It is a way of observing the mind without analysis – a way to observe the whole content of consciousness.

(d) The Art of Learning: Learning through experience is accumulation of information. It adds to our conditioning, it brings about fragmentation, narrowness and specialisation. It destroys human relationships and brings out isolation and division. Learning does not take place if the motivation is of acquiring, owning and possessing knowledge. The psychological mode of mind cannot learn anything in the true sense of the word.

Learning is possible when one is attentive. Attentivity implies awareness of the fact, the whole body must be aware of the fact, the wholeness. Attentivity is possible when one is not tuned to the chatter of the brain, but tuned to the set of frequencies of the fact. Learning is possible by encoding the information as wholes, it is possible when there is perception of the fact.

(e) The Art of Direct Action: Understanding and learning blossom to an action without an idea. It is born in a non-dual and non-conceptual frame of reference in the Mind Continuum. It is born of the timelessness. There is no time gap between understanding and action, it is instantaneous. The mind is really silent, so it brings regeneration or transformation or mutation of the human mind-brain-body system. The action without an idea is at a higher quantum energy potential than action with an idea. The action without an idea is direct, direct perception. It is in communion with nature.

These arts when stabilised over a long chronological time, bestow true relationship between man and man. Modern science suggests: (a) physical evolution, biological evolution, and trans-psychological dimension of Consciousness in man are harmonious and ordered; (b) relation of man to man is "harmony and order", "nonviolence and peace", and "love and compassion"; and (c) nature's benediction is ever flowing in every event and organism. This is also supported by intuitive findings of saints and seers of East and West, over uncountable aeons. The 'seers' have confirmed that mind-identified state of violence and aggression is totally eliminated, and man revels in harmony, nonviolence and peace. This is called "Creativity in Science", or "Symbiosis of Science and Spirituality".

New Relations Between Man and Nature

(a) You cannot call a tsunami as nature's wrath or fury, as this concept is disharmonious and disordered, and it adds to human misery, suffering and chaos. This is contrary to the established scientific facts of physical evolution, biological evolution and trans-psychological dimensions of Consciousness of man. The only possible alternative is that these catastrophes are constructs of human conditioned psyche. Expand your awareness to include all great calamities, catastrophes, earthquakes, tsunamis as a signal or message of Cosmic Consciousness that your living is antagonistic to the immutable laws of nature, and you need urgent steps for radical transformation of your conditioned psyche. This blossoms to new relations between man and the universe, and between man and man.

(b) A Model of the Emergence of Earthquakes and Tsunamis: Billions and billions of people are throwing the energy of violence throughout the waking state and the dream state of mind in the environment, and the energy of violence gets stored in the environments. When it reaches criticality, nature, out of its benediction, causes the energy of violence (violence-Unmanifest-violence packets) to go deep in the bowels in the earth and it gets stored there. If energy of violence reaches a critical mass in the earth, it explodes as an earthquake (which may take the shape of tsunami if faults in the earth are present). This dissipates all violence stored in the earth.

It may take man to build violence packets for uncountable aeons, but the release of this energy as a earthquakes takes fraction of a second. Nature invariably expresses itself as fact in a non-dual and non-conceptual frame of reference of Mind Continuum. It encodes a message on the superluminal channel (faster-than-light). Henry Stapp wrote, "Everything we know about nature is in accord with the idea that the fundamental processes of nature lie outside space-time but generate events that can be located in space-time".[18] In 1975, Jack Sarfatti, a physicist, postulated that faster-than-light connections exist in nature. It is superluminal transfer of negentropy (another term for order) without signals.[19]

Physicists have shown rationally that "our rational ideas about the world in which we live are profoundly deficient".[20] It appears that something very exciting is happening. When the packet of energy (violence-Unmanifest-violence) reaches a critical level, it explodes (information transfer) in the realm outside space-time and this is governed by superluminal channel; but this release is located in the space-time matrices. This release of energy in the space-time matrices appears as an earthquake.

The mind (self, violence) cannot interact directly with matter,[21] as the mind is at a higher quantum energy potential than matter of the earth. When violence attains a critical level, it explodes. The explosion first appears in the timeless realm, it encodes a message on the superluminal channel without signals, and then it generates events in space-time. The whole process takes a few microseconds. The answer to this paradox is that the symbolic-dualistic frame of reference must jettison the energy of violence and make a quantum jump to non-dual and non-conceptual frame which is free from any conflict, violence, greed, and so on.

To sum up, *all droughts, floods, catastrophes, calamities, earthquakes, tsunamis are not due to nature's wrath against mankind, but are constructs of the conditioned mind-brain-body system or inheritance of man over uncountable aeons.* An understanding gives you the feeling that man divorced from nature could not be happy and man's power to resist nature is hopelessly puny. This attitude is different from the tenets of scientific civilization to conquer nature. It reminds man that the age-old tide of human ambition and aggression must be jettisoned. And naturally these can be ended when the human mind-brain-body system is purified. An understanding of relationship between man and nature enables you to see God (Life Field, Cosmic Consciousness) face to face. No one can have true relationship without being 'intuitively' convinced that God alone is real and that all else is an illusion.

In Love, or Truth, or Beauty, or Reality, there is superluminal transfer of negentropy. In nature everything happens simultaneously and everything is interrelated. It is integrated and self-emerging system, a totality that exists. Life is mysterious wholeness. It is non-fragmentary, indivisible whole. It is homogeneous, undifferentiated and dimensionless. It is timeless, eternal and infinite. Life is immeasurable and unnamable.

(c) Ending of Earthquakes/Tsunamis: The answer to this paradox is at two levels:

(i) Individual Level: The word 'individual' means one who is not divisible who is whole – not fragmented. The individual is one who stepped out of stream. The conflicts, contradictions, challenges, psychological imbalances, ecological crisis have to be resolved and ended at the individual level.

(ii) Collective Level: These are calamities, catastrophes and tsunamis. When the individual resolves the individual challenges, he acts as a boat to ferry others across the turbulent river to the other shore of timelessness. The purity of the mind is the Mind of the universe – Mind Continuum, the Immense; the mind of the individual is the conditioned psyche – the mind spectrum, the successive consciousness. All this is a pointer to Oneness of Mankind, Brotherhood; and not distinctions between man and man.

The task of resolving the paradox is a difficult one because of the large gap between the superhuman task and the frailties of human endeavours and further the task of turning the age-old tide of human ambition and aggression.

The challenge of ending physical calamities is addressed to the restructured dimension (science-education-health) of the wholeness of life. This bestows understanding and living of life directly. This flowers to new set of values, which in turn can create a new civilisation and a new society.

(d) Peace, Security and Survival: Mr Kofi Annan, Secretary-General of the United Nations nominated sixteen eminent personalities mainly "to examine current challenges to peace and security" on September 23, 2003. We submitted a monograph to "initiate a global dialogue with an open mind" on July 5, 2004. An 'action plan' was also presented.[23]

(e) Intuition versus Intellect: In this article, what I have proposed to my colleagues – who live in the common waking state of mind or symbolic-dualistic frame of reference of mind spectrum – may seem too radical or even too utopian to merit their serious interest. But we have to appreciate that this proposal is based on deep intuition or insight which cannot be understood at the purely intellectual level. We have to appreciate that at present there is a race between our survival and continued degeneration toward disaster, and we cannot afford to lose this race. We have to make a quantum jump from a linear (successive consciousness) to a global Consciousness (Simultaneous or Cosmic Consciousness).

Modern science supports intuitive (insights) and rejects intellectual approach. Insight is fundamental and intellect is a derivative. Henry Stapp wrote in 1971:

> .... Human inquiry can continue indefinitely to yield important new truths.[24]

Max Planck, the father of quantum mechanics, wrote:

> Science .... means unresting endeavor and continually progressing development toward an aim which the poetic intuition may apprehend, but which the intellect can never fully grasp.[25]

G.F. Chew, in reference to a theory of particle physics, remarked:

> Our current struggle with [certain aspects of advanced physics] may thus be only a foretaste of a completely new form of human intellectual endeavor, one that will not only lie outside physics but will not even be describable as 'scientific'.[26]

(f) Our Ignorance: We are not aware of the immutable laws of Cosmic Consciousness or Supreme Intelligence. The cosmos is a handiwork of an Oceanic Intelligence, with an infinite power of creation. The mankind is living in a mysterious universe governed by Cosmic Consciousness.

Concluding Comment

Some scientists/philosophers/educationists call the earthquake/tsunami as constructs by nature due to its wrath or fury or tyrant behaviour. Weisskopf wrote that "this attitude is different from the tenets of scientific civilization which pretends to have conquered our impotence in the face of nature in the raw". This concept contradicts nature's benevolence of "love and compassion", and "harmony and order", and "non-violence and peace". It appears that catastrophes may be constructs of human conditioned psyche. It is evident that "the human truth becomes the truth of nature".

If man transcends the conditioning of the psyche, and lives in the non-dual and non-conceptual frame of reference of Mind Continuum, then all earthquakes/tsunamis will end in a long chronological time. Man then lives in realm of "harmony and order", and "love and compassion".

The new relationship of the great calamities, including earthquake and tsunami, is that instead of rupture and opposition between man and nature, there is a growing coherence of our insight (intuition) between man and nature. The aim of the physical calamity is to remind the human race that it must jettison the antagonistic attitude and learn to live in "harmony and order", "non-violence and peace", and "love and compassion" dimensions of life. Man has to understand, learn and live in equanimous awareness.

Every event/organism is governed by the immutable laws of Life, by Cosmic Consciousness, by Life Field. Everything is prearranged and everything is an effect. The truth of man and the truth of nature is the 'Oneness of Life'. Every expression of nature is interwoven and interlinked with the true nature of man. Man is the miniaturised cosmos.

References

1. Paul Davies, *Superforce*, New York: Simon and Schuster, 1984, pp. 243, 232.
2. Fred Hoyle, in Paul Davies, *ibid.*, p. 223.
3. Ruth Nanda Anshen, in Jonas Salk, *Man Unfolding*, New Delhi: Allied Publishers Private Ltd., 1972.
4. Victor F. Weisskopf, in *Introduction*, Hideki Yukawa, *Creativity and Intuition*, Kodansha International Ltd., 1973, pp. 9, 13

5. Ilya Prigogine and Isabelle Stengers, *Order Out of Chaos*, New York: Bantam Books, 1984, p. 312.
6. *Harijan*, 24.01.1934 & 02.02.1934.
7. *Ibid.*, 16.02.1934.
8. *Ibid.*, 20.04.1934.
9. Sri Aurobindo, *A Message on August 15, 1947*, in Tribune, August 15, 1997.
10. Vimala Thakar, *Contact with Vimala Thakar*, Nr. 20, November 1986, p. 22.
11. Lama Govinda, in *The Odyssey of Science, Culture and Consciousness*, New Delhi: Abhinav Publications, 1990, p. 37.
12. John Selby, *Seven Masters, One Path*, San Francisco: Harper, 2003, pp. 174-76, 179.
13. Tolle Eckhart, *Stillness Speaks*, in Tribune, 15.02.2005.
14. Buddha, in D.S. Kothari, *The Evolution of Consciousness*, New Delhi: National Publishing House, 1983, p. 239.
15. *Ibid.*, in B.M. Khanna, *The Odyssey of Science, Culture and Consciousness*, *op. cit.*, p. 165.
16. R. P. Kaushik, *Towards a New Consciousness*, Delhi: Darshan Yoga Sansthan, January 1974, p. 63.
17. B.M. Khanna, in *The Odyssey of Science, Culture and Consciousness*, *op. cit.*, pp. 160-61.
18. Henry Stapp, in Gary Zukav, *The Dancing Wu Li Masters*, Fontana/Collins, 1982, p. 310.
19. Jack Sarfatti, in Gary Zukav, *ibid.*, pp. 310-11.
20. Gary Zukav, *ibid.*, p. 309.
21. D.S. Kothari, in *The Evolution of Consciousness*, *op. cit.*, p. 236.
22. Sampooran Singh, in *Science, Human Values, Value Education and Peace*, New Delhi: NM & NWM, 1993, pp. 25, 86.
23. Sampooran Singh and Dr (Mrs) Kanwaljit Kaur, *Peace, Security and Survival: A Scientific Quest*, July 5, 2004.
24. Henry Stapp, *The Copenhagen Interpretation and the Nature of Space-Time*, American Journal of Physics, 40, 1972, 1098.
25. Max Planck, *The Philosophy of Physics*, New York: Norton, 1936, p. 83.
26. G.F. Chew, in Gary Zukav, *op. cit.*, p. 331.

28
Epilogue

The basic datum of existence is Consciousness which is of the nature of blissfulness. That is why our ancient Rishis have described the ultimate Reality as Sat-Chit-Anand : Existence-Consciousness-Bliss absolute.

When this Consciousness progressively manifests itself through the medium of Nature, it is called evolution of life and creation of the world. This life upon the Earth and in the Universe has many grades and levels from most rudimentary form bordering on inert matter to the yet most highly developed form of divinity transcending the human. And the fact is that there is practically no end to the process of evolution, so we do not know what will be its last stage, if there ever would be one such thing. We only know the latest stage in this endless process.

The latest form of life that is evolving, resulting from the emergence of a holistic and integral Consciousness – which the scientists call as the non-dual and non-conceptual frame of reference in the Mind Continuum – is integrated in nature, that is, in it all the faculties of human mind and all the aspects of life are fully developed and harmoniously interrelated to make it (life) an Organic Whole. This Organic Whole covers the vast range of life from spirituality to physicality, individuality to universatility and from silence to cosmic creativity. In practical terms, it is "Symbiosis of Science and Spirituality". The alternative titles are "Generation of Innovation in Science for Human Survival", or "Symbiosis of Science, Education, Ethics, and Health", or "An Overview of Human Destiny". These are various fields and functions of this integrated life structure.

For the past 10,000 years, or more, man has been trying hard to resolve his psychological imbalances and other challenges through thought structures, beliefs, concepts, ideals, and so on, but all have proved to be sterile in transforming the human psyche. The Union of Concerned Scientists (UCS) – consisting of more than 1670 scientists, including 104 Nobel laureates – issued a "World Scientists' Warning to Humanity" on April 1993, wrote:

> We must no longer allow it (earth) to be ravaged. This ethic must motivate a great movement, convincing reluctant leaders and reluctant governments and reluctant peoples themselves to effect the needed changes.

To solve all the problems of the world in every sphere it is imperative that the life and the man must be integrated. Without this integration no solution is at all possible. Today we are the victims of a self-contradiction: on the one hand, we are desperately trying to solve the problems of war, violence, exploitation, oppression, tension, etc. to make our human future safe and secure, and on the other, keeping the root cause of these problems, the human/ mental fragmentation intact, rather increasing it. This way we are heading towards a sure self-destruction. We must realize this folly and mend our ways to ensure our survival on this planet-earth.

A healthy and holistic life of necessity must contain all these aspects simultaneously. If, on the other hand, anyone of them is missing, it would mean that the life is not healthy and the structure is fragmented. This unhealthy and fragmented structure would create all the problems in the world which actually is the case today and has been through the millennia.

Dr. Sampooran Singh, in a personal communication in April 2005, stated that he had a strong intuition (insight) on "Symbiosis of Science and Spirituality". I neither doubt nor accept his claim especially in the light of Edwin Hubble, a renowned physicist's words in "*The Nature of Science and Other Lectures*". He stated:

> And sometimes, through the strangely compelling experience of mystical insight, a man knows beyond the shadow of a doubt that he has been in touch with a reality that lies behind mere phenomena. He himself is completely convinced, but he cannot communicate the certainty. It is a private revelation. He may be right, but unless we share his ecstasy, we cannot know.

The intuition is in the realm of non-dual and non-conceptual frame of reference, or timeless moment, or direct perception. The communication of intuitive truth is at the relative field of life or reason or intellect level, and this is expressed as under:

(a) Psychological Frame of Reference: The Unmanifest, through "objective observing and understanding", expresses itself as manifest; the perception recalls the past memory and grafts itself on pure perception resulting into distorted perception. The movement of thought structure is within me, it is independent of the movement of ego, the "me". It leads to conflict, chaos, chemical pressure and nervous tension.

(b) Objective Science: If objective observation is continued for a long chronological time, it reveals the objective fact or objective truth. There is no transference of energy between the object and the subject. It moves from *fact* to *fact*. It is a movement from the known to the modified known, so it constructs artifacts with higher sensitivity. It leads to fabrication of war gadgets capable of destruction. It is leading mankind to self-annihilation.

(c) Subjective Science (World): We have to restructure science so that it takes in its fold both the objective and the subjective dimensions of life. The restructured science has the dimensions:

[Science-Education-Ethics-Health]

The ethics implies that it takes in its ambience the sensual and trans-sensual dimensions of life. A living in the *present*, the *now* heals and living in the *past* and *duality* suppresses man's perceptive sensitivity and enhances disease. The former leads to order, harmony, nonviolence, peace and bliss.

The aim of this monograph is "to initiate an open dialogue with an open mind", and call for responses of eminent scientists. It is suggested to call for responses of scientists of ISRO, CSIR, DRDO, DST, Universities, Engineering and Medical Research Institutes. It may perhaps be useful to call for responses of other scientists, working elsewhere in the world.

It means that we must integrate the human being and human life to establish harmonious relationship in the world and thereby solve its problems. Again, to integrate life, an integral and holistic Consciousness must be discovered and established in earth-life. That is exactly what Dr Sampooran Singh as one man Mission, is endeavouring to do and he has outlined his plan of action. In fact, it is a revolution in life itself through the introduction of a new and holistic principle of Consciousness.

Dr Singh is a pioneer with the head of a scientist and the heart of a saint. I pray for the success of his mission as therein lies the security of our future: the survival of human and the planetary life.

May God glory Thee! My Pranams,

Swami Om Poorna Swatantra

02.10.2005

President, New World Movement, New Delhi
Ashram, Sthirpali Badi, Rajasthan 331305

Index

F

G

H

I

J

O

P

U

V